Solid Waste Landfill

Engineering And Design

Edward A. McBean
Professor, Department of Civil Engineering, University of Waterloo

Frank A. Rovers
President, Conestoga-Rovers Associates, Waterloo, Ontario

Grahame J. Farquhar
Professor, Department of Civil Engineering, University of Waterloo

Prentice Hall PTR, Upper Saddle River, NJ 07458

Library of Congress Cataloging-in-Publication Data

McBean, Edward A.
    Solid waste landfill engineering and design / Edward A. McBean, Frank A. Rovers,
    Grahame J. Farquhar.
        p.   cm.
    Includes bibliographical references and index.
    ISBN 0-13-079187-3
    1. Sanitary landfills—Design and construction.   I. Rovers, Frank
A.   II. Farquhar, Grahame J.   III. Title.
TD795.7.M38   1995
628.4'4564--dc20

94-10676
CIP

Acquisitions editor: Mike Hays
Cover designer: Tommy Boy Graphics
Production editors: Raymond F. Pajek and John Morgan
Art production manager: Gail Cocker-Bogusz
Proofreader: Elizabeth Jolin

Illustrations by Conestoga-Rovers and Associates

© 1995 Prentice Hall PTR
Prentice-Hall, Inc.
A Pearson Education Company
Upper Saddle River, NJ 07458

Transferred to digital print on demand, 2002
Printed and bound by Antony Rowe Ltd, Eastbourne

10   9   8   7   6   5   4   3

ISBN 0-13-079187-3

Prentice-Hall International (UK) Limited,London
Prentice-Hall of Australia Pty. Limited, Sydney
Prentice-Hall Canada Inc., Toronto
Prentice-Hall Hispanoamericana, S.A., Mexico
Prentice-Hall of India Private Limited, New Delhi
Prentice-Hall of Japan, Inc., Tokyo
Pearson Education Asia Pte. Ltd., Singapore
Editora Prentice-Hall do Brasil, Ltda., Rio de Janeiro

This book is dedicated to
Parker, Mabel, William, and Gerrarda

Contents

CHAPTER 12   DESIGN OF NATURAL ATTENUATION SITES   355

CHAPTER 13   LANDFILL GAS MIGRATION   379

Preface

Philosophy of the Book

The legacy of poorly-disposed of wastes is readily apparent to all levels of society. As a consequence, there have been enormous pressures from both the public and governmental agencies to improve the disposal practices associated with solid wastes. The result has been a continuing evolution in design and operational requirements for landfills.

In response, the contents of this book have been structured to describe the principles that govern waste decomposition in landfills and to describe the current practices utilized. Attention is given to both municipal and industrial waste disposal in landfills. Insights into the implications of the 3R's (reduce, reuse, and recycle) and the methodologies currently utilized in landfill site selection are also provided in the book. To accomplish these tasks, the book is organized into five broad categories as follows:

Part I Chapters 2 and 3 detail background information relevant to the engineering and design of solid waste landfills. Chapter 2 characterizes the composition of refuse being disposed of in today's landfills. This assessment thus provides the basis for recent historical and present-day disposal quantities and composition. However, the pressure for the 3R's(reduce, reuse, and recycle) is changing the character of the waste stream and, thus the latter portions of the chapter provide insight into how the waste stream can be expected to change which, in turn, will influence landfilling practices of the future.

Chapter 3 describes the methodologies currently being employed in the site selection of landfills. The intent in this chapter is, in part, to indicate the types of features considered in the selection of landfill sites which in turn influences the design and operational features at a landfill.

Part II This portion of the book, consisting of Chapters 4, 5 and 6, develops the fundamentals that will be utilized throughout the remaining chapters. Chapter 4 characterizes the decompositional features of the refuse, by considering the interacting principles. Chapter 5 develops the principles of Chapter 4 into preliminary quantification of magnitudes, addressing such questions as the time needed for decomposition to occur, and the quantity of gas that will be generated, per unit of refuse. Finally, Chapter 6 emphasizes the hydrogeologic principles that will be utilized throughout the remainder of the text.

It is highly recommended that a good understanding of Chapters 4 through 6 be obtained prior to proceeding onto the subsequent chapters, since an understanding of the principles will make the later discussions much more comprehensible.

Part III The third portion of the book, consisting of Chapters 7 through 12, focuses on leachate generation in terms of quantity an l quality of leachate and the variabilities of both over time. Emphasis is given to description of quantitative tools for estimating the volumes of leachate, descriptions of leachate collection procedures and leachate treatment alternatives, and finally, the components of natural attenuation in the event of leachate migration beyond the landfill. It will be apparent that tradeoffs need to be made regarding leachate. It is feasible, for example, to minimize leachate quantities by expending additional during the implementation of the landfill cover. The associated costs and effort must, however, be traded off against the slower rates of decomposition which will occur within the refuse, and the costs of treatment and disposal of the leachate. Numerous design decisions are implicit in this tradeoff. Again, the intent in these chapters is to provide the design principles necessary to allow these assessments to be made, and to indicate some useful information sources for pertinent data for specific applications.

Part IV The fourth portion of the book, Chapters 13 and 14, focuses on the generation rates, migration and control of landfill gas, and recovery of the gas as an energy source. The modeling principles are presented, along with the reported field experience, as a means of assisting future designs. In addition, indications are provided on gas recovery operations in this rapidly-evolving field.

Part V The last portion of the book details current operational features utilized on landfills, including such aspects as equipment, fencing needs and noise and odor controls. As well, guidance is provided on the need for monitoring and statistical analyses of the resulting monitoring data.

The book was prepared for a diverse spectrum of individuals involved in solid waste concerns. Thus, the intended audience includes senior undergraduate and graduate engineering students at universities. The utilization of the book for the teaching profession is enhanced through use of example design principles utilized by practitioners, use of case studies, and the inclusion of possible assignment problems at the end of individual chapters.

By an organized presentation of the theory and design principles, the book is also intended as a reference document for professionals in both engineering and planning in governmental positions and private companies. The book is not intended as a comprehensive design manual on landfills since detailed design considerations change from one region to the next and are continuing to evolve over time. Instead, the focus is intended to be on the principles of design; by being aware of the principles, an improved understanding of appropriate precautions and expectations is a natural outgrowth and will keep the text relevant for many years to come.

Extensive use is made of references for additional reading. These references are provided as directions to those readers wishing to explore specific concerns in greater depth.

Authors of the Book

The authors of this book represent a combination of professors and engineering practitioners. The intent by use of this mix is to ensure the book meets the needs of the educational profession while also responding to the needs of regulatory agencies and practicing engineers and planners.

Edward A. McBean (B.A.Sc. from the University of British Columbia and S.M., C.E., and Ph.D. from the Massachusetts Institute of Technology) is a Professor of Civil Engineering at the University of Waterloo. Ed has held academic appointments at Cornell University, the Massachusetts Institute of Technology, the University of California at Davis and the University of Waterloo. His professional employment experience includes work experience with Meta Systems Inc., Acres Consulting Services Ltd., UNDP and UNESCO. He currently serves as a senior technical advisor at Conestoga-Rovers and Associates. Ed has edited four books and is the author of 130 articles in the refereed technical engineering journals.

Frank Rovers (B.A.Sc. and M.A.Sc. from the University of Waterloo) is president of Conestoga-Rovers and Associates Ltd. Frank has been extensively involved in consulting engineering practice for twenty years and is the author of many technical reports and papers. Conestoga-Rovers and Associates (CRA) is an environmental engineering company with 450 employees with headquarters in Waterloo, Ontario and branch offices in cities including Niagara Falls, N.Y., Wappingers Falls, N.Y., Chicago, Illinois, Atlanta, Georgia, St.Paul, Minnesota, Detroit, Michigan, and Stockton, California.

Grahame Farquhar (B.A.Sc. and M.A.Sc. from the University of Waterloo, and Ph.D. from the University of Wisconsin-Madison) is a Professor of Civil Engineering at the University of Waterloo. Grahame has been active for more than twenty years with numerous consulting engineering companies in the area of waste disposal. Grahame has edited one book and is the author of 45 papers in the referred technical journals.

All of the senior authors have been heavily involved in the teaching of professional development courses, including those at the University of Wisconsin-Madison, University of Toronto, Nova Scotia Technical College, and UCLA.

Acknowledgments

We are under no delusion that the work reported in this book represents "our work". Clearly the material is the product of many people. Our intent in preparing the book was to assemble and to organize the considerable array of experience and understanding in the field of landfilling.

In preparing this book, we have drawn upon the experience and efforts of many, and for this assistance we are very grateful. During the years leading up to the writing of this book, the authors worked closely with many colleagues on various aspects of landfilling. The contributors are many, including:

- The employees of CRA who so generously provided examples and technical literature references. The advice and assistance of many is acknowledged, with special mention of Bruce McConnell, Klaus Schmidtlee, Peter Dimmell, Tony Crutcher, and Jim Yardley.
- The contents of the book very much reflect the ideas and the inspiration that the people from the "Wisconsin connection" provided, namely Joe Fluet, Bob Ham, Pete Kmet, Ron Lofy, Phil O'Leary, and Phil Stecker. They will find their contributions liberally dispersed throughout the work.
- Drafts of the book have been used during the teaching of graduate-level courses at the Univeristy of Waterloo and the University of California at Davis. The many useful comments by the students are greatly appreciated.
- The willingness of various municipalities who provided the opportunity for taking photographs of the landfills under their jurisdiction, including Metropolitan Toronto and Yolo County is gratefully acknowledged.
- The secretarial staff at CRA who so obligingly "revised the last revision". In this respect, special acknowledgment must be given to Jeannie Harris who continued to remain cheerful in the face of the numerous rewrites, and Diana Fedy who was instrumental in choreographing the entire volume,
- vi) The drafting department at CRA and in particular, Zoli Pillar, who, like the secretarial staff, responded to challenges to take our sketches and turned them into high quality drawings is also gratefully acknowledged.

To all of the above, we owe our sincere thanks for their assistance.

In an undertaking of the magnitude of this text, it is not possible to avoid errors, and for this we apologize in advance. Any corrections, criticisms, or suggestions for improvements will be greatly appreciated by the authors. We would also welcome any additional information and data that would make future editions of the book more complete.

CHAPTER 1

INTRODUCTION

1.1 HISTORY OF DEVELOPMENT OF WASTE DISPOSAL

Economists have defined waste as a material that is cheaper to throw away than to use. The ramifications of this definition, when it is translated into action by modern society, are apparent when the enormous quantities of materials that are discarded are considered. Thus, although personal attitudes are changing, there are still very significant quantities of solid wastes that must be managed now and in the foreseeable future.

Solid waste management, in its broadest sense, is concerned with the generation, on-site storage, collection, transfer, transportation, process and recovery, and disposal of solid wastes. Only in very recent years have annual per capita waste generation rates begun to demonstrate decline. Such declines have occurred partly because the public is more informed (awareness of reduce, reuse, and recycle), and partly because disposal costs have escalated.

The concerns with waste disposal and the concept of a "conserver" society are becoming apparent. There is increasing recognition that the world has limited resources and that the costs of waste management are going to continue to escalate. The increasing sophistication of waste management technologies will ensure that overall costs will rise. Because of siting difficulties and the increasing stringency of the containment technologies being required, landfilling is becoming increasingly expensive and is forcing the reexamination of options of dealing with solid wastes.

Perhaps the event that galvanized our society into an awareness of the problems that can result from mismanaged wastes was the Love Canal incident. Subsequent media attention surrounding that incident and other similar ones have made it very difficult to educate the public about, technologies now being utilized in the design of landfill sites as required by current standards of practice. Thus, municipal solid waste management is caught in the backlash of public hysteria over hazardous waste management. Nevertheless, a lack of public awareness and appreciation is only one of a number of issues with which professionals involved in landfill design and operation must contend.

As a result of the concern of the public and the realization by the political regulators of the economic damages and environmental effects that can result from substandard refuse disposal methods, there has been a rapid evolution in the technologies utilized in land disposal practices. The exploration of the evolving environmentally secure disposal technologies is the major focus of this text. Much of the discussion will relate to Municipal Solid Waste (MSW) because important principles and degrees of uniformity in disposal practices throughout North America can be detailed. Also, many of the principles associated with hazardous waste disposal in landfills are described, although a comprehensive examination of hazardous waste disposal practices is not attempted, since they are very problem specific. However, there is an increasing convergence of practices for disposal of MSW and hazardous wastes, and these will be discussed.

1.1.1 Historical Disposal Practices

To understand why there is a poor public image associated with solid waste disposal, it is necessary to review some unsuitable disposal practices utilized in the past.

One of the earliest known forms of waste management existed in fifth century b.c. in Greece, where the individual was responsible for collecting his/her own garbage and bringing it to the town dump (Kelly, 1973). The first garbage collection service was established in the Roman empire (Kelly, 1973). People tossed their garbage into the streets, and it was shoveled into a horse-drawn wagon by appointed garbagemen who then took the garbage to an open pit, often centrally located in the community. Bodies of dead people and animals were deposited in pits outside of town due to the odor.

The semiorganized system of garbage collection lasted only as long as the Roman empire. From the Dark Ages through the Renaissance, there was no organized method of disposing of wastes. However, a survey done in 1880 of garbage collection in major U.S. cities showed that 43 percent of cities provided some minimal sort of garbage collection. By 1915, 50 percent of the major cities had garbage collection, and by the late 1930s, this number had risen to 100 percent (Blumberg and Gotlieb, 1989).

Until the 1900s, solid waste was dumped directly on the land. Around 1910, methods for creating sanitary landfills were developed. One of the simpler methods, burying, was being used in the United States in 1904 (Public Administration Service, 1970).

Until the 1950s, municipal refuse disposal consisted of careless dumping operations, without consideration of proper planning and engineering procedures to maximize the usefulness of the disposal facilities and provide environmental protection. Open-pit dumping was standard practice (California, 1954), and the wastes were often burned in situ. Spontaneous combustion sometimes occurred, and at other times controlled burning practices were followed for volume control. Considerable problems of odors, noise, seagulls, and smoke were obvious and immediate environmental impacts. Unfortunately, many of these practices remain in some remote communities, even under the new environmental standards that regulate waste disposal practices.

The changes that occur beneath the mantle of a controlled tip (i.e., landfill) were investigated as early as 1932 by Bertram J. Jones, Director of Public Cleansing, and Frederick Owen, Chief Chemist, City of Manchester, England. Their work is published under the title "Some Notes on the Scientific Aspects of Controlled Tipping." This investigation did much to promote early recognition of controlled tipping as a safe and satisfactory method of solid waste disposal.

There was a gradual transformation to improved landfill practices, although it took considerable time for proper planning, engineering, management, and staff training to be applied to these issues. By late 1948 the conversion to "cut and cover" landfilling was essentially completed. By 1954 there were probably no more than 50 urban communities in the United States that were still operating the unsightly, rat-infested open dumps that continue to be described in great detail by landfill siting opponents. About that same time, the need to analyze groundwater downgradient from landfill sites was recognized, as a result of a growing recognition of the potential impact on groundwater quality by escaping contaminants. By 1959 the sanitary landfill was the major method of solid waste disposal for communities in the United States. (ASCE, 1959).

Even in the 1950s many municipal sites were designed and constructed with little or no engineering input. When such landfills were completed, they were generally covered with a minimum of soil, and surface vegetation was encouraged. There was limited recognition that surface water infiltrated a capped landfill and mixed with, and became contaminated by, the wastes. Eventually it was realized that a liner across the base of landfills with an engineered cap system to reduce long-term leachate generation and the escape of leachates to the environment were minimum requirements. Parallel to, and in many respects an instigating force behind these increased precautions against contaminating the environment, were changes in legislation.

The concept of controlled tipping, in which the refuse is sealed in cells formed from earth or other cover material, was implemented with the objective of changing the physical characteristics of the buried solid wastes to make it relatively free from odor, less attractive to vermin, and less dangerous to health. These and other required disposal practices indicate their increasing sophistication. Nevertheless, the "dump" stigma still remains in the minds of the public. The prevalent public attitude to the siting of solid waste management facilities is one of Not In My BackYard (NIMBY). The NIMBY phenomenon has developed due to the increased awareness and affluence of residents, public education, and media exploitation. The results have been increased understanding of the processes involved and the need for additional levels of protection in such respects as controls, backup systems, and improved operations. There have also been tremendous improvements in the ability to detect contaminants with better technology and instrumentation. Awareness of the dangers of chemical exposures and the understanding of transport mechanisms of leachates have also increased significantly.

Developed in this text are the principles and, to a more limited extent, the practices associated with landfill gas, leachate management, and environmentally secure operating procedures for solid waste management.

1.1.2 Landfilling as a Primary Disposal Means

Although landfilling has lost (appropriately) some of its share in the array of solid waste management technologies to those of the 3Rs (reduce, reuse, recycle), it is still the primary means of disposal of both residual municipal solid waste and many hazardous wastes and will continue to be so for the foreseeable future. It is to be expected that increased societal efforts will be directed toward the 3Rs technologies; nevertheless, there will remain significant quantities of unrecyclable wastes that must continue to be directed to landfills. An understanding of the impact of 3Rs separation programs on the character of the waste to be landfilled will be an important consideration in future landfill design. For example, recycling of waste paper will reduce the organic content of the refuse, which will impact landfill gas production and leachate generation in future years. The principles related to the changes in the wastes being disposed of and how they will affect landfilling practices will be discussed at various appropriate points within the text.

1.2 OBJECTIVES

Our intent is to provide a synthesis of existing knowledge on solid waste landfilling. It is necessary to be selective in what can be covered in a discourse of this nature. As a result, we focus on the principles underlying the issues. Thus, we will see how evolving standards of practice and the changes that will occur in solid waste management will affect designs and disposal practices and concerns.

Landfills designed to state-of-the-art technology offer a relatively straightforward and affordable option for the disposal of MSW. However, it must be realized that landfills present a number of practical challenges that must be adequately handled in the planning, design, operation, and closure phases to ensure that the problems commonly associated with open dumping do not occur.

References

ASCE Committee On Sanitary Engineering Research. 1959. Refuse Volume Reduction in a Sanitary Landfill. ASCE-*Journal of the Sanitary Engineering Division*, Vol. 85, No. SA6, November, pp. 37–50.

Blumberg, L., and Gottlieb, J. 1989. *War on Waste*. Island Press, Washington, D.C.

CALIFORNIA STATE WATER POLLUTION CONTROL BOARD. 1954. *Investigation of Leaching of a Sanitary Landfill*. Sacramento, California.

Kelly, K. 1973. *Garbage*. Doubleday, Toronto.

Public Administration Service. 1970. *Municipal Refuse Disposal* Interstate Publishers and Printers, Illinois.

OPPORTUNITIES
FOR REDUCTION, REUSE,
AND RECYCLING

2.1 WASTE QUANTITIES IN THE MODERN ERA

2.1.1 Quantity Projections

The quantities of solid wastes produced by the developed nations of the world are large and are increasing along with a growing affluence and improved standard of living. Every year Americans discard, directly and indirectly (by including commercial and industrial waste quantities), an amount of waste equal in weight to the Statue of Liberty (O'Leary et al., 1988), or about 160 million tons of refuse per year (Chemecology, 1989(b)). Current levels are approaching 1 ton per person per year, and the annual generation is expected to reach 190 million tons in the United States by the year 2000. In addition, nearly one-half of the 6000 U.S. landfills are expected to be filled by the mid1990s. Refuse quantities, in conjunction with the fact that many landfills are reaching capacity, indicate very severe impending problems. To make problems worse, some existing landfills and incinerators will not be able to meet increasingly stringent federal, state, and local environmental regulations and will be forced to close.

To put these quantities into perspective, New York City generates 27,000 tons of garbage per day. At an in situ density of 710 kg/m^3 (1200 lb/yd^3) and a 4:1 ratio of waste to cover soil, a landfill space requirement of 41,315 m^3 is required each day for New York City alone. That volume corresponds approximately to a football field covered to a depth of 10 m every day.

It is interesting to compare the purchase price of a major newspaper with its disposal cost. For example, $1.25 for purchase, $0.10 for tipping fee (at $100/tonne).

To understand solid waste generation rates and the opportunity for creating any changes in what is currently being disposed of in landfills, it is essential first to examine the components of the refuse. Solid wastes are frequently categorized into residential or municipal, and industrial, commercial, and institutional.

Residential/Municipal

Residential solid wastes are defined as those wastes generated and discharged from single- and multifamily dwellings and include wastes generated from within the dwelling, from the yard, and from activities outside the dwelling. The wastes themselves can be classified in many ways; common categories include food waste, garden waste, paper products, metals, glass and ceramics, plastics, rubber and leather, textiles, wood and rocks, dirt, and ashes. (Estimates of quantities within each of these categories of residential waste are provided in Section 2.2.)

In the early 1970s, publications routinely quoted the quantity of municipal solid waste generation rate in the United States as 2.4 kg per capita, per day. This value was derived from Black et al., (1968). Subsequent studies (e.g., National Center for Resource Recovery (NCRR), 1973; Applied Management Sciences, 1973; and Smith, 1975b) indicated that this number was high. Smith (1975a) found the national average daily per capita generation of residential solid waste in the United States to be in the range of 1.08 to 1.22 kg.

The Wisconsin Department of Natural Resources (1981) suggested that more accurate waste forecasts arise if residential generation multipliers based on the size of the community are used, as follows:

Population	Per Capita Generation Rate Per Day (kg)
< 2500	0.91
2500–10,000	1.22
10,000–30,000	1.45
> 30,000	1.63

Many other investigators have also developed estimates of per capita generation rates. As an example, Table 2.1 gives the residential solid waste quantities summarized by types of household. Such breakdowns are useful because the number of households in each of the indicated categories is typically easily available from municipal governments (CRA, 1991).

Some of the variations in estimates of solid wastes arise because of the different timings of studies and the related changes in character of wastes disposed of over time. Others occur because of sampling difficulties, seasonal changes

Table 2.1 WASTE GENERATION COEFFICIENTS FOR HOUSEHOLDS (KG/YR/HOUSEHOLD)

Dwelling Type	Domestic Wastes	Yard Wastes	Total Wastes
Detached	1025	249	1274
Mixed	840	125	965
Multiple	342	0	342

Source: Adapted from CRA (1991).

(e.g., whether or not yard wastes were included), and fluctuations in the economy that translate into variable waste generation rates.

Industrial

Industrial wastes encompass all discarded materials resulting from industrial operations, including processing, plant facilities, packaging, shipping, office, and cafeteria wastes. There is little similarity in the types of wastes discarded by different industries; however, it can generally be assumed that wastes from industries within the same Standard Industrial Classification (SIC) sector will produce similar wastes. Accurately predicting industrial waste productions is more difficult than predicting residential waste; less information is available because wastes are frequently handled either by the industries themselves or by private disposal contractors and not by municipal or public sources. In addition, waste generation is a function of economic activity either in terms of production output or in terms of the number of employees. Although production output is commonly acknowledged to be a better indicator of activity, this information is frequently proprietary. The number of employees is typically available through state employment agencies.

Industrial waste generation coefficients for different industrial sector categories, as reported in different studies, are provided in Table 2.2. The last two columns in Table 2.2 give the arithmetic mean and the geometric mean of the data. There is considerable merit in using the geometric mean, because it decreases the dependence of the result on individual high values.

Commercial and Institutional Wastes

Commercial solid wastes are defined as wastes generated and discarded from a wide variety of activities that can be categorized as retail, wholesale, and service establishments. Institutional solid wastes are those generated at facilities such as hospitals, schools, and municipal office buildings.

There is considerable difficulty in assigning solid waste generation rates to a number of the businesses included in the Industrial/Commercial/Institutional (ICI) categories, since few data are available. CRA (1991) reported the list summarized in Table 2.3.

Table 2.2 INDUSTRIAL WASTE GENERATION COEFFICIENTS

Industrial Sector (1970 SIC Code)[a]	Coefficients Reported by Rhyner and Green (1988) (tonnes/employee/year)								
	Niessen and Alsobrook (1972)	Weston (1971)	Goleuke and McGauhey (1970)	Wisconsin (1974)	Mylnarski and Northrop (1975)	Steiker (1973)	Washington State[b]	Arithmetic Mean	Geometric Mean
Food and Beverage (10)	7.57	5.26	4.37	4.42	3.01	14.14	0.24	5.57	3.6
Rubber and Plastic (16)	8.92	2.36	1.4	1.01	0.35	4.69	1.56	2.9	1.86
Leather and Related Products (17)	8.15	0.15	2.26	0.18	0.34	1.27	7.61	2.85	1.07
Textile Industries (18)	3.33	0.24	0.48	0.76	0.35	—	0.22	0.9	0.53
Knitting Mills, Clothing (24, 23)	1.99	0.28	0.48	0.21	0.35	1.02	0.65	0.71	0.54
Wood Products (25)	7.74	9.34	19.67	14.73	15.51	20.68	2.21	12.84	10.58
Furniture and Fixtures (26)	2.52	0.47	18.28	1.13	0.35	5.24	1.46	4.21	1.82
Paper and Allied Products (27)	3.68	13.06	11.37	13.52	5.96	9.77	8.02	9.34	8.59
Printing and Publishing (28)	5.29	0.44	12.07	1.03	0.88	2.5	0.6	3.26	1.68
Primary Metal Products (29)	2.89	21.77	6.1	6.09	13.59	22.26	1.78	10.64	7.43
Metal Fabricating (30)	8.25	1.54	6.1	3.38	0.97	3.3	1.52	3.58	2.78
Machinery (31)	5.03	2.36	3.79	3.29	0.83	3.23	0.43	2.71	2.08
Transportation Equipment (32)	2.44	1.18	3.08	1.18	0.87	6.07	1.12	2.28	1.8
Electrical Products (33)	2.67	1.54	2.7	2.43	0.59	3.08	0.54	1.94	1.6
Non-Metallic Mineral Products (35)	5.84	2.18	16.43	20.69	44.36	18.69	0.66	15.55	8.14
Petroleum and Coal (36)	1.45	—	—	26.35	—	—	1.77	9.86	4.07
Chemicals and Chemical Products (37)	13.43	0.57	7.45	7.45	2.35	3.75	0.66	5.09	3.05
Miscellaneous Manufacturing (39)	1.45	0.13	2.26	1.09	0.34	2.17	0.67	1.16	0.81

[a]No coefficients were available for two sectors—tobacco products (15), and primary industries (00–09). The value for miscellaneous manufacturing was used for these.

[b]Matrix Management Group (1988).

Table 2.3   INDUSTRIAL/COMMERCIAL WASTE GENERATION COEFFICIENTS

Sector (1970 SIC Code)	Coefficient (tonnes/employee/year)	Reference
Construction (40–42)	1.09	5
Transportation (50–51)	1.12	2
Storage Services (52)	0.68	4,6
Communications (54)	0.68	4,6
Electric and Other Utilities (57)	0.68	4,6
Wholesale Trade (60–52)	1.70	4
Retail Food (63)	2.66	1
General Merchandise (64)	1.54	1
Other Retail (65–69)	1.54	1
Finance, Insurance, Real Estate (70–73)	0.08	1
Education and Related Services (80)	0.06	5
Health and Welfare (82)	1.14	4
Recreation Services (84)	0.76	1
Accommodation and Food (88)	1.50	1
Other Services (83, 85–86, 89)	0.68	4
Public Administration (90–99)	0.54	3

References:
1. Based on Gore and Storrie, 1991.
2. Matrix Management Group, 1989.
3. Matrix Management Group, 1988.
4. Rhyner and Green, 1988.
5. CRA, 1991.
6. Value for miscellaneous services used.

Conversion Factor:

1 tonne = a metric ton or 1000 kg

Alternatively, 1 tonne = 2204 lb = 1.102 ton

Table 2.4 gives the range and typical values of per capita solid waste generation rates for various waste production sectors.

Typical densities for solid waste as a function of source are indicated in Table 2.5. Residential waste has a typical curbside density of 150 kg/m^3 (250 lb/yd^3); in a compaction vehicle, typical density is 300 kg/m^3 (500 lb/yd^3), with a range of 180 to 415 kg/m^3, and good sanitary landfill operation practices will achieve 590 to 830 kg/m^3 (1000 to 1400 lb/yd^3). Typical increases in density at varying stages of disposal are listed in Table 2.6.

Table 2.4 TYPICAL PER CAPITA SOLID WASTE GENERATION RATES (BY SOURCE)

Source	Typical		Range	
	(kg/day)	(lb/day)	(kg/day)	(lb/day)
Residential	1.1	2.5	0.9 – 2.3	2.0 – 5.0
Industrial/Commercial	0.7	1.5	0.45 – 1.6	1.0 – 3.5
Demolition	0.27	0.6	0.05 – .36	0.1 – 0.8
Other municipal	0.18	0.4	0.05 – .27	0.1 – 0.6

Table 2.5 TYPICAL DENSITIES OF MUNICIPAL SOLID WASTES BY SOURCE

Source	Density			
	Range		Typical	
	(kg/m^3)	(lb/yd^3)	(kg/m^3)	(lb/yd^3)
Residential (uncompacted)	90 - 180	150 - 300	150	250
Residential (compacted)	180 - 360	500 - 1000	300	750
Commercial/Industrial	175 - 350	300 - 600	445	500

Note: 1 lb/yd^3 x 0.5933 = 1 kg/m^3

Table 2.6 TYPICAL DENSITIES OF REFUSE AT VARYING STAGES OF DISPOSAL

	Density	
	(kg/m^3)	(lb/yd^3)
From noncompacting garbage trucks	395	500
From compactor garbage trucks	355 – 415	600 – 700
After compaction on landfill site	445 – 505	750 – 850
After extra compaction	595	1000
Maximum compaction likely on landfill site (before long-term settlement)	710	1200

Source: Lutton et al. (1979)

2.1.2 Variability in Disposal Quantities

At the insistence of state regulators, landfill operators are now recording refuse disposal quantities; however, the accuracy of such records, even for very recent years, is variable. For example, Figure 2.1 shows the history of waste generation for the Regional Municipality of Waterloo, Ontario, Canada. The graph shows growth characteristics of landfill quantities over a 14-year period. Although the historical trends of these data are significant, the division into residential and ICI wastes from data such as those depicted is uncertain. Record-keeping practices are improving, which will allow greater accuracy in establishing this separation in the future.

The growth in quantities of solid wastes for disposal will determine the lifespan of landfill sites. Assumptions about how the waste stream will change over time due to, for example, population growth, per capita waste generation rates, and diversion to the 3Rs will affect projections of the remaining site life. For example, due to a steep increase in tipping fees, the implementation of 3Rs programs, and the banning of certain recyclables at the Keele Valley Landfill (operated by the Municipality of Metropolitan Toronto, Canada) the quantities of solid waste delivered to the site fell 32 percent between September 1990 and September 1991. This increased the landfill site life projections by 5 years. What is not so obvious is how much of the decrease in delivered tonnages was the result of diversion of wastes to other sites with lower tipping fees.

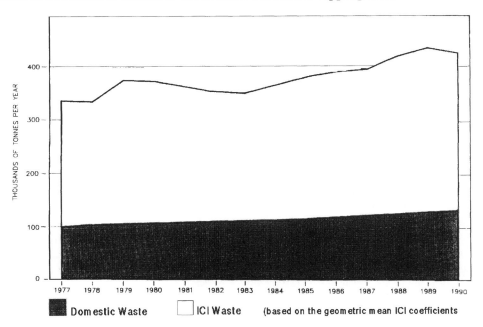

Fig. 2.1 Estimated total waste generation by source. (Conestoga–Rovers & Associates)

Despite the relative lack of confidence in landfill quantity estimates, the following major factors are apparent throughout North America:

1. There is an increasing amount of refuse to be disposed of, although the rate of increase is decreasing.
2. The sources of solid wastes and the composition of the solid waste stream are becoming better understood.
3. Waste tonnages will decrease due to increased tipping fees, the apparent landfill capacity crisis, and increased recycling and reuse efforts.

2.2   COMPOSITION OF THE WASTE STREAM

Municipal solid wastes (MSW) generate leachate through processes of decomposition and percolation through the waste. Knowledge of MSW decomposition processes and leachate production and composition has evolved considerably over the last 20 years, however, much remains to be learned.

Good measures of the waste stream composition are hard to obtain, in part because the opening of bags to determine the wastes present is an onerous task. Also, people are reluctant to have their garbage sorted. Additionally, seasonal trends relating to yard wastes, spring cleanup, ashes, and the like, as well as the need to collect data over a large number of households to ensure a representative sample, complicate the problem of determining refuse composition. Nevertheless, the validity of this technical literature is improving because, despite the difficulties, many researchers have undertaken the enormous task of item-by-item segregation and weighing of refuse to estimate solid waste composition.

Household and commercial wastes are composed of a variety of components but are surprisingly similar in overall composition (McGinley and Kmet, 1984). This is not the case for industrial wastes. Many different component characterizations have been developed, including those provided in Tables 2.7 through 2.11. In addition, Ham et al., (1979) present a breakdown of the composition of an average bag of municipal refuse: paper accounts for 42 percent, yard waste such as wood, grass, brush, greens, and leaves account for 14 percent, and food wastes account for 12 percent. Of the remaining 32 percent, noncombustibles such as metals, glass, ceramics, and ashes account for 24 percent, and other rubbish such as leather, rubber, and plastic for the remaining 8 percent. These figures reveal that the organic fraction of the waste stream (paper, yard and food wastes) amounts to approximately 68 percent of the total waste stream.

The data in Tables 2.7 to 2.11 clearly show that a significant portion of the waste stream comprises items that have the potential for being recycled or reduced, including food and garden wastes, paper wastes, plastics, metal, and glass.

Table 2.7 REFUSE COMPOSITION: WASTE COMPOSITION ANALYSIS (RESIDENTIAL, COMMERCIAL AND INDUSTRIAL WASTE STREAM) PERCENTAGE BY WEIGHT

	Franklin (1986) for year 1984	Franklin (1986) for year 2000
Food	8.1	6.8
Paper, cardboard	37.1	41.
Yard wastes	17.9	15.3
Wood	3.8	3.8
Plastics	7.2	9.8
Ferrous metals	9.6	9.0
Glass	9.7	7.6
Textiles, rubber	4.6	4.6
Ceramics, rubble, other	2.	2.1
Total	100	100

Table 2.8 MUNICIPAL REFUSE COMPOSITION (% WET WEIGHT)

Component	Mean Percentage	Standard Deviation
Food	12.1	5.6
Garden wastes	18.0	17.3
Paper	43.6	9.4
Plastics/rubber	2.7	1.6
Textiles	1.9	1.4
Wood	2.1	0.9
Metal	8.9	2.1
Glass and ceramics	8.1	2.6
Ash and soil (other)	2.6	—

Sources: Chian and deWalle (1977); Emcon (1980); Rovers and Farquhar (1973)

Table 2.9 PERCENT COMPOSITION OF REFUSE (% BY WEIGHT)

Source	Paper	Food, Garden	Plastic Rubber Leather	Glass, Ceramic	Rags	Dirt, Ash, Rock	Metal	Wood	Other
Wigh & Brunner (1981)	41.8	20.6	6.9	7.8	4.1	6.1	8.3	1.8	5.6
Jones & Malone (1982)	44.8	1.4	9.0	7.6	3.1	20.2	10.8	0.5	2.7
Walsh & Kinman (1979)	42.6	14.3	8.7	12.2	a	6.1	12.2	2.6	1.3
Fungaroli & Steiner (1979)	55.0	15.0	3.0	10.0	3.0	2.0	10.0	2.0	a
Fungaroli & Steiner (1979)	53.5	13.0	1.5	7.7	0.8	1.5	6.7	2.3	9.8
Kemper & Smith (1981)	48.6	12.8	5.5	11.9	5.3	5.6	7.3	2.3	0.7
Emcon Associates (1975)	35.5	19.6	4.2	9.1	1.1	21.2	8.0	1.3	a
Cameron (1975)	47.6	21.6	5.4	7.0	3.6	1.4	8.7	4.7	a
Arithmetic average	46.2	14.8	5.5	9.2	3.0	8.0	9.0	2.2	5.2
Standard deviation	6.4	6.4	2.6	2.1	1.6	8.1	1.9	1.2	3.4

[a]Refuse was not categorized according to this heading.

Table 2.10 UNIT WEIGHT OF UNCOMPACTED AND COMPACTED REFUSE COMPONENTS

Waste Component	Uncompacted Unit Weight (lb/ft^3)	Water Content (% of dry weight)	Ratio of Compacted to Uncompacted Weight	
			Normally Compacted	Well Compacted
Food waste	8–30	50–80	2.9	3.0
Paper and paper board	2–8	4–10	4.5	6.2
Plastics	2–8	1–4	6.7	10
Textiles	2–6	6–15	5.6	6.7
Rubber and leather	6–16	1–12	3.3	3.3
Yard waste	4–14	30–80	4.0	5.0
Wood	8–20	15–40	3.3	3.3
Glass	10–30	1–4	1.7	2.5
Metals	3–70	2–6	4.3	5.3
Ash, brick, dirt	20–60	6–12	1.2	1.3

Source: After Tchobanoglous et al. (1977)

Table 2.11 COLLECTED AND COMPACTED MSW DENSITIES (per 100 kg of refuse)

Waste Component	Weight (kg)	Volume (m^3)	Compaction Factor in Landfill	Volume in Landfill (m^3)
Food wastes	15	0.51	0.33	0.168
Paper	40	4.83	0.15	0.725
Cardboard	4	0.80	0.18	0.144
Plastics	3	0.46	0.10	0.046
Textiles	2	0.31	0.15	0.047
Rubber	0.5	0.04	0.30	0.012
Leather	0.5	0.03	0.30	0.009
Garden waste	12	1.14	0.20	0.228
Wood	2	0.08	0.30	0.024
Glass	8	0.41	0.40	0.164
Tin cans	6	0.67	0.15	0.100
Nonferrous	1	0.06	0.15	0.009
Ferrous metals	2	0.06	0.30	0.018
Dirt, ash, brick	4	0.08	0.75	0.060
Σ	$\Sigma = 100$			

Chemecology 1989(a) profiled 40 U.S. waste-characterization studies and concluded the following about the average composition of solid wastes from household and commercial sources (prior to recycling):

1. Nearly one-half consists of paper products, including newsprint and cardboard.
2. Nearly one-third is made up of various organic wastes, including yard clippings and food wastes, plastic, rubber, textiles, wood, and disposable diapers.
3. .Seasonal variations in yard trimmings add to the waste in the summer, whereas more ashes are disposed of in winter. Grass clippings may comprise 20 percent of residential refuse during the growing season.
4. The balance is composed of inorganic wastes such as metals, glass, ceramics, rocks, ash, and dirt.

In examining waste stream composition data it is most important to distinguish between descriptions of the residential waste stream and those of the combined residential/ICI waste stream.

2.3   DISPOSAL ALTERNATIVES

2.3.1 Basic Strategies

During the last twenty years, many of the industrialized countries have been regulating the generation, disposal, and management of waste; however, in the United States and Canada, the primary emphasis has been on waste disposal rather than on waste reduction. In 1986 the U.S. Environmental Protection Agency (EPA) determined that more than three-quarters of all municipal wastes were deposited in the nation's 6000 municipal landfills (O'Leary et al, 1988). This situation is changing in that regulatory agencies are becoming increasingly critical of landfills, often requiring two and three levels of safety in the approval process for landfill siting. However, the need for landfills continues.

The question typically raised by residents living in the vicinity of a proposed landfill is, Why do we need another landfill? We should stop being so wasteful. The result is tremendous public resistance to the siting of solid waste management facilities, in general, and landfill sites, in particular. To respond to this type of question, it is necessary to examine what other options exist to manage solid wastes.

The standard of living of North Americans is highly correlated to the generation of solid wastes. Some segments of society promote wastefulness and rapid obsolescence. Likewise, it is difficult to change habits, especially when increased consumption is glamorized. And, historically, solid waste management has had a tradition of low cost, and residents have not been charged in accord with their use. In economic terms, this is referred to as a "public good": People are not encouraged to use less because the costs of their individual usage do not determine their billings.

Nevertheless, changes in the management of solid wastes are going to occur. Of interest are the changes that have occurred in the last few years and what might realistically be expected in the next few decades. There is no question that adjustment will take place. Legislated and/or proposed regulations have greatly improved standards of landfill design, performance, and monitoring to protect people and the environment. Costs of solid waste management are increasing, as are societal pressures for better handling of solid wastes.

Future solid waste management plans will include resource-conservation and separation programs to reduce the creation of waste, resource-recycling programs to reprocess wastes into useful goods, incinerator technologies to reduce the volume of waste, and new landfill design and operation technologies to dispose of residential waste in an environmentally sound manner (O'Leary et al., 1988).

Current methods of disposal include direct landfilling in regions with no land shortage, use of highly sophisticated incinerators, and landfills that include bailing and shredding in other locations. Disposal never means total disappearance but only transfer from an inconvenient state to a more convenient state for the short term.

Alternatives to landfilling depend on the form of the refuse. To minimize reliance on landfills, regulating agencies, with strong support from the public, are now stressing the need for the 3Rs programs. For example, some of the more readily biodegradable organic fraction (food waste, paper, cardboard and yard waste) may be collected separately from other wastes and composted or sold to be recycled. Paper is largely recyclable and is a good fuel for waste-to-energy plants. Many yard wastes can be composted. In addition, certain waste components, such as plastic, glass, and metals, can be collected for recycling. Source separation of refuse into its components facilitates different types of reuse and recycling. The remainder of municipal wastes may be commingled wastes (textiles, rubber, leather, wood and inorganic wastes, crockery, and soil).

Another concern arising from solid waste generation is the presence of hazardous materials in the waste stream. In 1989 alkaline batteries accounted for 88 percent of the mercury found in municipal refuse. Manufacturers have removed most of the mercury from alkaline batteries and plan to eliminate it completely by 1998 (Austin, 1991).

Hazardous wastes are those materials that pose a potential hazard to human health or living organisms. Hazardous wastes arise from numerous residential, industrial, and commercial sources. Residential sources include bleaches, cleaning fluids, insecticides, gasoline, used oil, paints, and paint thinners. Commercial sources include inks from print shops, solvents from cleaning establishments and cleaning solvents from auto repair shops. This variety of hazardous wastes means that a range of alternatives for waste management must be considered. Some of these wastes may be used as raw material providing the opportunity for recycling, but the matching of wastes to raw materials is a significant timing, transformation, transportation, and storage problem.

Table 2.12 EXAMPLES OF SOURCE REDUCTION STRATEGIES

- ❖ Purchase products with minimal packaging.
- ❖ Reduce the amount and/or toxicity of the wastes that are now generated.
- ❖ Reduce the quantity and cost associated with its handling and environmental impact.
- ❖ Reduce waste by designing, manufacturing, and packaging products with minimum toxic content, minimum volume of material and/or a longer useful life.
- ❖ Develop and use products with greater durability and repairability.
- ❖ Substitute reusable products for disposable single-use products.
- ❖ Use fewer resources (e.g., make two-sided photocopies).
- ❖ Increase the recycled material content of products.
- ❖ Develop rate structures that encourage generators to produce fewer wastes.
- ❖ Maintain a compost pile.

Five basic strategies of solid waste management are summarized in the following list.

1. *Source Reduction*   Reducing the quantity of material requiring disposal can take a variety of forms, a number which are listed in Table 2.12. The intent is to avoid having to deal with the waste by not generating it. Up to a 25 percent reduction of the per capita waste generation rate may be achievable through increased awareness and commitment by the public and industries. However, once solid wastes are generated, society must deal with the material. This is the focus of the remaining four strategies.
2. *Reuse*   Reuse means finding another or similar use for a product rather than discarding it. Examples include refurbishing of old appliances for resale, reupholstering old furniture, and repairing old automobiles.
3. *Recycling*   Recycling involves the changing of the material into another usable form. Resource recycling is the collection, separation, and reclamation or composting of wastes. Examples include the use of de-inked newsprint in paper manufacturing, and the reclamation of glass or the recovery of metal and plastic containers for the secondary materials market. More than 75 percent of municipal solid waste in the United States is potentially recyclable. As will be apparent from the paragraphs to follow, the best place to separate out materials for recycling and reuse is at the source, before the solid wastes are mixed or commingled.
4. *Energy-from-Waste Incineration*   Capturing the heat energy from incineration of refuse is a subset of (3), but there is not necessarily any attempt to

separate the material prior to mass burning. A combination of recycling and incineration is feasible (if some of the material is separated out prior to burning). However, there is also a conflict between the two strategies: commercial recycling pays for plastics and newspapers, yet they can also be burned.

   5. *Landfilling*   Regardless of other options, wastes that cannot be resold as recyclables, incineration ashes, and the like will ultimately be disposed of in a landfill. The characteristics of the refuse stream vary from municipality to municipality due to many factors including the character of the municipality, the affluence of the area, and the type of industry present, but landfills will continue to be needed.

   Clearly, no single strategy will be used in isolation. The solid waste management field is seeing gradual changes in the mix of the strategies. Wastes will be handled differently in the future than they have been in the past. There is an increasing recognition of the need to involve the population and have them willing to participate. Some of this change is being mandated by legislation, and some is a result of environmental concern. The challenge is to change a solid waste management system now largely dependent on landfills to one in which integrated resource recovery is the predominant technology.

   The focus of this text is on landfilling. In this chapter we examine some of the alternatives that exist for decreasing the volume of material being landfilled and how changes in the quantities will affect landfilling practices. We will look at the changes that are occurring in solid waste management thinking. What is happening now? What will happen as a natural outgrowth of current attitudes? What might happen in the future? An example of alternative product flows is illustrated in Figure 2.2.

2.3.2 Examples of the Utilization of the 3Rs

   The 3Rs approach is directed toward achieving a lower per capita waste disposal rate by various reduction strategies. The goal for waste recycling in Ontario, Canada, by the year 2000 has been set at 10 percent reduction, 15 percent reuse and 25 percent recovery (Crutcher and Yardley, 1991).

   The recent trend in the waste disposal industry is being fueled by different viewpoints which have the following goals:

   • Reduce the amount of waste material produced.
   • Reduce the quantity of natural resources that are utilized by the manufacturing sector.
   • Reduce the volume of material to be landfilled.

   The first two goals are the basis for the 3Rs approach. The third goal and possibly the second are the basis for the bulk waste volume reduction initiatives. However, what opportunities exist?

The concept of complete recycling is desirable but not feasible. There are different approaches to making partial recycling simpler. For example, separating municipal wastes at the source is feasible to a limited extent. Alternatively, separation can take place after collection. The difference between these two approaches has substantial implications. Once the materials are mixed together, it is difficult to separate them and end up with a clean or noncontaminated product that can be sold. When the materials are presorted at pickup, they can be utilized more efficiently, but separating the materials at the source and maintaining them have associated high costs (e.g., in the form of residents' time and initiative and in terms of collection vehicle costs, which are approximately

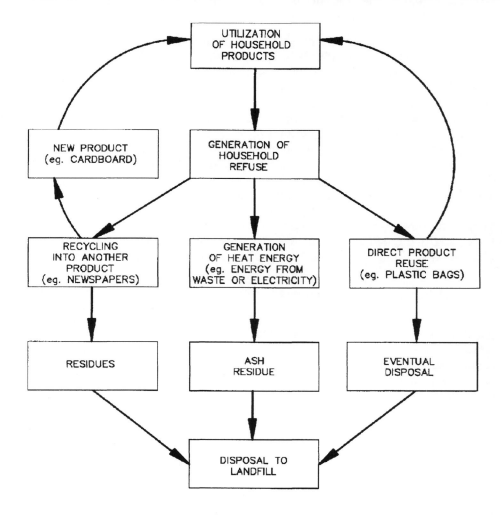

Fig. 2.2 Schematic of example product flows.

$75,000/vehicle for multiresidential collection).

Various means of source separation are utilized. Source separation may involve curbside collection, drop-off, and buyback centers. Residents may sort recyclable materials into color-coded containers. Garbage collectors then put the recyclable materials into a special collection vehicle with appropriate bins for each of the materials. In another option residents separate their refuse into wet compostable and dry recyclable components.

Separation of solid waste components at the household level has had considerable success. Separating wastepaper, cardboard, aluminum cans, glass, and plastic containers at the source of generation is proving to be one of the most positive and effective ways of achieving the recovery and reuse of materials. However, there are still people who, due to the inconvenience, simply mix the recyclables with their garbage. The need for continued public awareness programs to change the attitudes of these people is evident. In some jurisdictions, separation is compulsory; otherwise there is no garbage pickup.

In some recycling programs residents use only one container to store recyclable materials and the operator of the collection vehicle separates them into the appropriate bins. A variety of specialized collection vehicles have been developed to facilitate this curbside separation. Alternatively, homeowners may drop off the materials at collection centers. These facilities range from single-material collection points (e.g., igloo containers) to sophisticated facilities that receive multiple goods. The approaches involving greater resident participation have historically had low participation rates.

Examples of materials commonly being separated by one or more of these methods are listed in Table 2.13. The recycling opportunities for various types of plastics are described in Table 2.14. A resin-coding system has been designed to assist in separating the plastics.

Table 2.15 gives examples of ongoing initiatives for waste reduction. From many of the initiatives included in Table 2.15 it is apparent that a fundamental decision is needed at the municipal level to advance materials recovery. Such decisions can include the option to require individual homeowners to separate their own refuse (for curbside pickup or drop-off and/or buyback facilities) or, alternatively, to forego source separation and use materials recovery facilities (MRFs) to separate the commingled wastes.

The concept of MRFs or comprehensive recycling of garbage involves processing the refuse through a complex line of separating machines or hand sorting. The quality of the recovered material is lower than that recovered at the source because of contamination by refuse, especially when the sorting is done mechanically. With commingled wastes, a high recovery percentage requires multiple sifting through the refuse and complete opening of all garbage bags. Table 2.16 lists some of the problems of MRFs. A third option is a combination involving the separation of some solid waste components including wastepaper, cardboard, aluminum cans, glass, and plastic containers at the source of generation followed up by a MRF. The sorting of the dry fraction of the waste stream is best accomplished in a modified MRF. Here some manual sorting is required but the majority of the separation functions are accomplished by mechanical means

Table 2.13 EXAMPLES OF MATERIALS CURRENTLY BEING SEPARATED

- ❖ Aluminum — aluminum cans and secondary aluminum (window frames, storm doors). The demand for recycled aluminum is high, as it takes 95 percent less energy to produce an aluminum can from an existing can than from ore

- ❖ Paper — old newspaper, cardboard, high- and low-grade paper, kraft paper, box board

- ❖ Plastics — of two types
  clean commercial-grade scrap and postconsumer scrap, e.g., polyethylene terephthalate (PET)—used for the manufacture of soft drink bottles
  high-density polyethylene—used for milk and water containers and detergent bottles. Currently less than 5 percent of available scrap plastic is recycled.

- ❖ Glass — container glass (for food and beverage packing), flat glass (e.g., for windows) and amber and green glass

- ❖ Ferrous metals — e.g., from cars and appliances. Steel cans can easily be separated from mixed recyclables

- ❖ Yardwastes — leaves and grass clippings can be used for composting and mulching

- ❖ Construction & demolition wastes — e.g., scrap lumber can be used to create wood chips for use as a fuel in combustion facilities (Recycling of this class of materials is increasing as tipping fees increase.)

- ❖ Wood wastes — e.g., wood pallets from industry can be shredded and burned

Table 2.14 RECYCLING OPPORTUNITIES FOR VARIOUS PLASTICS

Plastic Type	Example
Polyethylene terephthalate (PET, PETE)	Soft-drink bottles; frequently recycled
High-density polyethylene (HDPE)	Milk and detergent jugs; frequently recycled
Polyvinyl chloride (PVC)	Some shampoos and other similar products; rarely recycled; burning produces toxic gases
Low-density polyethylene (LDPE)	Plastic film and wrap; rarely recycled
Polypropylene (PP)	Food lids and containers; rarely recycled
Polystyrene (PS)	Food containers and foam burger boxes, hot drink cups; occasionally recycled from schools, cafeterias; and restaurants
Mixed Resins	All other plastic resins; rarely recycled

Table 2.15 EXAMPLES OF ONGOING INITIATIVES FOR WASTE REDUCTION

❖ Use of recyclable products by the public and industrial consumers, including reuse of items such as plastic containers, shopping bags, and gift wrap

❖ Recycling of paper, plastic, glass, and metal through local landfills and recycling centers. Seventy-five thousand trees are used to produce one week's edition of a major Sunday newspaper. Glass accounts for approximately 8 percent of municipal refuse. Glass containers are very recyclable. Recycling one glass jar saves enough energy to light a 100-W bulb for 4 hours. Plastics make up approximately 7 percent of our refuse and can be recycled or are good fuels for energy-from-waste plants

❖ Arranging for industrial wastes from one plant to be used as raw materials for a process in another plant

❖ Concentrating on industrial recycling, because industry generates approximately 60 percent of the total waste. To reduce, or generate less waste, plants must usually change their production process, modify equipment, or improve waste management practices. For example, modernization prompted by stringent federal air and water pollution control has reduced incidents of mercury poisoning. In the wake of regulations, engineers develop a new process to eliminate the hazardous wastes. It may involve substitution of other, less hazardous material

❖ Reducing waste by having, developing, and using products with greater durability and repairability; substituting reusable products for disposable single-use products

❖ Increasing the recycled materials content of products

❖ Developing rate structures that encourage waste generators to produce less waste

❖ Finding ways to recycle many metals, which account for approximately 9 percent of municipal refuse; some contain toxics such as lead and cadmium that are troublesome in landfills or incinerators. More than one-half of aluminum cans are currently recycled. Aluminum can crushers reduce their volume, thus reducing the handling and shipping costs

❖ Developing initiatives to recycle the food, rubber, leather, textiles, wood, and other wastes that account for approximately 17 percent of the total waste and are not easily recycled (e.g., Skajaa (1989) and Spencer (1989))

❖ Eliminating hazardous components of waste. Many of the hazardous inorganic constituents present in refuse, such as mercury, arsenic, selenium, and lead can be converted biologically into a variety of compounds, some of which are toxic (e.g., under anaerobic conditions ethylmercury, dimethylarsine, nitrosamines are produced). The most effective way to eliminate the small quantities of hazardous waste is to separate them at the point of generation of the refuse. These hazardous constituents are now being collected by many municipalities on special collection days

❖ Decreasing the amount of residential solid wastes. Although residential solid wastes account for a relatively small proportion of the total wastes generated in the United States (10 to 15 percent), they are perhaps the most important because they are generated in areas with limited storage space that is rapidly infested with vermin and insects. If stored for lengthy times, containers with tight lids should be used and food waste should be ground and discharged to the sewer system. Food wastes decreased from 14% of solid wastes to 9% from 1960 to 1992 because of garbage grinders. In some states, composting of leaves is legislated. In approximately 1 year, bacterial and fungal decomposition occurs, and the leaves can become a useful soil amendment

❖ The projections of the saving by composting leaves are 2 percent of curbside wastes. Excluding yard wastes and putrescibles from the refuse (as a result of composting) may be of little significance in humid regions, where the majority of moisture entering the landfill is a result of precipitation. In arid areas, however, excluding such wastes may have a significant impact upon the moisture levels present within the landfill

Table 2.16 CONSIDERATIONS IN MATERIALS RECOVERY FACILITIES

❖   Facilities have encountered considerable escalation in cost and technological capabilities.

❖   The technical problems are being overcome in many cases, and the cost picture may improve.

❖   Often the more central (industrial) location of MRFs saves haulage cost and energy in comparison with a landfill.

(shredding, trowelling, separation, screening, etc.).

There is also considerable merit in source separation of household hazardous wastes prior to collection, since hazardous wastes can significantly detract from the success of a MRF. Although small in quantity, household hazardous wastes can contaminate commingled wastes. The key to diverting these wastes is ensuring that these products are kept for specific collection days. The question then becomes, Will the householder store the material until that time? If not, separation of these materials from the commingled waste is considerably more difficult. Major waste components may be contaminated by small amounts of wastes such as motor oils, household cleaners, and paints. This contamination reduces the value of the individual components for recycling.

2.3.3 Composting

A rapidly expanding part of resource recycling is the composting of food wastes and yard wastes (approximately 18 percent of domestic refuse can be composted). Yard and food wastes are collected and put into piles or windrows, where they decompose aerobically and are greatly reduced in volume. Composting can potentially reduce the organic content of the refuse to be disposed of (Kashimanian and Taylor, 1989), and the product can be used as an additive for lawns and gardens.

When quantities are small, composting by static piles can be undertaken without significant environmental control to regulate decomposition rates. In larger, more complex operations, under-air is sucked into the composting material to optimize the aerobic digestion process. In both cases, water and nutrients are added to promote the composting process.

In-vessel composting involves the establishment of a highly regulated environment for the composting biomass in a totally enclosed vessel or simply a concrete trough with the top of the composting material exposed. This composting process is highly regulated with the addition of air and regular turning cycles utilizing specially designed mountings and turning equipment.

The production of compost by windrows or static piles can take from 6 months to 1 year depending on the environmental controls applied and the composition of the organic fraction. Compost can be produced from an in-vessel unit in as little as 2 weeks. The steps involved in composting residential organic wastes are summarized in Table 2.17.

Table 2.17 COMPONENTS OF YARD WASTE PROGRAM

Step 1   Leave grass clippings on the lawn. Avoid the inconvenience and cost of bagging grass clippings. Mow the lawn regularly, cutting off only one-third of the grass blade height

Step 2   Compost yard waste. This diverts yard waste from landfills while creating a rich fertilizer for gardens.

Step 3   Have leafy yard waste picked up separately for composting. If leaves are not used for home composting or mulching, rake them to curbside to be picked up for central composting.

Table 2.18 FEATURES OF COMPOSTING

❖   Composting returns to the land in an environmentally acceptable form that which has been taken from the land (i.e., raw materials).

❖   Composting does not remove or alter those elements in refuse such as heavy metals created by manufacturing processes that under certain conditions can be harmful to health. Discriminating use of compost is therefore necessary.

❖   Compost is a soil improver, not a fertilizer, such as manures and chemical fertilizers.

Some of the favorable attributes of composting are listed in Table 2.18. Compostable yard wastes are generally too moist to burn efficiently; thus, it is best to keep them out of the refuse if the refuse is to be used in mass burning.

The trend to composting has been strengthened in eight states in the United States that have imposed a ban, varying in extent, on the disposal of yard waste at disposal facilities. Yard waste materials may be taken to disposal facilities only if they are destined for use as daily cover (Glenn 1989, 1990).

In terms of impact on the landfills of the future, the removal or at least reduction of yard wastes in landfilled refuse will decrease leachate strength and landfill gas volumes produced in the early years, because decomposable organic yard wastes break down rapidly. The more resistant materials generally decompose over a much longer period of time; therefore, this biodegradation component will not change much. The potential for contaminants in this compost will continue to be a concern.

2.3.4 Incineration

Incineration, with or without energy recovery, is the second largest means of waste disposal in the world. The main problems created by the incineration of municipal wastes are air pollution and the disposal of ash residues. A century of experience has been collected, although not all this experience has been desirable. Originally, incinerators exhausted all combustion gases and particles directly to the atmosphere, but the air pollution laws of the 1970s in North America curtailed those practices. As a result, many older incinerators have been shut down. Most experience with incinerators has been gained in western Europe (where 23 percent of the refuse is incinerated (Brussef and Rocherolles, 1979) and Japan (where 60 percent of the refuse is incinerated (Sugito, 1979)). In the European Community, over 80 percent of the 500-plus incinerators have resource recovery facilities. Switzerland burns more than 70 percent of its MSW, and Sweden has 25 energy-from-waste incinerators, which handle over 80 percent of its municipal wastes (Warmer, 1990). In the United States there are energy-from-waste facilities operating in 28 states, managing 29 million tons, or 16 percent, of the country's MSW (Austin, 1991). These plants produce 2100 MW of electricity. Another 100 plants are in various stages of planning or construction.

In modern refuse-derived fuel plants, incineration occurs at extremely high temperatures (1800 to 2000°F) for long enough to minimize the release of pollutants. A study by Oak Ridge National Laboratories showed that 99.3 percent of the intake of dioxins and furans by individuals residing close to an incinerator was from background sources, and not the MSW incineration facility (Council on Plastics, 1989).

The combustion of unprocessed solid waste releases 10,000 to 17,000 J/g. (Coal releases about 23,000 J/g per gram.) It is noteworthy that recycling reduces the energy from the waste to about 70 percent of these levels (due primarily to the recycling of cardboard, which has high energy recovery).

Refuse may be thought of as consisting of moisture, dry combustibles, and noncombustibles. The major origin of the dry combustible portion is plant life. The dry combustible in such items as paper, wood, natural textiles, vegetable wastes, brush, and leaves is largely cellulose ($C_6H_{10}O_5$). Cellulose has a calorific value of 17,400 J/g. The second major class of dry combustibles consists of hydrocarbons, fats, oils, waxes, resins, synthetics (plastics and textiles), rubber, etc. They have high calorific values ranging up to 44,000 J/g and averaging about 37,000 J/g. Moisture contributes no heat units but absorbs extensive heat on evaporation in the furnace. Food waste and greens are high in moisture—about 75 percent when fresh.

Many recyclable materials, including glass bottles and aluminum cans, do not burn well. Incineration for resource recovery does not compete for compostable yard wastes either, since they are too moist to burn efficiently.

An alternative practice in some locations involves separation of the solid waste into noncombustible and combustible components. The noncombustible materials are sent to a landfill and the combustibles, referred to as refuse-derived fuel (RDF), are sent to refuse-derived fuel plants (as opposed to mass burn plants, where waste is burned in bulk). Here the waste is shredded or processed into a uniform fuel after the recyclables are removed. This process creates a fuel with an energy value similar to that of high-grade lignite coal. The heating value of components of municipal solid waste in terms of potential for energy recovery are given in Table 2.19.

Table 2.19 TYPICAL ENERGY CONTENT OF MSW COMPONENTS

Constituent	Btu/lb Constituent
Food waste	
Garbage	6,484
Fats	16,700
Rubbish	
Paper	7,572
Leaves	7,096
Grass	7,693
Street sweepings	6,000
Wood	8,613
Brush	7,900
Greens	7,077
Dirt	3,790
Oils, paints	13,400
Plastics	14,368
Rubber	11,330
Rags	7,652
Leather	8,850

Source: MITRE (1975)

Modern incinerators operate more cleanly and more productively than ever before. Semi-suspension burning is the most efficient method of incinerating wastes. The pulverized refuse is swept by air across the boiler, burning as it falls toward the bottom gate. Natural gas is used to preheat the boiler air to approximately 450°F (Kennedy and Palombella, 1979).

The incinerators can reduce large quantities (volumes) of garbage to ash while producing steam or electric energy. Dry ash is roughly 20 to 25 percent by weight of the unburned waste. Incineration is desirable in that it conserves landfill capacity by volume reduction. Burning of refuse obviously reduces but does not eliminate the need for landfill space (garbage once burned is reduced in volume to about 40 percent of its precombustion state). Incineration extends the estimated landfill life by a factor of 3 to 12 (Legrand and Chynoweth, 1989). The leftovers from incineration include waste in the form of fly ash and bottom ash. The ash that settles on the bottom grate of the incinerator chamber is called bottom ash, comprising 90 percent of incinerator ash. The remaining 10 percent is fly ash, which is captured by the emissions controls that filter the flue gases.

Incinerators may also produce a lot of other organic compounds as byproducts, including dioxins, furans, PCBs, PAHs, chlorinated benzenes, and chlorinated phenols. Because of the sorption characteristics of fly ash, many of these compounds are absorbed by the fly ash. Fly ash contains a much greater concentration of metals in a mobile form (leachable by water) than bottom ash. Thus, air pollution controls are essential, and battles over health risks and site location exist and will continue. Air pollution persists as the major concern for incinerator technologies (O'Leary et al., 1988).

For decades, nearly all the fly ash and bottom ash was buried at "secured" (hazardous waste) landfills to minimize the hazards of salts and heavy metals abundant in the material. Recently, however, the technology has become available to stabilize and possibly reuse incinerator ash as a road base aggregate. Options for ash stabilization include vitrification, in which ash is melted, cooled into a glasslike substance, and spun into insulating material, and solidification, in which an adhesive material is added to the ash to produce a concretelike substance. Solidification, the least expensive and simplest ash stabilization technique, essentially locks in trace metals present in the ash, resulting in an inert, stable solid (Woods, 1991).

Another potential use for incinerator ash is in road construction. As early as 1975, ash aggregate was approved for use in Houston, Texas, in highway construction. The material, called "littercrete," consists of 89 percent ash aggregate, 9 percent asphalt, and 2 percent lime. Topped with asphalt, the roadbed has exceeded performance standards for conventional road materials over the past 16 years (Woods, 1991).

One last comment about incineration is relevant in regard to greenhouse gases. The methane gas produced by anaerobic decomposition of wastes in a landfill (to be more fully discussed in Chapters 4 and 5) is 30 times more harmful in terms of the greenhouse effect than the carbon dioxide released by the incineration of wastes.

2.4 LEGISLATION

At the federal level in the United States, the Resource Conservation and Recovery Act of 1976 (RCRA) amended by the Solid Waste Disposal Act Amendments of 1980 and the Hazardous and Solid Waste Amendment of 1984 (HSWA) are the primary sources of solid waste legislation. Subtitle D of RCRA deals with nonhazardous solid waste disposal and requires the development of a state comprehensive solid waste management program that outlines the authorities of local, state, and regional agencies. State and local legislation is continually evolving and varies from state to state and thus will not be addressed in a formal sense in this text.

The transition from a consumer society to a conserver society is not an easy feat. Although in general the public feels obliged and even eager to contribute to the solutions to excessive solid waste production, a regulatory impetus is often required to initiate a positive response. The need for legislation exists because voluntary programs have often failed to achieve long-term participation, including both citizen participation and the ability to create permanent change. For example, many citizens have complained that the consumer/citizen did not ask for more packaging, or more disposable products, or lower quality goods that need replacing more frequently. As a result of these types of considerations, legislation is being imposed to force the adoption of the 3Rs. Examples of regulatory methods to promote reduction, reuse, and recycling are listed in Table 2.20.

Regulations in response to identified needs and uniformity are still evolving. In the past, government regulations have sometimes deterred waste generators from reducing waste by making treatment and disposal options more attractive. However, more recent legislation has been directed specifically at waste reduction. For example, in California the Integrated Waste Management Act mandates a 25 percent diversion from the waste stream by 1994 and 50 percent by the year 2000.

Many mandatory and voluntary programs divide the responsibility for sorting and collecting recyclable materials between homeowners and municipalities. In addition, enforcement bans are being invoked at different landfill sites, for example, against corrugated cardboard, wood waste, asphalt, tires, and wood pallets. A major concern is that mandatory programs will achieve greater participation but will also result in public antagonism and illegal private dumping.

2.5   ECONOMIC ASSESSMENTS

As an example of the changes that are taking place in solid waste management, two decades ago solid waste collection costs amounted to approximately 60 percent of the total annual cost of solid waste management. However, with increasing costs of the 3Rs and landfilling, this percentage is rapidly changing. Of interest is how these increased costs will be paid for.

Unlike many features of modern society, solid waste disposal costs have not been charged in accord with use. For example, many communities charge for water costs in accord with usage by metering consumption—such an approach encourages conservation. However, weighing solid wastes at curbside is time consuming and has not been employed. Some attempts have been made to restrict the number of bags of refuse that will be picked up, but the restriction on quantities has frequently been met by substantial increases in illegal dumping.

Some administrators have assumed that recycling programs will pay for themselves with the revenue generated by the sale of materials. Unfortunately, the demand for recycled materials changes monthly, and the revenue is not covering the costs of collection and processing. The market for recycled goods is generally unstable; the prices are elastic, meaning that the prices change in response to supply. Nevertheless, to pay a processing facility to accept the material may be less costly than the cost of landfilling. In the northeast United States the costs are $30 to $75/tonne for incineration, whereas the tipping fees may

Table 2.20   EXAMPLES OF REGULATORY METHODS USED TO PROMOTE REDUCTION, REUSE, AND RECYCLING

Product recycling	Where recycling opportunities are in place, product recycling at the landfill has been effected to direct that segment of the waste stream to recycling (i.e., cardboard, tires, glass).
Packaging regulations	Although excessive packaging can be considered a controversial or even worldwide problem, some authorities have initiated strict packaging regulations that affect not only internal packaging but imports and exports as well.
Mandatory solid waste management plans	There is a growing focus on the ICI responsibility for waste reduction. For example, the Province of Ontario has recently implemented a regulation requiring all larger industries and commercial enterprises to institute solid waste management audits.
Recognition of environmental responsibility	To recognize and reward efforts of manufacturers to reduce the environmental impact of their products, the government of Canada has instituted a "Green Seal" that is affixed by manufacturers to their products following rigorous approval procedures.
Purchase guidelines	To promote the purchase of products incorporating recycled materials, many larger purchasing agencies have stipulated a minimum recycled content in certain product specifications.
Payment for collection and disposal services	Although most solid waste collection and disposal costs are recovered through the municipal tax levies, some jurisdictions are attempting to reduce the curbside waste stream by instituting a levy on each refuse bag collected. This is done through sale of the garbage bags or tags. The revenue from the sale of the bags or tags is used to ensure program cost recovery.
Excessive tipping fees	To encourage ICI participation in waste stream diversion from landfill activities, some municipalities have instituted extraordinary tipping fee schedules. Although this effort usually has the desired effect in the jurisdiction in question, it often means diversion of the waste stream to other landfills outside the region.
Container deposit programs	These programs, common in the soft drink and brewing industries, are being expanded in some jurisdictions to include a wide range of containers. This movement is opposed by the soft drink and distilling industries.

Fig. 2.3 Wood chipper in use at Yolo County landfill.
(From R. Allan Freeze/John A. Cherry, GROUNDWATER, ©1979. Reprinted by permission of Prentice Hall, Englewood Cliffs, New Jersey.

reach $125/tonne at a landfill.

One approach being utilized at some landfill sites is the use of higher tipping fees to divert wastes from landfilling. Tipping fees have also become a tool to discourage disposal of refuse from outside the collection area. In many cases, fees are appreciably higher than the actual cost of providing the sanitary landfilling to encourage the 3Rs, but unless the tipping fee is uniformly widespread, the result may be simply a diversion of wastes to other sites.

The volume of garbage could be considerably reduced if industries such as the plastics industry, newspapers and newsprint suppliers, and grocery products manufacturers and distributors adopted a returnable deposit concept similar to that used in the soft drink industries. However, the administration of these additional programs is not likely to be simple.

Industrial recycling is much different from residential recycling: wastes are typically much more uniform, and costs of management of the solid wastes are much easier to monitor. As a result, it is often possible to arrange for recycling (e.g., haul loads of wood to wood grinders). Wood chippers such as that depicted in Figure 2.3 are used to shred large pieces of wood (e.g., large branches, yard wastes, wood pallets) into chips that can be used as fuel, composted, or used as daily cover material. The tub grinder shown has a revolving upper section and a stationary lower section containing a hammermill.

2.6 IMPLICATIONS FOR FUTURE LANDFILLING

If the residents of the United States and Canada were not part of an advanced industrial society, it is doubtful that either the government or the public could be induced to pay much attention to pollution. Concern with pollution is a luxury in the sense that a nation or an individual who is preoccupied with obtaining sufficient food, clothing, and shelter for survival will have neither the time nor the inclination to worry about pollution, except in those cases where it is an obvious and imminent threat to public health. As important as the pollution problem may be, it is less important than the obvious prerequisites for survival.

On the other hand, the conditions of life in an affluent society tend to contribute to a concern with pollution. The greater amount of leisure time enjoyed by the population leads to a greater demand for recreational resources and aesthetic satisfaction, and the higher level of education enables people to better comprehend the dangers and dynamics of pollution.

Due to the affluence of society in the United States and Canada, there is a recognition by society of the need for good disposal practices. Thus, the initiatives discussed in the previous sections of this chapter are beginning to change the characteristics of the landfill waste stream, which will affect the composition of the material within the landfill (Crutcher and Yardley, 1991). The changes in the organic and inorganic fraction of the refuse will influence the design and operation of landfills.

In this chapter we established the basis of the existing components of refuse and indicated the changes in composition and quantities that may occur in the short-term future. As we stated in Chapter 1, this text focuses on the implications of landfilling of solid wastes. However, as apparent from the trends discussed, the nature of future refuse will be changing over time. Therefore, in the chapters to follow we focus on the principles and practices of current landfilling and discuss the implications of future changes in the refuse stream.

As will become apparent in the chapters to follow, modern landfills and energy-from-waste plants have little in common with the open dumps of the past. Waste management considerations are beginning to be captured prominently in product design, as public pressure and the substantially increasing costs of solid waste management force industry to invest in environmentally friendly products.

The principles developed herein are intended to assist in designing new sites and remediating old sites. There will always be a residue that is nonrecyclable, noncombustible, and noncompostable. Landfilling will continue to be essential, but increased societal efforts will be directed toward implementation of the 3Rs. It is expected that through the 3Rs programs there will be a significant reduction in waste quantities produced as well as a significant fraction of waste being diverted from disposal at landfills.

References

Applied Management Services. 1973. *The Private Sector in Solid Waste Management: A Profile of its Resources and Contribution to Collection and Disposal.* U.S. Environmental Protection Agency Report SW-51d.1. Washington, D.C.

Austin, T. 1991. Waste to Energy? The Burning Question. *J. Civil Engineering,* October, pp. 35-38.

Black, R.J., Muhich, A.J., Klee, A.J., Hickman, H.C., and Vaughan, R. 1968. *The National Solid Wastes Survey: An Interim Report. NTIS Report PB260102.* U.S. Department of Commerce, NTIS Report. PB260102, Washington, D.C.

Brussef et Rocherolles. 1979. Faculté des Sciences de l'Université de Paris.

Chemecology. June 1989(a). *Household Waste,* A Chemical Manufacturers Association Publication. Vol. 18, No. 5, Washington D.C.

Chemecology September 1989(b). *Our Garbage Problem Won't Go Away By Itself.* A Chemical Manufacturers Association Publication, Vol 18, No. 7, Washington D.C.

Chian, E., and Dewalle, F.B., 1977. *Evaluation of Leachate Treatment.* U.S. Environmental Protection Agency Report EPA 600/2-77/186a and b, Vols. 1 & 7. Cincinnati, Ohio.

COUNCIL ON PLASTICS AND PACKAGING IN THE ENVIRONMENT. 1989. *Incineration and Energy Recovery: An Environmentally Sound Approach to Solid Waste Problems.* April. Washington, D.C.

CRA Ltd. 1991. *Refuse Volume Forecast for the Region of Waterloo.* Report to the Regional Municipality of Waterloo, Ontario, Canada.

Crutcher, A., and Yardley, J. 1991. *Implications of Changing Refuse Quantities and Characteristics on Future Landfill Design and Operations.* Ed. by M. Haight, Municipal Solid Waste Management. University of Waterloo Press, Waterloo, Canada.

EMCON Associates. 1975. *Sonoma County Solid Waste Stabilization Study.* EPA/530-SW-65d.1. U.S. EPA Office of Solid Waste Management Programs, Washington, D.C.

EMCON Associates. 1980. *Methane Generation and Recovery from Landfills.* Ann Arbor Science Publishers, Michigan.

Franklin, M. 1986. *Characterization of Municipal Solid Waste in the United States, 1960–2000.* U.S. EPA Report. July.

Fungaroli, A., and Steiner, R. 1979. *Investigation of Sanitary Landfill Behavior* Volume 1, Final Report. U.S. Environmental Protection Agency Report - 600/2-79-053a.

Glenn, J. 1989. Regulating Yard Waste Composting. *J. Bio Cycle,* 30(12), December, pp. 38–41.

Glenn, J. 1990. Impact of Statewide Yard Waste Disposal Ban. *J. Bio Cycle, 31(8),* pp. 32–35, August.

Goleuke, C.G., Mcgauhey, P.H. 1970. *Comprehensive Studies of Solid Waste Management.* NTIS Report PB218265. U.S. Department of Commerce, Washington, D.C.

Gore and Storrie LTD. 1991. *Ontario Waste Composition Study.* Report to the Ontario Ministry of the Environment, Waste Management Branch. Vol. 1, Toronto, Canada.

Ham, R., et al., 1979. *Recovery, Processing and Utilization of Gas from Sanitary Landfills.* U.S. Environmental Protection Agency Department EPA-600/2-79-00, February.

Kashimanian, R.M., and Taylor, A.C. 1989. Costs of Composting Yard Wastes vs. Landfilling. *J. Bio Cycle, 30(10),* October, pp. 60–63.

Kennedy, J.W., and Palombella, R.F. 1979. "Blueprint for the Future." Paper presented at Energy from Waste Conference. Toronto, Ontario. November pp. 4–8

Legrand, R., and Chynoweth, D. 1989. "Digestion as a Solid Waste Management Tool." 12th Annual Madison Waste Conference, Madison, Wisconsin. September 20–21

Lutton, R.j., Regan, G.l. And Jones, L.W. 1979. *Design and Construction of Covers for Solid Waste Landfills.* U.S. EPA Report 600/2-79-165, August.

Matrix Management Group. 1989. *Waste Stream Composition Study 1988–89*, Final Report. Dept. of Engineering, Solid Waste Utility, Seattle. June.

Matrix Management Group, R.W. Beck & Associates, and Gilmore Research Group. 1988. *Best Management Practices Analysis for Solid Waste,* Vol. 1, 1987. *Recycling and Waste Stream Survey,* Second Edition. Washington Dept. of Ecology, Washington. December.

Mcginley, P., and Kmet, P. 1984. *Formation, Characteristics, Treatment and Disposal of Leachate from Municipal Solid Waste Landfills.* Special Report, Bureau of Solid Waste Management, Wisconsin Dept. of Natural Resources, Madison.

Mitre Corporation. 1975. *Energy Conservation Waste Utilization Research and Development Plan.* Report MTR-3063. July.

Mlynarski, H.D., and Northrop, G.M. 1975. *Derivation of Residual Coefficients for Typical Polluting Industries in New England,* NTIS Report PB258996. Prepared for the Science Foundation, Research Applied to National Needs Program, U.S. Dept. of Commerce, Washington, D.C.

National Center For Resource Recovery. 1973. *Municipal Solid Waste; its Volume, Composition and Value,* National Center for Resource Recovery Bulletin, Vol. 3, No. 2, Washington, D.C.

Niessen, W., and Alsobrook, A.F. 1972. "Municipal and Industrial Refuse Composition and Rates." *Proceedings: 1970 National Incinerator Conference, American Society of Mechanical Engineers,* New York, pp. 319–337.

O'Leary, P.R., Walsh, P.W., and Ham, R.K. 1988. Managing Solid Waste, *Scientific American,* Vol. 259, No. 6, December, pp. 36–42.

Rhyner, C.R., and Green, B.D. 1988. The Predictive Accuracy of Published Solid Waste Generation Factors. *Waste Management and Research,* Vol. 6, pp. 329–338.

Rovers, F.A., and Farquhar, G.J. 1973. Infiltration and Landfill Behaviour. *ASCE—J. Environmental Engineering Division,* 99, EE-9, pp. 671–690.

Skajaa, J. 1989. Food Waste Recycling in Denmark. *J. Bio Cycle, 70,* November, pp. 70–73.

Smith, F. 1975a. *Comparative Estimates of Post Consumer Wastes,* NIIS Report PB 256491. U.S. Department of Commerce, Washington, D.C.

Smith, F. 1975b. *A Solid Waste Estimation Procedure: Materials Flows Approach,* U.S. Environmental Protection Agency Report. EPA SW-147, Washington, D.C.

Spencer, R. 1989. Organic Waste Recycling Facility Launched. *J. Bio Cycle,* 70, November, pp. 42–45.

Steiker, G. 1973. "Solid Waste Generation Coefficients: Manufacturing Sectors", Regional Science Research Institute Discussion Paper Series, No. 70. *Regional Science Research Institute,* P.O. Box 8776, Philadelphia, Pennsylvania 19101.

Sugito, D. 1979. "Solid Waste Management in Japan." *Proceedings of Recycling Congress,* Vol. 1, pp. 165-169, Berlin.

Tchobanoglous, G., Theisen, H., and Eliassen, R. 1977. *Solid Waste Engineering Principles and Management Issues.* McGraw-Hill, New York.

Walsh, J.j., and Kinman, R.N. 1979. "Leachate and Gas Production Under Controlled Moisture Conditions." *Symp. Municipal Solid Waste: Land Disposal—Proceedings of the 5th Annual Research Symposium.* EPA-600/9-79-023a, March 26–28, pp. 41–61.

Warmer Campaign, The. 1990. *Warmer Factsheet: Waste Incineration.* January. Kent, United Kingdom.

Weston, R.F., INC. 1971. *New York Solid Waste Management Plan: Status Report 1970.* NTIS Report PB213557, U.S. Department of Commerce, Washington.

Wigh, R., and Brunner, D. 1981. "Summary of Landfill Research, Boone County Field Site." Presented at *7th Annual Research Symposium, Land Disposal: Municipal Solid Waste,* EPA-600/9-81-002a, pp. 209–242.

Wisconsin Department of Natural Resources. 1974. *Report on the State of Wisconsin Solid Waste Management Plan,* Madison.

Wisconsin Department of Natural Resources. 1981. *The State of Wisconsin Solid Waste Management Plan.* Madison.

Woods, R. 1991. Ashes To Ashes?*Waste Age* Vol. 22, No. 11, Nov., pp. 46–52.

2.7 PROBLEMS

2.1 Estimate the landfill volume needed for a community of 150,000 people. Utilize typical coefficient data from the various tables within the chapter. Assuming a compacted density within the site of 800 lb/yd^3 and an average depth of wastes of 20 ft, estimate the surface area needed over a 25-year time period.

2.2 Define the 3Rs program and identify the preferred order of each R. Give three examples of each R.

2.3 In addition to the main purpose of waste diversion, what other benefit does a recycling program bring?

2.4 Detail the five major components in current refuse being delivered to an MSW landfill site and indicate how these components are likely to change over the next two decades.

2.5 Indicate the seasonal and business cyclical variations in refuse composition that likely occur in your community.

2.6 What factors were responsible for the high rate of recycling of aluminum cans? What similar initiatives might be successful in establishing recycling of other components of the waste stream (if any)?

2.7 The 3Rs will impact the quantities of refuse from the residential, commercial, and industrial sectors. Develop a list of the changes that you would expect to observe within the next 10 years.

2.8 Develop a checklist of hazardous constituents leaving the typical residence. Comment on how that list has changed over the last 20 years.

2.9 A city of 100,000 people has an existing landfill capacity of 1.6×10^6 m^3. Determine the remaining site life. If the rate of waste production increases to 120 percent of current disposal rates, by how much will the site life be decreased?

If recycling decreases the refuse generation rate to 50 percent of current disposal rates, by how much will the site life be extended?

2.10 Design a recycling strategy employing source separation that might be useful in your community.

What is the likely success rate of the source separation program? Consider the number of vehicles that would be required in the collection activities.

2.11 Determine the volume and weight of refuse collection in a community of 100,000 people. Estimate the depth of refuse collected over a year if the refuse is placed on a football field, at its curbside density. Determine the depth of refuse following compaction.

2.12 List several factors that will determine the maximum period between garbage collections.

2.13 As a consultant, you are to design an effective source separation system for your community. Identify a minimum of five aspects of your proposed program that will be essential to its success. Considerations should include citizen participation, cost, and disposal of residual materials. What reduction of refuse would you expect from your program?

2.14 Using the information on resource recovery and reuse, estimate the percentages of the various components of refuse that will be landfilled in the future.

Information Commons
Level 0

Customer name: Kaisaier Aisikaer
Customer ID: ********

Title: Waste treatment and disposal /
ID: 200567621
Due: 19/05/2016 23:59

Title: Solid waste landfill engineering and
design /
ID: 201042766
Due: 19/05/2016 23:59

Total items: 2
12/05/2016 00:48

Thank you for using the Information Commons.

Problems with your library account?
Phone 0114 2227200 or email
library@sheffield.ac.uk

SITE-SELECTION METHODOLOGIES

3.1 INTRODUCTION

Not many years ago, refuse disposal facilities were located in sites such as old gravel pits, in areas close to wetlands and other low-lying areas, or at previous dump sites. Proximity to the generation area of the refuse was considered extremely important, so available, low-cost disposal space was critical. These sites were often located in regions where groundwater aquifers were recharged with water from precipitation. Initially, little regard was given to the potential for causing environmental contamination.

Because of reactions to identified problems, the open dump has rapidly evolved into the modern sanitary landfill. Examples of the problems and subsequent resolutions include smoke and odor problems, which led to the decision to stop burning solid wastes at the landfill site; problems with rodents, insects, and aesthetics, which led to the use of a daily soil cover; differential and overall settling plus the fire potential, which led to increased efforts at compaction of the waste; and the potential for groundwater pollution, which led to daily cover operations, berming, liners and leachate collection systems. The net result of these changes and many others that will be detailed later is that a significant evolution has taken place in the disposal practices at solid waste sites.

Although the management of solid wastes has improved, historical mismanagement of solid waste disposal has alienated the public. Residents are influenced by the legacy of earlier landfill-related problems. Landfill design and operations have changed markedly over the last few decades, yet people do not necessarily differentiate between past and present design practices. The net result is that a landfill, whether it is a municipal site or a hazardous waste disposal site, is a negative facility in the eyes of the public. People's attitudes toward landfills are particularly negative if the site is in the vicinity of where they are living or working. Although conscious of the need for an acceptable disposal system for wastes, citizens are probably neither sufficiently informed nor sufficiently motivated to seek it. The terms Not In My Back Yard (NIMBY), Build Absolutely Nothing Anywhere Near Anyone (BANANA), and Not On Planet Earth (NOPE) are convenient titles for a complex series of issues, fears, and forces influencing today's communities and residents. Neighbors of landfills object to the noise, dust, odors, and scavenger bird nuisances. Opposition to a landfill siting does not stop in the immediate environment, as complaints of traffic safety, litter, and groundwater contamination can arise from large distances around a landfill operation.

Everybody wants his or her garbage dealt with properly, but not close by. Once the garbage truck has disappeared around the bend of the road, the average citizen assumes (and hopes) that its contents are no longer his or her problem. As a result, municipalities are discovering the considerable power of organized public opposition to waste management planning for both municipal and hazardous wastes. Such opposition has in some cases become an insurmountable force.

Whether a person is cynical or optimistic about the role of the public in waste management planning, it is unrealistic to expect that negotiation, cooperation, or compromise will remove all opposition. Decision makers now need a sophisticated understanding of why the public objects to many of the choices, and they need a defensible, thorough structure or methodology that will demonstrate why a particular site is selected as a disposal site. They also must recognize at the outset of the site-selection process that there are no ideal sites, and conflicting issues will arise. Tradeoffs must be made.

Due to (normally) increasing opposition, the high cost of confirming a site's attributes, and the limited number of locations that will meet the criteria for disposal, procedures for identifying and ultimately selecting a site often involve considerable expense. Disposal costs are skyrocketing. In this chapter we examine a series of the available methodologies that have been utilized in the site-selection process. The methodologies can be equally applied to the various components of a solid waste management system, including transfer stations, municipal waste landfills, and hazardous waste landfills; however, for ease of presentation, we have assumed that the site to be selected is a sanitary landfill.

An important aspect of the methodologies is the role of the public. The basic purpose of public participation is to promote the productive use of inputs and perceptions from private citizens and public interest groups to improve the quality of environmental decision making. Without valid efforts to incorporate public opinion into the siting and operation of sites, it is unlikely that the sites will be approved.

3.2   SITE-SELECTION CRITERIA

To minimize haulage costs, a site should be as close to the generator as possible. If haulage cost was the only consideration, the optimal disposal site would be the centroid (as weighted by refuse generation quantities) of the contributing sources. However, the site-selection criteria for many siting problems have progressed far beyond the questions of economics of haulage cost.

A major difficulty in selecting a preferred site for a component of the solid waste management system (e.g., a landfill or a transfer station) from a number of alternative sites, is that differing environmental impacts for each site cannot be simply quantified. For example, some environmental impacts associated with a specific site are visual aspects, public health aspects, and costs, to specify only a few. The environmental impacts or criteria are thus referred to as being non-commensurable (they are not additive).

In assessing a site as a possible location for landfilling, many criteria could be employed. Nevertheless, there are several fundamental criteria that will serve to exclude a site from consideration:

1. The site must be structurally sound and free from potential problems such as landslides, subsidence, and flooding.
2. The effect on the neighborhood of heavy, large earthmoving equipment and significant traffic flow associated with the site operation must be assessed, as well as the need for additional facilities (e.g., access roads) to be constructed.
3. The extent to which the landfill site affects the quality (and perhaps quantity) of groundwater and surface water in the vicinity of the site must be assessed.

The existence of significant concerns associated with the first criterion will generally preclude a potential site from further consideration. The types of concerns implied by (2) and (3) in turn can be translated into an array of considerations such as those indicated in Table 3.1. Table 3.1 is not meant to be exhaustive, but it does indicate the diversity and number of features that will be scrutinized as part of the site-selection process. It should also be apparent that the various features are not measurable in similar units, which makes the comparisons between sites particularly challenging. Because there are many features that need to be considered, and because the issues are diverse and collecting the data to carry out the evaluations at different sites is expensive, it is essential to have a site-selection methodology that organizes the search for the best site. The approach must be able to rank and apply the site-selection factors in a logical and defensible way.

The first stage in the site selection process is the identification of a number of potential sites, each of which is somewhat less than perfect. Analyses must then be carried out to determine which of the deficiencies can be overcome. It is generally best to have the natural environment mitigate as many concerns as feasible. However, it is highly unlikely that a site will be identified that will have all such concerns mitigated.

The site-selection process must then choose out of the array of possible sites that which best satisfies the goals involved. A set of goals involved with site selection could include the following:

1. The risk to public health is minimized
2. The site minimizes the impact on the environment

Table 3.1 SOME QUESTIONS RELEVANT TO SITE SELECTION

❖ What are the zoning, planning, or existing land use commitments at the proposed site?

❖ What governmental controls are in place on allowed land uses at the site?

❖ What is the availability of the land area (who owns it)?

❖ What are the economic considerations related to haul costs, capital and operating costs and devalued land prices?

❖ What are the soil types and conditions at the site?

❖ What streams, rivers, lakes and reservoirs exist in the vicinity that may be affected by surface and subsurface runoff ?

❖ What floodplains and floodways are in the vicinity?

❖ What parks, open spaces, and recreation areas are in the vicinity?

❖ What subsurface, hydrogeologic, and geologic conditions are present? Are there any instabilities in the subbase?

❖ Does the soil have a high cation exchange capacity to attenuate contaminants?

❖ Are there sole-source aquifers in the vicinity?

❖ Are there features of archaeological or historical significance at the site?

❖ Are there endangered species of plants or animals on, or near, the site?

❖ Are there wetlands in the vicinity that will be affected?

❖ Will traffic patterns be capable of handling the anticipated increment without creating excessive local conflicts? What is the proximity to major roadways? Are there load limits on roadways?

❖ What are the aesthetic considerations associated with odor, noise, and dust to nearby residents?

❖ Are there airports in the vicinity that will be affected by the bird populations associated with landfills?

❖ Are there confined or unconfined aquifers at shallow depths?

❖ What potential ultimate uses exist for the completed site?

3. The site maximizes the level of service to the facility users

4. The site minimizes the cost to facility users.

However, these goals must be translated into specific features that can be evaluated based on the characteristics of the individual sites. If the situation occurs where a particular site is worse in terms of all the various types of impacts, in comparison with another site, the latter site is clearly the preferred site, and the selection process is straightforward. However, in most situations this is not the case. Instead, one site is preferred to another with regard to some factors, whereas another site is preferred for other reasons. As a consequence, some means of identifying the more important impacts must be utilized to aid in the site-selection process (i.e., the attributes must be prioritized by some procedure).

An example of the translation of goals to specific features as a first step in establishing priority rankings is given in Table 3.2. In this table, five overall priority groups are identified that characterize overall concerns. These are only suggested priority groups; a specific application may identify additional priorities or, alternatively, modified classifications. However, those noted in Table 3.2 will be utilized for purposes of the following discussion.

The priority groups must then be subclassified into discrimination features, such as those listed in Table 3.2. For example, different sites will take different tracts of agricultural land out of active production—a site that utilizes less agricultural land would be preferred to an alternative that requires more agricultural land or, more acres of high-quality agricultural land. The list of discriminating features will be problem-specific; however, the intent is to quantify the various discriminating features. The same measure can then be used

Table 3.2 PRIORITY GROUP RANKINGS AND ASSESSMENT FEATURES EMPLOYED IN DISCRIMINATION BETWEEN SITES

No.	Priority Group	Assessment Features
1	Public Health and Safety	Hydrology/Hydrogeology Traffic Safety/Traffic Service and Operations
2	Natural Environment	Biophysical Agriculture
3	Social Environment	Population Impact Community Facilities Dust/Odor Noise Visual Impact Land Use Compatibility
4	Cultural Environment	Heritage Features Archaeology
5	Economic Cost	Dollars

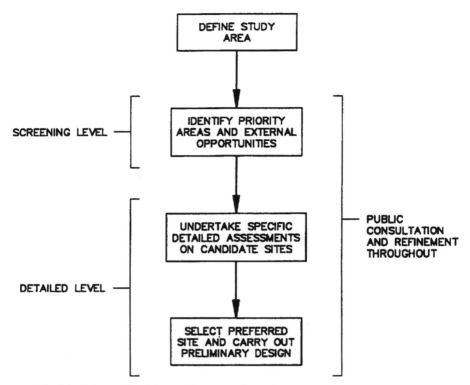

Fig. 3.4 Schematic of layered approach to site assessment.

across the range of sites being compared to establish their relative merits and classify and rank them. The decision-making framework must be systematic and utilize the information available at various points in time as the level of detail increases and the geographic focus narrows.

There are two major approaches to addressing the potential risks:

1. Prudent siting will minimize the most significant risks
2. The residual or remaining risks can be made small by design, operation, monitoring, mitigation, and contingency systems.

However, it is highly probable that any site-selection task will involve tradeoffs between important requirements, since no one site will be the most desirable in all respects. The result is that real systems are characterized by multiple objectives and goals that often conflict or compete with one another. In determining which tradeoffs are acceptable, the site-selection goals (risk, impact, service, cost) are used to rank the factors and rationalize the decision. To make matters more complicated, each of the decision-makers involved has his or her own set of priorities in terms of the features being assessed. Thus, the tradeoffs acceptable to one decision maker will differ from those of other decision makers.

Among procedures within the overall site-selection process, some are much better at screening out those alternatives worthy of more detailed examination as opposed to being utilized for the detailed examinations themselves. There is considerable merit in a "layered" approach for site assessment such as depicted in Figure 3.1.

3.3   EXAMINATION OF ALTERNATIVE METHODOLOGIES

A number of different evaluation methods have been developed to aid in the evaluation and assessment process. The method to apply in a specific situation is in part a function of the stage of the assessment (e.g., a preliminary screening as opposed to detailed assessment) and in part a function of the availability of data characterizing the site and its possible effects on the surrounding environment. We briefly describe the methodologies before discussing them in detail:

- *Ad hoc methods* compare alternatives in narrative terms without using any explicitly stated method to order the preferences. These methods involve "professional judgment," are difficult to explain to the public, and are not likely to be acceptable in today's environment.
- *Checklist methods* compare and evaluate alternatives against a specified set of criteria with no compensatory rules or tradeoffs.
- *Economic methods*—use economic procedures and principles to translate noncommensurable units into monetary units.
- *Cartographic methods*—compare and evaluate alternative sites using maps
- *Pairwise comparison methods* use the sequential comparison of alternatives in pairs as a basis for subsequent ordering of preferences.
- *Matrix methods* use a matrix for the summary, comparison, and evaluation of criteria and alternatives. Descriptive matrices use only the display properties and rely on "professional judgment" to order the preferences and thus are simply a minor extension of the ad hoc methods. Mathematical matrices include additive models and use mathematical operations to order preferences and to enable tradeoffs to be developed between the attributes. The matrix procedures are particularly useful in evaluating the sensitivity of the selection procedure to uncertainty in the assumptions or data.

3.3.1 Ad Hoc Methods

The ad hoc methods are based on professional judgment and describe impacts in narrative terms, without explicitly specifying criteria, ratings, or weights. Thus, it is difficult for persons other than the initiator to follow the logic of the reasoning. The problems with this procedure include the following:

- There is no assurance that different alternatives are evaluated in the same manner.

- There is no assurance that the full dimensions of the environment are considered.
- There are no assurances that the public concerns are addressed.
- The method is often not traceable and lacks accountability.
- The method cannot be easily applied.

For example, one might employ the ad hoc method in the selection of a landfill site by the following decision sequence: Site A is better than Site B because it has more available capacity, but Site A has the potential for increasing traffic safety concerns. Site C is on prime agricultural land, whereas Site B is visible from a nearby highway. Site C is the best site, since it costs the least for development.

It is obvious that there are major problems with this approach in that there is little technical structure and no application of consistent criteria. This procedure is unlikely to be accepted today but has been presented because it serves to emphasize the need for a methodical structure for the evaluation process.

3.3.2 Checklist Methods

Checklist methods compare and evaluate alternatives against a specified set of criteria (or list) with no tradeoffs. The results are expressed as simple yes/no responses. Checklist methods have widely varying characteristics and degrees of complexity. Examples of checklists include the following:

Evaluation Checklists

- *Unordered checklists of criteria/impacts* are lists of unranked criteria on which evaluation of proposals is based. As a result, all criteria are considered equal in importance. This assessment is based on "dominant" alternatives. Alternative A is considered to be dominant to alternative B if A is better than B in at least one respect and no worse than B in any other respect. The result is that Alternative B can be eliminated from further evaluation.
- *Scaling-weighting checklists* are an expansion of the unordered list of criteria. Scaling-weighting lists rank and weight the criteria to reflect the importance of each. This procedure assigns numerical values to the criteria for each proposed site. The ranking weight values are then multiplied by the numerical values. Sites with total values that rank highest are considered further.
- *Satisficing checklists* specify certain specific conditions that an alternative must satisfy before it can be considered acceptable. This evaluation method is an effective means of eliminating unacceptable alternatives. An example of such a criterion might be a requirement or a regulated limit. Any alternative that cannot meet the regulation can be rejected, regardless of any positive attributes it may possess.

Impact Identification Checklists

- *Simple checklists* consist of a catalog of environmental, economic, and social factors. The checklist is able only to identify impacts and attempts to ensure that impacts are not overlooked. It does not determine their relative importance.
- *Questionnaire checklists* consider a series of questions relating to the impacts of a landfill site. The method determines the amount of data and information available.

The checklist approaches are utilized during the screening of the alternatives. There is not sufficient information generated to allow a detailed evaluation of the sites. The magnitudes of impacts are not determined, only whether the impacts are positive or negative. The checklist (regardless of the form selected) must be comprehensive to ensure that all impacts are identified.

3.3.3 Economic Evaluation Methods

Economic evaluation methods attempt to represent all aspects of a project in monetary terms. Costs and benefits are expressed in terms of the individual's willingness to pay for the benefits (the maximum amount individuals affected by the project would be willing to pay for project benefits). The willingness to accept the cost incurred is measured in terms of the maximum amount individuals would be willing to accept regarding the costs of the project.

Market prices, where they are available, are used as the reference points for estimating the willingness-to-pay and willingness-to-accept values. The estimation of externalities (effects on third parties), such as noise and loss of natural areas, that are not accounted for in the marketplace is an important and difficult aspect of cost-benefit analyses. Methods of determining willingness-to-pay values in the absence of market prices typically include direct estimation of consumers' preferences using polls, surveys, bidding games, questionnaires, and voting.

By their nature, many of the impacts of a landfill site are not easily measured in monetary units for each of the time periods, nor are they easily assigned values via polls, survey, and the like. The net result is that the implementation of this procedure is very difficult and hard to defend in a public forum.

3.3.4 Cartographic Methods

Cartographic methods of site assessment compare and evaluate alternatives using maps and are used at the site-identification stages. The two main forms of cartographic evaluation methods are constraint mapping and overlay mapping.

Constraint Mapping   Constraint mapping is a common siting technique that defines characteristics that are undesirable for sites and then systematically eliminates areas that possess these constraints. This approach utilizes maps on which unacceptable characteristics for a site are identified, and their geographical locations are mapped. Maps of all unacceptable characteristics are overlaid, and only unconstrained geographical areas remaining are considered for sites or corridors. If no sites remain, the need for the site or the definition of the study area needs to be reconsidered, or the constraints relaxed. The negative aspects of the constraint mapping approach include the following:

- Considerable time and expense are incurred in gathering details to prove conclusively that areas or sites are not suitable. These detailed data are subsequently of little use once the site is excluded.
- It is inherently assumed that areas or sites that remain following constraint mapping are acceptable without definition of the positive attributes that make them acceptable. The procedure does not inherently allow for ranking or comparison of remaining areas or sites.

Nevertheless, the procedure can be effective for moving from large geographical areas to smaller, "target" areas.

Overlay Mapping   Overlay mapping is similar to constraint mapping but with the addition of weighting and rating factors that are combined into shaded map overlays. The weighting factors denote the degree of shading (constraint) of each specific evaluation component. All overlays are superimposed, and the relative suitability of geographical areas for the proposed site are indicated by the intensity of shading on the composite map. The lighter the shading, the more suitable the area.

In summary, the cartographic methods have particular value during the preliminary site-selection process, in which the sites deserving of further analysis are identified from a very large number of potential sites. However, once they have been narrowed down to a short list of candidate sites, the utility of the cartographic procedures diminishes.

3.3.5 Pairwise Comparison

The pairwise comparison procedure considers alternative sites in a pairwise fashion to establish the relative importance or attractiveness of each alternative. In its simplest form, the procedure examines individual impacts associated with two of the sites being considered to develop a measure of how frequently each site is superior to the other. Values are estimated by comparing each alternative against other alternatives for various criteria. These indications of superiority are then informally weighted in terms of the importance of the criteria.

E X A M P L E

Site-A is preferred to Site-B in that it utilizes less agricultural land, but Alternative B is preferred to A in that it is more aesthetically pleasing. If these two criteria were all that there were to choose between the two sites, then a preference that weighted loss of agricultural land as less desirable than aesthetics would identify Site A as the preferred site to Site B.

The informal preference weighting is a major detraction of this simple procedure. In addition, if many features are to be considered, the number of components to be compared (e.g., loss of agricultural land, visual aesthetics, or potential for groundwater contamination) makes this pairwise comparison procedure very cumbersome.

Fuzzy-Set Procedures   The pairwise comparison procedure is improved by using *fuzzy-set procedures*. Fuzzy-set theory (after Zadeh and Tanaka, 1975) is based on subjective interpolation: by comparing evaluation factors, an inefficient alternative (one that is dominated by other alternatives) is identified. The best alternative is then selected by identifying the extent of dominance of one alternative over another. It is important to note that dominance testing is not an easy step, since it is unlikely that one alternative will be dominant in all the evaluation factors. Consequently, subjective assessments are required to rank the evaluation factors, thus prioritizing the dominance findings. McBean and Zukovs (1983) examine the prioritization concerns in greater detail.

The fuzzy-set information is amenable to illustration using a matrix summary, which provides an overall visual assessment of alternatives. In Table 3.3. six alternative sites are ranked using values of 1 through 6. In terms of desirability, a value of 1 is assigned to the best site, and 6 to the worst. Thus, insofar as the hydrology/hydrogeology component is concerned, Site D is the preferred site, Sites A and E are the second most preferred, and so on. Where there is a virtual tie in terms of ranking, the average rank is utilized.

Numerically summing the rankings is then an approach to selecting the preferred alternative (the lower the sum of the ranks, the more preferred the site). However, there are dangers in selecting the preferred site on this basis alone, among which are the following:

1. The ranking procedure provides no relative value information: Is Site D very much different (better) than Sites A and/or E insofar as hydrology/hydrogeology, public health and safety? Is the difference between F and B of similar magnitude to that between B and C with regard to agricultural impact?

2. The visual impact of the ranking procedure can be significantly biased by inclusion of varying numbers of discriminating features. For example, public health and safety may be more important than social environment, yet the latter is described by six features in Table 3.3, as opposed to two features for the former. Counts of the number of times that Site D dominates Site A and vice versa are summarized in Table 3.4. Simply counting the numbers of dominant features indicates that Site A is preferred (10 occa-

Table 3.3 NONPARAMETRIC RANKING OF ALTERNATIVES WITH RESPECT TO A SERIES
OF FEATURES

No.	Priority Group	Site Plan Area Alternatives					
		A	B	C	D	E	F
1	*Public Health and Safety*						
	Hydrology/Hydrogeology	2.5	5.5	5.5	1	2.5	4
	Traffic safety/Traffic service and operations	3	3	3	1	6	5
2	*Natural Environment*						
	Biophysical	1	4.5	3	2	4.5	6
	Agriculture	4	2	3	5	6	1
3	*Social Environmental*						
	Population impacts	4.5	3	6	4.5	1	2
	Community facilities	1	2	6	3	4	5
	Dust/Odor	1.5	6	4	5	3	1.5
	Noise	2	5	6	4	3	1
	Visual impact	1	2	3	6	5	4
	Land use compatability	1	5.5	5.5	3.5	3.5	2
4	*Cultural Environmental*						
	Heritage features	2	4	6	5	3	1
	Archaeology	4	2	4	6	4	1
5	*Costs*						
	Costs	1	5.5	5.5	2.5	2.5	4

Notes:
The lower the value, the more preferred the site.
When equal ranks for different alternatives are identified, the average rank is
assigned to all those alternatives judged as equal.

sions of dominance versus 2), but if public health and safety is considered to
be much more important, then perhaps the 2 features in which Site D is
preferred to Site A are more important than the 10 occasions of dominance
of Site A.
3. Testing of the sensitivity of measurements is difficult in the ranking procedures.

Table 3.4   FREQUENCY OF DOMINANCE FOR SITES A AND D (From Table 3.3)

No.	Priority Group	Occurrences of Site A Preferred to D	Occurrences of Site D Preferred to A
1	*Public Health and Safety*	0	2
2	*Natural Environment*	2	0
3	*Social Environmental*	5	0
4	*Cultural Environmental*	2	0
5	*Costs*	1	0

Parametric Ranking   The assignment of ranks, one through the number of alternative sites (i.e., a *nonparametric* ranking system), loses proximity information. For example, Sites A and D might be very similar (but not precisely equal) with respect to hydrology/hydrogeology, yet the next best site might be considerably different. Therefore, an alternative parametric or expanded-range ranking approach assigns the value of unity to the best site for each criterion and a much higher number, for example, 10, to the worst. For those sites in between, individual values are assigned commensurate with their proximity to the best and to the worst, for example, 3.0, 6.4, and 9.8. In other words, the matrix may now appear as indicated in Table 3.5. The values retain their magnitude and are not forced into being ranked.

Note that the parametric approach still suffers from the remaining difficulties associated with the pairwise comparison. Also, the introduction of a completely spurious site can change the rank assignments. This problem is particularly relevant when the spurious site is really bad in only a few of the priority groupings. For instance, consider a fictitious Site G that is catastrophic regarding hydrology/hydrogeology. The first line of entries in Table 3.6 might then be adjusted to the following values:

	Site Plan Area Alternatives						
	A	B	C	D	E	F	G
Hydrology/Hydrogeology	1.1	1.3	1.3	1	1.1	1.2	10

The relative rankings of Sites A through F, due to the consideration of Site G, make this feature (hydrology/hydrogeology) become nondiscriminating—there is virtually no difference between Sites A through F insofar as this feature is concerned.

If some of the features of dominance are more important than others, it is possible to weight each of the discrimination features. Such a *weighting scheme* may be via category assignment such as employed by McBean and Zukovs (1983) or it may involve a specific weight assignment as described in the next section.

3.3.6 Matrix Methods

The last group of procedures to be examined employ matrices for comparison and evaluation. Mathematical matrices assume additive models and use mathematical operations to order preferences and determine the relative importance of each. The most commonly used matrices employ two opposing axes (e.g., criterion and alternative).

Leopold et al. (1971) first proposed a matrix procedure that incorporated a list of project activities in addition to a checklist of potentially impacted environmental characteristics. An impact is identified by an action and its environmental consequences. The magnitude of an impact is described by assigning a numerical value from 1 to 10, with 10 representing the largest magnitude, based on an objective evaluation of the facts. In addition, the consequences of an impact are assessed and rated on a scale of 1 to 10. A multiplication of the mag-

Table 3.5 PARAMETRIC RANKING OF ALTERNATIVES WITH RESPECT TO A SERIES OF FEATURES

No.	Priority Group	Site Plan Area Alternatives					
		A	B	C	D	E	F
1	*Public Health and Safety*						
	Hydrology/Hydrogeology	3	8	8	1	3	6
	Traffic safety/Traffic service and operations	2	2	2	1	6	5
2	*Natural Environment*						
	Biophysical	1	5	4	2	5	8
	Agriculture	6	3	4	8	10	1
3	*Social Environmental*						
	Population impacts	5	4	8	5	1	3
	Community facilities	1	2	7	4	5	6
	Dust/Odor	1	5	3	4	2	1
	Noise	3	7	8	5	4	2
	Visual impact	1	2	3	8	6	7
	Land Use Compatability	1	6	6	4	4	2
4	*Cultural Environmental*						
	Heritage features	3	5	8	6	4	1
	Archaeology	4	3	4	6	4	1
5	*Costs*						
	Costs	1	10	10	2	2	5

nitude and importance values provides a measure of the response to an alternative, a measure that can be compared with the responses of other alternatives to select the best alternative.

A major concern with the method of Leopold et al. (1971) and to a large extent with subsequent variations and extensions (e.g., Chase, 1973; Dee et al., 1973; Canter, 1977) is the difficulty of assigning values on a scale of 1 to 10 to characterize the magnitude and importance of impacts. On the positive side, this type of procedure and its variations are highly amenable to sensitivity analyses.

The models based on Leopold et al. (1971) are additive models in that they reduce the assessment and evaluation exercise to one in which each alternative is assigned a numerical value (index or score) intended to represent the utility and attractiveness of the project. Values are assigned to the criteria and applied to the various project alternatives (each plan). The results are then weighted by a preference to determine the project with the most desirable overall attributes.

The weight assignments are of two types. First, the discrimination features are weighted within the priority groups (e.g., biophysical versus agricultural). Second, the priority groups themselves are weighted (e.g., public health and safety versus the natural environment). These assignments embody preference information—how people feel about the relative importance of the different types of impacts of a landfill. The usual requirement is that the sum of the weights equal unity, to retain relative values. Thus, a common means of assignment is to request the decision makers to assign weights to the priority groups. The values listed in Table 3.6 were derived by such a method. Table 3.6 gives the individual weights assigned to the discriminating features within each of the priority groups.

The weight assignments may then be utilized in combination with the rank assignments from Section 3.3.5 to reflect preference functions, as the tabular array in Table 3.7 indicates. The calculations may be carried out using a spreadsheet program. For the example depicted in Table 3.8, Site A is the preferred site.

The matrix calculation procedure can be easily applied to the other pairwise comparison procedure indicated, namely, the expanded-range ranking system. An important feature of the matrix approach is its usefulness for carrying out sensitivity analyses in a simple manner— since the weight assignments are the preferences of decision makers, it is easy to explore the degree to which the selection of the best site is sensitive to the value assignments. Utilization of a spreadsheet program makes the sensitivity testing a relatively simple task.

Table 3.6   WEIGHT ASSIGNMENTS OF PRIORITY GROUPS

No.	Priority Group	Weight
1	*Public Health and Safety*	33.4
2	*Natural Environment*	20.4
3	*Social Environment*	15.5
4	*Cultural Environment*	15.4
5	*Cost*	15.3
		100

Note: If all the priority groups were considered equal, each would be given a weight of 20.

Table 3.7   WEIGHT ASSIGNMENTS OF DISCRIMINATING FEATURES

Priority Group	Discriminating Feature		Weight
Public Health and Safety	Hydrology/Hydrogeology		68
	Traffic safety		32
		Sum =	100
Natural Environment	Biophysical		46
	Agricultural		54
		Sum =	100
Social Environment	Population Impacts		25
	Community facilities		12
	Dust/Odor		19
	Noise		15
	Visual impact		16
	Land use compatability		14
		Sum =	100
Cultural Environment	Heritage features		54
	Archaeology		46
		Sum =	100

Note: If the features within the public health and safety priority group were considered equal, each would be assigned a weight of 50.

Table 3.8 SITE SCORING EVALUATION

Priority Group	Factor Weight	Factor Ranking Site						Weighted Factor Site						Priority Group Weight	Weighted Factor Site					
		A	B	C	D	E	F	A	B	C	D	E	F		A	B	C	D	E	F
Public Health and Safety																				
Hydrology/Hydrogeology	0.675	3.00	8.00	8.00	1.00	3.00	6.00	2.03	5.40	5.40	0.68	2.03	4.05							
Traffic safety/Traffic service and operations	0.325	2.00	2.00	2.00	1.00	6.00	5.00	0.65	0.65	0.65	0.33	1.95	1.63							
								2.68	6.05	6.05	1.00	3.98	5.68	0.334	0.89	2.02	2.02	0.33	1.33	1.90
Natural Environment																				
Biophysical	0.460	1.00	5.00	4.00	2.00	5.00	8.00	0.46	2.30	1.84	0.92	2.30	3.68							
Agriculture	0.540	6.00	3.00	4.00	8.00	10.0	1.00	3.24	1.62	2.16	4.32	5.40	0.54							
								3.70	3.92	4.00	5.24	7.70	4.22	0.204	0.75	0.8	0.82	1.07	1.57	0.86
Social Environmental																				
Population impacts	0.246	5.00	4.00	8.00	5.00	1.00	3.00	1.23	0.98	1.97	1.23	0.25	0.74							
Community facilities	0.119	1.00	2.00	7.00	4.00	5.00	6.00	0.12	0.24	0.83	0.48	0.60	0.71							
Dust/Odor	0.187	1.00	5.00	3.00	4.00	2.00	1.00	0.19	0.94	0.56	0.75	0.37	0.19							
Noise	0.147	3.00	7.00	8.00	5.00	4.00	2.00	0.44	1.03	1.18	0.74	0.59	0.29							
Visual impact	0.160	1.00	2.00	3.00	8.00	6.00	7.00	0.16	0.32	0.48	1.28	0.96	1.12							
Land use compatability	0.143	1.00	6.00	6.00	4.00	4.00	2.00	0.14	0.86	0.86	0.57	0.57	0.29							
								2.28	4.36	5.88	5.04	3.34	3.34	0.155	0.35	0.68	0.91	0.78	0.52	0.52
Cultural Environmental																				
Heritage features	0.535	3.00	5.00	8.00	6.00	4.00	1.00	1.61	2.68	4.28	3.21	2.14	0.54							
Archaeology	0.465	4.00	3.00	4.00	6.00	4.00	1.00	1.86	1.40	1.86	2.79	1.86	0.47							
								3.47	4.07	6.14	6.00	4.00	1.00	0.154	0.53	0.63	0.95	0.92	0.62	0.15
Costs																				
Costs	1.000	1.00	10.0	10.0	2.00	2.00	5.00	1.00	10.0	10.0	2.00	2.00	5.00							
								1.00	5.00	6.00	2.00	3.00	4.00	0.153	0.15	0.77	0.92	0.31	0.46	0.61

Final Site Score

	A	B	C	D	E	F
	2.69	4.89	5.61	3.41	4.49	4.04

Notes:
FW=factor weight for each priority group (sum=1.00)
PGW=priority group weight (sum=1.00)

3.4   SITE END USES

Part of the acceptability of a site for use as a landfill is how the completed site will relate to the surrounding land uses.

Completed landfill sites have been successfully utilized for a range of subsequent land uses: parks and recreation areas, such as ski slopes, toboggan runs, coasting hills, ball fields, golf courses, amphitheaters, and playgrounds; botanical gardens; residential and industrial development; and parking areas. The use of a completed sanitary landfill as a green area is very common; however, for reasons that will become more obvious in later chapters, caution must be used in selecting potential land uses. Small, light buildings such as concession stands, sanitary facilities, and equipment storage sheds are usually required at recreational areas. These must be constructed to minimize settlement and gas problems. Other concerns include ponding, cracking, and erosion of the cover material. The cost of designing, constructing and maintaining buildings is considerably higher than it is for those erected on well-compacted earth fill or on undisturbed soil.

3.5   SUMMARY

All the procedures outlined in this chapter must be structured toward the development of an information-dissemination program to develop support for proposed siting of the solid waste landfill. The program must begin early in the long-range planning stages; thus, a process involving several levels of analyses is frequently employed.

The selection of the best site involves both factual and subjective information. Because there are many diverse features to be considered, data collection is costly and involves a great deal of effort. The factual information also must be quantified, since it is multidimensional in character and must be distilled into a form that is comprehensible to nontechnical decision makers.

The assignment of preference functions is a difficult task that must be left to decision makers. The values assigned are necessarily uncertain; thus, being able to utilize sensitivity analysis effectively is a positive attribute.

Due to the problems outlined in this chapter, it is generally recommended that several of the procedures discussed be employed. If the findings from different procedures are consistent, then the selection of the best site is probably relatively straightforward. If the findings are not consistent, identification of the reasons for the inconsistencies will direct additional information collection efforts to the problem areas. In addition, removal of any obviously inferior candidates followed by a reevaluation of the remaining alternatives may allow a better resolution of the necessary tradeoffs.

Quantification of numerous site-assessment features and of the values within the various procedures will be developed in the remaining chapters.

REFERENCES

Canter, L. 1977. *Environmental Impact Assessment*. McGraw-Hill, New York.

Chase, G. 1973. *Matrix Analyses in Environmental Impact Assessment*. Engineering Foundation Conference on Preparing Environmental Impact Statements. Henniker, N. H., July.

Dee, N., et al., 1973. An Environmental Evaluation System for Water Resources Planning. *Water Resources Research* 9 (3): 523–536.

Leopold, L. B., Clarke, F. E., Hanshaw, B. B., And Balsley, J. R. 1971. *A Procedure for Evaluating Environmental Impact*. U.S. Geological Survey, 645, Washington, D.C.

McBean, E., And Zukovs, G. 1983. A Decision Making Analysis Methodology for Pollution Control Strategy Formulation. *Canadian Water Resources Journal* 8 (2): 64–87.

Zadeh, L., Fu, K., And Tanaka, K. 1975. *Fuzzy Sets and Their Applications to Cognitive and Decision Processes*. Academic Press, New York.

3.6 PROBLEMS

3.1 Five potential landfill sites were evaluated using the priority groups and discrimination features listed in Table 3.2. The following nonparametric and parametric rankings were obtained:

	Non-parametric/Parametric				
	U	V	W	X	Y
Priority 1					
Hydro	4/1.2	2/1.2	1/1	3/1.6	5/10
Traffic	3/1.1	4/1.2	1/1	2/1.1	5/10
Priority 2					
Biophysical	5/8	1/1	2/4	3/5	4/6
Agricultural	4/8	2/3	4/6	1/1	3/4
Priority 3					
Population impacts	2/3	5/9	4/7	3/4	1/1
Community facilities	2/2.5	3/3	4/7	5/8	1/2
Dust/Odor	4/5	2/2	1/1	3/6	5/10
Noise	2/3	5/6	1/2	4/5	3/5
Visual impact	4/8	1/1	3/2	5/9	2/10
Land use	3/6	4/7	2/2	5/8	1/1
Priority 4					
Heritage Features	5/9	2/3	3/5	4/5	1/2
Archaeology	3/7	5/8	4/8	2/1.5	1/1
Priority 5					
Costs	3/5	4/6	5/9	2/2	1/1

- Do a pairwise comparison using the frequency of dominance for sites U, V, W, and X for the nonparametric rankings and choose the best site (you should have six comparisons).
- How does your answer change if priorities 1 and 2 are considered extremely important as compared with priorities 3, 4, and 5.
- Do a pairwise comparison of your best alternative in (a) with alternative Y. How do they compare?

(a) Do a pairwise comparison using the frequency of dominance for sites U, V, W, and X for the nonparametric rankings and choose the best site (you should have six comparisons).

(b) How does your answer change if priorities 1 and 2 are considered extremely important as compared with priorities 3, 4, and 5.

(c) Do a pairwise comparison of your best alternative in (a) with alternative Y. How do they compare?

3.2 (a) Using the information in Problem 3.1 and employing the parametric rankings, determine the best site. Weight the discriminating features as well as the priority groups equally. (*Hint*: The weight of each of the five priority groups is 20).

(b) How does Site Y affect your results?

(c) Use the following weighting schemes with the parametric rankings to determine the best site.

Priority Groups		Discriminating Features		
1	35			
2	20			
3	15	1 {	A	65
4	10		B	35
5	20			
		2 {	A	50
			B	50
			A	20
			B	10
			C	25
		3 {	D	15
			E	15
			F	15
		4 {	A	45
			B	55

(d) How do your findings in (a) through (c) differ?

(e) Describe how the ranking scheme might be misused to bias toward an alternative.

(f) If Site Y is eliminated because of major health and safety reasons, how will your decision change if the parametric rankings change as follows:

	U	V	X	Y
Hydro	5	2	1	7
Traffic	3	7	1	3

3.3 Tabulate the important topographic, geographical, and hydrogeologic factors that must be considered when selecting sites for sanitary landfilling.

PRINCIPLES OF DECOMPOSITION IN LANDFILLS

4.1 INTRODUCTION

Solid wastes deposited in landfills decompose by a combination of chemical, physical, and biological processes. The decomposition produces solid, liquid, and gaseous byproducts, all of which may be of concern in the overall management of a landfill. The biological processes acting on the organic materials within the refuse commence soon after refuse placement. However, interdependencies among the three processes require that chemical and physical processes also be considered along with the biological processes. For example, the physical and chemical processes influence the availability of the nutrients essential for the biological action.

Physical decomposition of solid waste results from the breakdown or movement of the refuse components by physical degradation and by the rinsing and flushing action of water movement. Upon reaching field capacity (the moisture level beyond which any increases in moisture will drain by gravity), flow of dislodged refuse particles occurs as a result of pressure gradients, and diffusion as a result of concentration gradients. As the moisture level of the refuse increases, additional refuse particles are dislodged.

Chemical processes resulting in refuse decomposition include the hydrolysis, dissolution/precipitation, sorption/desorption, and ion exchange of refuse components. Chemical decomposition generally results in altered characteristics and greater mobility of refuse components, thereby enhancing the rate at which the landfill becomes more chemically uniform.

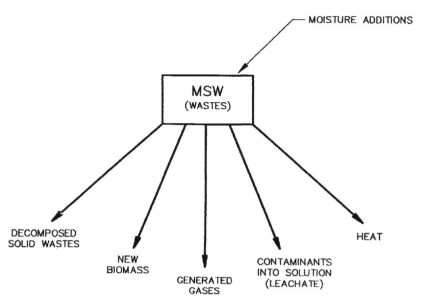

Fig. 4.1 Byproducts of solid waste decomposition.

Although both physical and chemical decomposition of refuse materials are important in landfill stabilization, biological decomposition is the most important process. Specifically, biological decomposition is the only process that produces methane gas.

Biological decomposition occurs with naturally present bacteria. It is a complex process within landfill sites, consisting of various biologically mediated sequential and parallel pathways by which refuse is decomposed to various end products.

The products of the physico-chemical and biological decomposition processes are depicted on Figure 4.1. In this chapter we focus on the fundamental principles that characterize these processes.

4.2   THE LANDFILL AS A BIOCHEMICAL REACTOR

As a result of the combination of processes referred to in Section 4.1, the landfill is a form of biochemical reactor, similar to an anaerobic digester in a wastewater treatment plant. Of course, there are potentially important limitations on the degree to which the landfill contents are mixed. The result is that variabilities in such features as moisture, refuse age, and composition in various locations within the refuse are of great importance to the degree and rate at which the refuse decomposes and to the byproducts produced. Thus, knowledge of the moisture content, leachate character, and migration of the gas within the refuse are essential to understanding the rate and current status of the decomposition processes.

Biological decomposition takes place in three stages, each of which has its own environmental and substrate requirements that result in characteristic end products.

4.2.1 Aerobic Decomposition

Aerobic processes require the presence of oxygen. Thus, aerobic decomposition occurs on initial placement of the refuse, while oxygen is still available. Aerobic decomposition may continue to occur on, and just below, the surface of the fill, as well. However, because of the finite amount of available oxygen buried within the refuse and the limitations on air transport into the fill, aerobic decomposition is responsible for only a small portion of the biological decomposition within the refuse.

During this first stage of decomposition, aerobic microorganisms degrade organic materials to carbon dioxide, water, partially degraded residual organics, and considerable heat. Aerobic decomposition is characteristically rapid, relative to subsequent anaerobic decomposition, and the oxygen demand of this refuse is high. A general relation for this decomposition is

$$\text{Degradable waste} + \text{oxygen} \rightarrow CO_2 + H_2O + \text{biomass}$$
$$+ \text{ heat} + \text{partially degraded materials}$$

An example of the aerobic decomposition of a general constituent $CH_aO_bN_c$ is

$$CH_aO_bN_c + \frac{1}{4}\left(4a - 2b - 3c\right)O_2 \rightarrow CO_2 + \frac{1}{2}\left(a - 3c\right)H_2O + cNH_3 \qquad (4.1)$$

The aerobic microorganisms produce carbon dioxide levels as high as 90 percent, and the temperature rises to as high as 70° C. The elevated carbon dioxide results in the formation of carbonic acid in the refuse, thus resulting in acidic pH levels in the leachate:

$$CO_2 + H_2O \rightarrow H_2CO_3 \text{ (carbonic acid)} \qquad (4.2)$$

Leachate is not usually produced in this decomposition cycle because the refuse has not reached field capacity in this early stage. Any leachate that is produced during this initial stage is generally the result of channelization through highly permeable pathways or voids in the refuse. The composition of this leachate is typically particulate matter entrained by the percolating water, soluble salts present in this refuse, and small amounts of soluble organic matter.

4.2.2 Acid-Phase Anaerobic Decomposition (Nonmethanogenic)

The second stage of refuse decomposition involves facultative microorganisms that become dominant as the oxygen is depleted. These microorganisms continue the decomposition processes. In this, the acid or acetogenic phase, high concentrations of organic acids, ammonia, hydrogen, and carbon dioxide are produced. Acid fermentation prevails, with characteristic end products being high levels of carbon dioxide, partially degraded organics (especially organic acids) and some heat, as described by the following equation:

Degradable wastes $\rightarrow CO_2 + H_2O$ + organism growth
$$+ \text{ partially degraded organics}$$

The production of carbon dioxide (high partial pressure) and large amounts of organic acids result in the lowering of the pH of the leachate to the range of 5.5 to 6.5, which in turn causes the dissolution of other organics and inorganics. The result is a chemically aggressive leachate with high specific conductance.

4.2.3 Anaerobic Decomposition (Methanogenic)

As the biodegradation of the refuse progresses, the oxygen becomes depleted, the redox potential is reduced, and the third stage of refuse decomposition involving the anaerobic methanogenic bacteria becomes dominant. These organisms produce carbon dioxide, methane, and water, along with some heat. Characteristically, these organisms work relatively slowly but efficiently over many years to decompose remaining organics.

The methanogenic bacteria utilize the products of the anaerobic acid stage, for example, hydrogen,

$$4H_2 + CO_2 \rightarrow CH_4 + 2H_2O \tag{4.3}$$

and acetic acid,

$$CH_3 COOH \rightarrow CH_4 + CO_2 \tag{4.4}$$

Consumption of the organic acids raises the pH of the leachate to the range of 7 to 8. Consequently, the leachate becomes less aggressive chemically and possesses a lower total organic strength. Organic acids that cannot be used directly by the bacteria are converted to methane by an intermediate step. Volatile fatty acids act as a substrate for methanogenic bacteria, but high concentrations inhibit the establishment of a methanogenic community and at very high concentrations are toxic.

The methane bacteria that function in the methanogenic stage obtain energy from two reactions: (1) the reduction of CO_2 through the addition of H_2 to form CH_4 and H_2O and (2) the cleavage of the $CH_3 COOH$ into CH_4 and CO_2. Although energy is captured by the microorganisms during this stage, very little synthesis of new cell material occurs (McCarty, 1963).

The gases N_2 and H_2S may also be produced during anaerobic decomposition. Nitrogen is produced from the microbial process of denitrification, in which the nitrate ion is reduced. Hydrogen sulfide is produced by sulfate-reducing microorganisms.

Hydrogen is produced during the nonmethanogenic stage but is consumed during the methanogenic stage. Toerien and Hattingh (1969) reported that the latter reaction proceeds at a much more rapid rate than the former; therefore, H_2 is generally not found in the presence of CH_4.

The time required for the methanogenic stage to commence may be from six months to several years after placement. The shorter time period is associated with situations of higher water content and flow rate. It is noteworthy, however, that instability in the system or rapid variations in water movement may inhibit the methanogenic bacteria.

The optimal pH for the methanogenic bacteria is in the range of 6.7 to 7.5. However, there is still some activity with pH in the range of 5.0 to 9.0. The best temperature conditions are within the range of 30 to 35° C for mesophilic bacteria, with a second optimum around 45° C for thermophilic bacteria. The temperature of the landfill dictates which class of bacteria (mesophiles or thermophiles) are functional. Most landfills operate in the mesophilic range. There is a significant decrease in methanogenic activity at temperatures below 10 to 15°C. The best ratio of carbon to nitrogen for the methanogenic bacteria is 16:1, and the more water present, the better high organic acid concentrations or the presence of any toxic material will dramatically interfere with methanogenic bacteria.

During the methanogenic phase, leachate characteristically has a near-neutral pH, low volatile fatty acid content, and low total dissolved solids (TDS). A small portion of the organic refuse, the ligand-type aromatic compounds, are slow to degrade anaerobically. These compounds are important factors in adsorption and complexation (Lu et al., 1984).

The methanogenic stage does not mark the end of the hydrolysis and fermentation that occurs in the acetogenic stage. These steps continue, but the methanogenic bacteria population grows to a level at which the bacterial rate of consumption of the acetic stage end products approaches the rate of production.

4.2.4 Cumulative Effect

Landfills actively receiving refuse normally undergo all three decomposition processes simultaneously, at various locations within the refuse. However, normally within the first few years following landfill closure, the anaerobic stages become dominant and remain so until all the available organics have been decomposed. As the landfill ages, the landfill gas production rates gradually decrease over time, although small amounts of methane continue to be produced decades after closure. Therefore, a great deal can be learned about the status of landfill decomposition by monitoring the gas and leachate.

FORMS OF DECOMPOSITION

PRIMARY
BYPRODUCTS

AEROBIC DECOMPOSITION

CO_2, H_2O, NITRATE AND NITRITE

ANAEROBIC DECOMPOSITION

CH_4, CO_2, H_2O, ORGANIC ACIDS,
NITROGEN, AMMONIA, FERROUS
AND MANGANOUS SALTS, AND
HYDROGEN SULFIDE

Fig. 4.2 Schematic of byproducts from alternative forms of biological decomposition.

As a result of the biological processes discussed and the relative absence of oxygen under conditions normally encountered in landfills, the predominant gaseous byproducts are methane and carbon dioxide and much lesser quantities of hydrogen sulfide, nitrogen, hydrogen, and ammonia. The byproduct information is summarized in schematic form in Figure 4.2.

The design of control and recovery systems for gas and leachate facilities must be based on an understanding of the fundamentals or principles of decomposition. We examine these principles in the remainder of this chapter. In Chapter 5 we provide some specific magnitudes and estimates of amounts and composition of gas and leachate.

4.3   BIODEGRADATION PATTERN FOR ORGANIC MATTER

The biodegradation of refuse follows a very predictable decomposition pattern. Certain compounds decompose aerobically. As the oxygen supply is exhausted, facultative and anaerobic microorganisms take over. The role of the carbon cycle as a fundamental component of the overall decomposition process, was elucidated by Zehnder et al. (1982). A modified form is diagrammed in Figure 4.3.

During the aerobic phase of decomposition, the organic matter is decomposed by the microorganisms to carbon dioxide and water, and oxygen is utilized.

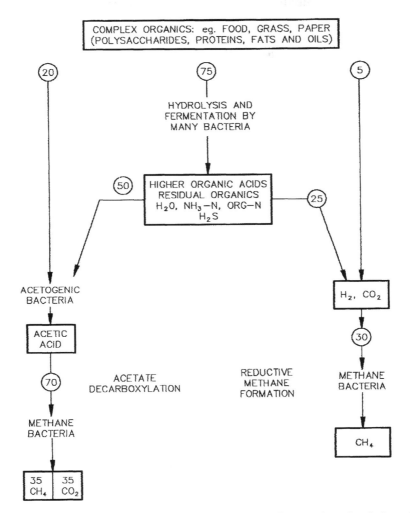

Fig. 4.3 Biodegradation pattern for organic matter. The numbers in circles give the percent of carbon.

An example of aerobic decomposition is the oxidation of carbohydrates to carbon dioxide and water:

$$C_6H_{12}O_6 + 6\,O_2 \rightarrow 6\,CO_2 + 6\,H_2O \tag{4.5}$$

One molar volume of CO_2 is produced for each molar volume of oxygen utilized. The carbohydrates and water are solid and liquid, respectively, and occupy a negligible volume in comparison with the gas evolved. The net result is the replacement of oxygen by CO_2 within the refuse mass.

Fatty materials are similarly oxidized, according to the following typical reaction of stearic acid:

$$C_{18}H_{36}O_2 + 26\,O_2 \rightarrow 18\,CO_2 + 18\,H_2O \tag{4.6}$$

Here, however, there is a contraction of 31 percent of the original volume of the oxygen present, corresponding to 6.2 percent of the air (21 percent oxygen in air) in the fill, and air is drawn into the fill. The net result is an enrichment of the air within the refuse mass, with respect to nitrogen.

In summary, aerobic decomposition is characterized as rapid, with CO_2, heat, and organics produced. The period of aerobic decomposition is of short duration because of the limited availability of oxygen within the refuse. The liquid waste products of microbial degradation include organic acids that in turn contribute to additional chemical activity within the landfill.

When these processes have exhausted all the oxygen, anaerobic activity commences, with the formation of CH_4 and CO_2 from the identical initial compounds in Equations 4.5 and 4.6:

$$\text{for carbohydrates:} \quad C_6H_{12}O_2 \rightarrow 3\,CO_2 + 3\,CH_4 \tag{4.7}$$

$$\text{for stearic acid:} \quad C_{18}H_{36}O_2 + 8\,H_2O \rightarrow 5\,CO_2 + 13\,CH_4 \tag{4.8}$$

As the equations demonstrate, solids and liquids are utilized, and 6 to 18 volumes of gas are produced. The net result is the development of increased pressures. The pressure buildup and the concentration gradients are responsible for advection and diffusion of landfill gas from the decomposing refuse.

As methane and carbon dioxide are produced, nitrogen is swept out, and the net result after lengthy periods of decomposition, as depicted schematically

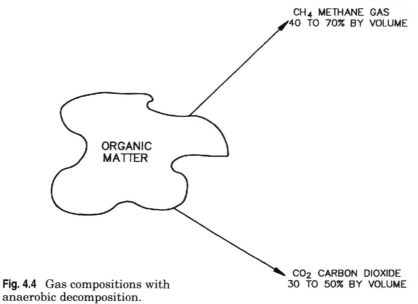

Fig. 4.4 Gas compositions with anaerobic decomposition.

on Figure 4.4, gas compositions between 40 and 70 percent methane, and between 30 and 60 percent carbon dioxide, with other gases at less than 1 percent by volume. The methanogenic stage produces CO_2 and CH_4, a neutral and rising pH, and a decrease of carboxylic acid concentrations. Lower leachate concentrations and lower temperatures occur.

Following lengthy periods (years), there is a deceleration in the generation rate of methane in response to decreasing levels of degradable organic matter in the refuse. However, the composition of the generated gas remains essentially constant, and the leachate reaches neutral pH levels with relatively low organic and inorganic concentrations. The generated gas constituents are primarily CO_2 and CH_4 with traces of mercaptans, solvents, hydrogen sulfide, and hydrogen (these various subconstituents total less than 5 percent by volume), plus water vapor and hydrocarbons.

4.4 LEACHATE PRODUCTION CHARACTERISTICS

4.4.1 Temporal Variations

The chemicals within the leachate vary over time depending on the physical, chemical, and biological activities occurring within the landfill. Figure 4.5 illustrates idealized leachate concentration production curves. These curves are theoretical in character but indicative of the temporal variations.

Leachate flows are delayed until field capacity is reached, although field capacity need be reached only in localized regions of the refuse, for leachate to develop. In addition, some leachate flows occur by short circuiting. Field capacity is generally reached after 1 to 2 years when lateral development of refuse placement is utilized, and longer if vertical development is used (because overlying refuse lowers infiltration/percolation). Once field capacity is reached, variations in leachate generation occur seasonally, because of varying infiltration/percolation rates in response to varying climatologic phenomena. Two effects of moisture migration are depicted in Figure 4.6.

The temporal variations in leachate generation are depicted in the lysimeter data of Fungaroli and Steiner (1979) in Figure 4.7. The data were collected under realistic water and temperature conditions. Moisture was added to simulate the seasonality of precipitation conditions. The graphs show that flow or percolate is delayed until local field capacity is reached, and thereafter percolate levels escalate until they parallel the moisture added.

Temporal variations in the chemical concentrations of leachate also occur. However, unlike the leachate production curves that demonstrate significant responses to variations in the infiltration/percolation rate, the concentration levels generally increase to a peak level, then gradually decline (though there may be modest seasonal fluctuations). Nevertheless, the peaks and fluctuations in concentration level out as a result of both aging and increasing size of a landfill due to spatial averaging. Rates of decline vary for different chemicals, as well. The readily soluble and biodegradable materials peak sooner and higher. For example, acetate, BOD, and chloride all peak higher and sooner than phenolics and zinc.

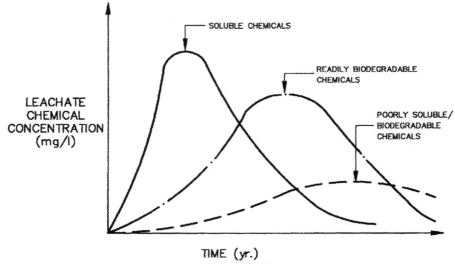

Fig. 4.5 Idealized chemical production curves.

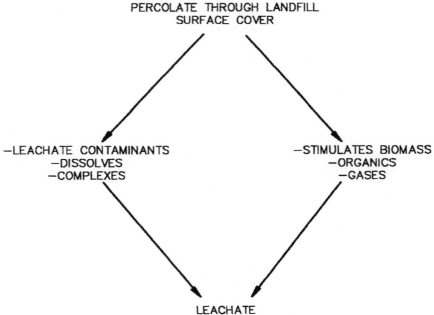

Fig. 4.6 Alternative moisture flux pathways.

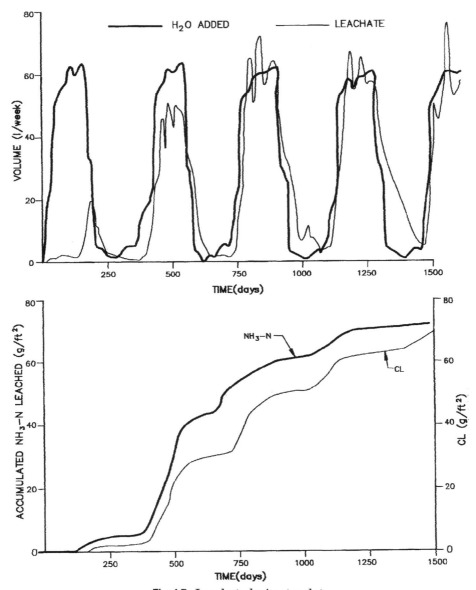

Fig. 4.7 Leachate lysimeter data.

Table 4.1   TYPICAL CHEMICAL CONCENTRATIONS IN YOUNG LANDFILL LEACHATE

Parameter	Leachate Concentration (mg/L)	Typical Sewage Concentration (mg/L)	Typical Groundwater Concentration (mg/L)
COD	20,000–40,000	350	20
BOD_5	10,000–20,000	250	0
TOC	9,000–15,000	100	5
Volatile fatty acids (as acetic)	9,000–25,000	50	0
NH_3–N	1,000–2,000	15	0
Org–N	500–1,000	10	0
NO_3–N	0	0	5

The source quantity of chemicals in the refuse is finite and is responsible for the attainment of a peak and then a decline as the residual source depletes. In addition, as depletion occurs, release rates tend to stabilize, with the fluctuations dampening as the site ages.

Although detailed discussion of leachate variations will be developed in Chapter 11, it is appropriate to examine the temporal variations here.

4.4.2 Young Leachates

In the first few years, leachates contain the readily biodegradable organic matter. Thus, these young leachates tend to be acidic due to the presence of volatile fatty acids. The pH is typically in the range of 6 to 7 and may be lower in stressed (dry) landfills. The young leachates are derived from processes such as the complex biodegradation of organics (e.g., cellulose) and simple dissolved organics (e.g., organic acids).

The result of these actions is that chemical concentrations are generally in the ranges indicated in Table 4.1. Typical values of sewage and groundwater concentrations are provided for comparison.

With time, leachates become simple dissolved organics (e.g., gases CH_4, CO_2, H_2, H_2O, and biomass) and decrease in strength.

4.4.3 Old Leachates

After 4 to 5 years, the pH increases to the range of 7 to 8. The changes occur as a result of depletion of the readily biodegradable organics and the production of gases. The poorly biodegradable organics remain. Typical chemical concentra-

Table 4.2 TYPICAL CHEMICAL CONCENTRATIONS
IN OLDER LEACHATE

Parameter	Concentration (mg/L)
COD	500–3000
BOD_5	50–100
TOC	100–1000
Volatile fatty acids such as acetic acid	50–100

tions are provided in Table 4.2.   Nitrogen levels are very useful as indicators of the age of the leachate. Ammonia nitrogen (NH_3–N) and organic nitrogen (Org–N) are produced by the decomposition of organics and are stable in the anaerobic environment; nitrate nitrogen (NO_3–N) is consumed in the anaerobic environment. These trends are demonstrated in Table 4.3.

4.5   CHARACTERISTICS OF LANDFILL GAS

4.5.1 Changes in Landfill Gas Production Over Time

Just as there are changes in leachate production rates over time, variations also occur in landfill gas production rates. Quantities and rates of gas production are a function of numerous factors, including density of refuse, leachate flow, moisture levels, and refuse age and composition.

Table 4.3 NITROGEN CONSTITUENT CONCENTRATIONS FOR VARIOUS SOURCES

Sample	Age (yr)	NH_3–N (mg/L)	Org–N (mg/L)	NO_3–N (mg/L)
Sewage	—	15	10	0
Groundwater	—	0	0	5
Young leachate	1	1000–2000	500–1000	0
Several sites (Germany)	12	1100	—	—
Du Page Co. (Illinois)	15	860	—	—
Rainham (U.K.)	24	17	—	—
Waterloo (Canada)	35	12	—	—

Figure 4.8 shows the four stages of gas production. The time scale depends on the factors cited.

Phase I: Aerobic   Oxygen entrained at the time of refuse placement is consumed during the aerobic phase. Therefore, because additional oxygen supply is limited after refuse placement, this aerobic phase of biodegradation takes only a few days (longer if the refuse is dry). The aerobic decomposition generates heat, with a typical temperature rise, ΔT, of 10 to 20°C above refuse placement tem-

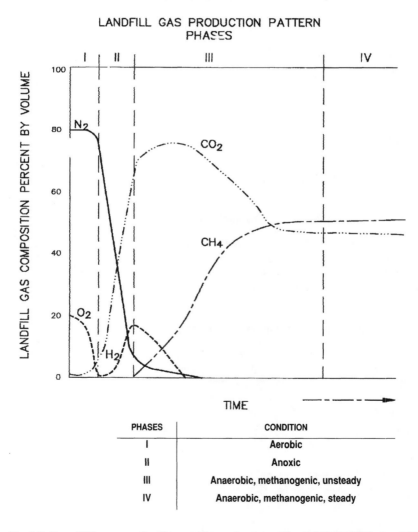

Fig. 4.8 Landfill gas production patterns (source: Farquhar and Rovers, 1973).

peratures. This temperature increase can be much higher if the moisture content is also high.

Phase II: Anoxic, Nonmethanogenic   During phase II, a CO_2 bloom occurs as a result of the acid fermentation discussed earlier. Some hydrogen gas is also produced. Indications of the duration of this phase were reported by Ludwig (1967) at the Azusa landfill. Using lysimeters, Chian and deWalle (1977) found also that, under saturated conditions, more than 70 percent CO_2 by volume was produced in 11 days.

Phase III: Anaerobic, Methanogenic, Unsteady   Methanogenesis begins in phase III. Indications of the time required to reach 50 percent of CH_4 by volume during this third phase range from 3 months in wet refuse to much longer, and perhaps never, in dry refuse.

Phase IV: Anaerobic, Methanogenic Steady   Throughout phase IV the production of CH_4 remains steady at 40 to 70 percent by volume. Eventually, the production of methane declines as refuse organics get depleted on a unit basis, but the slowly biodegradable organics yield CH_4 for decades (e.g., cellulosic organics such as wood and paper). There are many reports of CH_4 production from refuse older than 30 years, but the rate of production is low.

4.5.2 Factors Affecting Landfill Gas Generation

The rate at which landfill gas is generated depends on many factors. Continuing decomposition and gas production can be expected for up to 30 to 100 years, but these occur at a high level for a much shorter period of time. There is no simple equation or rate constant that can adequately describe the rate of decomposition in a landfill due to the existence of many types of decomposable matter. However, it is possible to at least characterize the importance of the various factors in qualitative terms. Figure 4.9 indicates relatively comprehensively the multitude of factors involved.

Water leaches substances from the waste, which in turn enter the aqueous phase and become accessible to the microorganisms. The existence of adequate moisture content is therefore essential for gas generation. The nutrient content of the aqueous phase is the second most important parameter because the growth of a bacterial population is essential for gas production and depends on nutrients for metabolism. Environmental parameters such as pH and temperature affect the activity of bacteria in the anaerobic environment. Two additional parameters believed to affect gas production rates are particle size and waste density.

Moisture Content   Moisture content is considered the most important parameter in refuse decomposition and gas production. It provides the aqueous environment necessary for gas production and also serves as a medium for

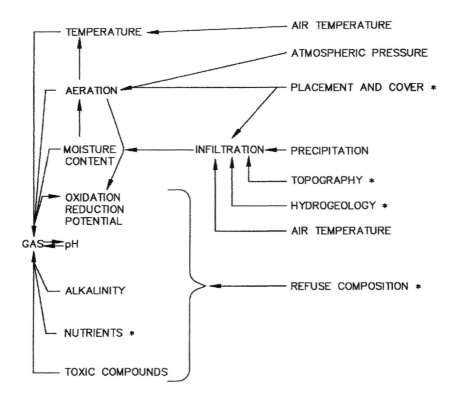

* FACTORS OVER WHICH SOME CONTROL MAY BE EXERTED
DURING SANITARY LANDFILL DESIGN AND OPERATION.

Fig. 4.9 Factors effecting gas production.

transporting nutrients and bacteria throughout the landfill. The subsistence moisture level required by methanogenic bacteria is very low and occurs even in the driest of landfills. Landfill gas is therefore produced at all landfills. Gas production is increased only moderately as moisture content increases up to field capacity because the nutrients, alkalinity, pH, and bacteria are not transferred readily within the landfill. If the moisture content in the refuse exceeds field capacity, however, the moving liquid carries nutrients, bacteria, and alkalinity to other areas within the landfill, creating an environment favorable for increased gas production.

The moisture content of solid wastes is expressed as the weight of moisture per unit weight of wet or dry material. Specifically, the wet-weight moisture content $M_{w/w}$ as a percentage is defined as:

$$M_{w/w} = \frac{w-d}{w}100 \qquad (4.9)$$

where: w =weight of the sample including water (kg)
        d =weight of the sample after drying at 105° C (kg)

The overall moisture content of refuse as received at a landfill ranges typi-cally from a low of 15 to 20 percent to a high of 30 to 40 percent on a wet weight basis. A typical average moisture content is 25 percent. Table 4.4 gives refuse moisture contents for various types of refuse as reported by Emcon (1975).

The moisture content can vary greatly in different zones of the landfill. Sur-face water infiltration through the landfill cover can progressively increase the moisture content, beginning in the upper zones. Figure 4.10 schematically depicts the moisture flux mechanisms. Refuse immediately below the cover must reach field capacity before leachate is released downward, unless channelization occurs. Groundwater infiltration through landfill sidewalls can saturate waste from the base of the landfill to the level of the surrounding groundwater and even somewhat higher as a result of capillary action. Placement of wetter wastes, such as sludge, in isolated areas within a landfill can also cause substan-tial variation in moisture content. Moisture levels can also be influenced by recy-cling the leachate (to be discussed in Chapter 11).

Moisture movement, in addition to moisture content, has been demon-strated as a significant variable affecting methane gas production. Moisture movement through decomposing refuse increases gas production by 25 to 50 percent over the production observed during minimal moisture movement, but at the same overall moisture content.

Nutrient Content   Bacteria in a landfill require various nutrients for growth, primarily carbon, hydrogen, oxygen, nitrogen, and phosphorus, but also small amounts of sodium, potassium, sulfur, calcium, magnesium, and other trace met-als. Certain nutrients are required not only in sufficient quantities but in certain ratios as well. The greater the quantity of easily "digested" nutrients, the greater the rate of gas generation; nutrients that are more difficult for the bacteria to utilize result in a lower rate of generation. Numerous toxic materials, such as heavy metals, can retard bacterial growth and consequently retard gas produc-tion.

Bacterial Content   The bacteria involved in aerobic biodegradation and meth-anogenesis exist in the refuse and soils. However, seeding the refuse with bacte-ria from another source can result in a faster rate of development of the bacteria population. Digested wastewater sludge and digester effluent can be sources of additional bacteria.

pH Level   The optimum pH range for anaerobic digestion is 6.7 to 7.5 or close to neutral. Within the optimum pH range, methanogens grow at a high

Table 4.4 REFUSE MOISTURE CONTENT

Refuse Component	Moisture Content (% of dry weight)			
	Sample 1	Sample 2	Sample 3	Sample 4
Food waste	151	133	118	122
Garden waste	67	99	102	91
Paper	29	28	38	36
Plastic, rubber, etc.	21	20	15	20
Textiles	38	28	28	25
Wood	13	17	22	18
Metals	6	7	4	4
Glass, ceramic	1	1	0	1
Ash, dirt, rock	10	26	15	13
Fires	47	47	47	51

Source: Emcon (1975)

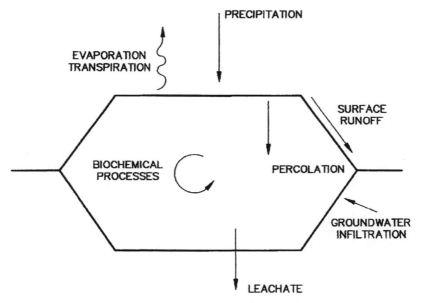

Fig. 4.10 Moisture flux at a landfill.

rate, so methane production is maximized. Outside the optimum range, below a pH of 6 or above 8, methane production is severely limited. The pH levels in a landfill may be influenced by the presence of industrial wastes, alkalinity, infil-tration of groundwater, and the relative rates of organic acid production and methane generation. Young leachates typically have a pH of less than 6 to 7 due to the presence of volatile fatty acids.

The pH of the refuse and leachate significantly influences chemical and bio-logical processes. An acidic pH increases the solubility of many constituents, decreases adsorption, and increases the ion exchange between the leachate and organic matter. An acidic pH is generally the result of the formation of organic acids during the initial stages of anaerobic decomposition. These acids become the substrate for the methanogenic bacteria. As these organics begin to prolifer-ate, the pH should rise as the acids are converted to methane. If the pH is too low, however, methanogenesis will be inhibited.

E X A M P L E

Estimate the pH of leachate in contact with landfill gas. Assume that the composi-tion of the landfill gas in contact with the leachate is 50 percent carbon dioxide and 50 percent methane and that the landfill gas is saturated with water vapor at 50°C and atmospheric pressure. The alkalinity of the leachate is 500 mg/L.

Solution

The saturation concentration of CO_2 for the indicated conditions is 379 mg/L. The pH of the leachate is found from the relation

$$\frac{[H^+][HCO_3^-]}{[H_2CO_3^*]} = k$$

where: $[H^+]$ = molar concentration of the hydrogen ion (mol/L),
      $[HCO_3^-]$ = molar concentration of the bicarbonate ion,
      $[H_2CO_3^*]$ = molar concentration of carbonic acid (mol/L) as
              $(CO_2)aq + H_2CO_3$.

For all practical purposes it can be assumed that the computed concentration value of $(CO_2)aq$ is equal to the term H_2CO_3 and that at the pH values encountered in landfills all the alkalinity is due to the carbonate ion.

The molar concentrations of HCO_3^- and $H_2CO_3^*$ are

$$HCO_3 = \frac{500 \text{ mg/L}}{50,000 \text{ mg/mol}} = 0.01 \text{ mol/L}$$

$$H_2CO_3^* \cong CO_2 \cong \frac{379 \text{ mg/L}}{44,000 \text{ mg/mol}} = 0.00861 \text{ mol/L}$$

The pH of the leachate [k at $50\infty\infty°C$ is 5.07×10^{-7}] is

$$\frac{[H^+]\,[0.01]}{[0.00861]} = 5.07 \times 10^{-7}$$

$$[H^+] = 4.37 \times 10^{-7} \qquad pH = 6.36$$

Temperature Temperature conditions within a landfill influence the type of bacteria that are predominant and the level of gas production. As mentioned previously, the optimum temperature range for mesophilic bacteria is 30°C to 35°C, whereas the optimum for thermophilic bacteria is 45°C to 65°C. Thermophiles generally produce higher gas generation rates; however, most landfills exist in the mesophilic range. Landfill temperatures often reach a maximum within 45 days after placement of wastes as a result of the aerobic microbial activity. Landfill temperatures then decrease once anaerobic conditions develop. Greater temperature fluctuations are typical in the upper zones of a landfill as a result of changing ambient air temperature. Figure 4.11 illustrates temperature fluctuations at various depths with the refuse at a shallow, relatively dry landfill. Smaller temperature fluctuations occur in the central and deeper zones because of the insulating effects of the overlying refuse mass. Landfill refuse at 15 m depth or greater is relatively unaffected by ambient air temperatures and has been observed with temperatures as high as 70°C. Isolated zones of higher temperature may exist within a landfill of generally lower temperature. These higher temperatures tend to appear at deep landfills (greater than 40 m) where sludge is added and/or leachate is recirculated. At shallow landfills, ambient temperatures can affect the refuse temperature.

Elevated gas temperatures within a landfill are a result of biological activity. Landfill gas temperatures are reported to be typically in the range of 30°C to 60°C (Emcon, 1980 and 1981). Optimum temperatures range from 30°C to 40°C, whereas temperatures below 15°C severely limit methanogenic activity. The actual temperatures that can be expected in a full-scale landfill are questionable; most published data refer to expected landfill temperatures and not actual measured temperatures for varying conditions. One publication indicates that a maximum temperature of 24°C to 46°C can be expected as a result of aerobic decomposition soon after landfilling (Ham et al., 1979).

Temperature also affects chemical solubility, because solubility increases with increasing temperatures.

The role of temperature on rate production, k, has been characterized in

Hartz et al. (1982) as:

$$k_{T_2} = k_{T_1} e^{\left(\dfrac{E_a\,(T_2 - T_1)}{R T_2 T_1} \right)}$$

(4.10)

$$k_{10^\circ C} = 0.17 k_{25^\circ C}$$

where: T = temperature (°K)
         E_a = 20 kcal/mol
         R = ideal gas constant = 1.987 kcal/°K/mol
         k_T = CH$_4$ production rate (M^3/d)

The rate of decomposition decreases with decreasing temperature.

Oxidation–Reduction Potential   The oxidation–reduction (or redox) potential is controlled by the microbial activity in the refuse and the introduction of oxygen

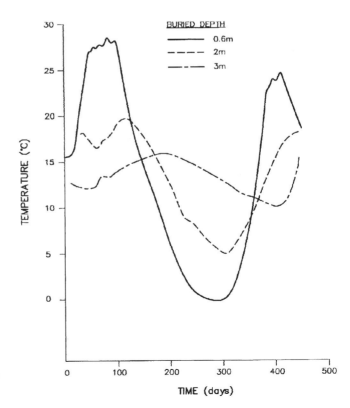

Fig. 4.11 Temperature variability as a function of time at various depths.

through rainfall and diffusion. Because oxygen is toxic to methane-forming bacteria, production of methane generally requires that the oxidation–reduction potential (ORP) be less than 330 mV (Fungaroli and Steiner, 1979; Zehnder, 1978).

As a direct result, high levels of infiltration may cause methane generation rates to decrease, especially within the upper zones of the landfill due to the high redox levels in the rainwater. A decreased redox potential results in the formation of sulfides, which tend to precipitate out the soluble metals.

Particle Size on Gas Production   Smaller particle sizes of shredded refuse are believed to have a beneficial effect on landfill gas production. A reduced particle size exposes a greater surface area of the refuse to the important parameters affecting gas production, including moisture, nutrients, and bacteria. A well-shredded mass of waste should result in increased microbial activity and transfer of nutrients, particularly if sufficient moisture is present.

Density on Gas Production   There are very few conclusive data available regarding the effect of density on landfill gas production. Within the typical density range of most landfills of 300 to 450 kg/m^3 in place, there does not appear to be a significant relationship between refuse density and gas production.

A landfill with baled wastes may be an exception, however. Bales themselves have a higher density, but when placed in a landfill, they have major air voids between them. Water infiltration through the landfill cover and percolating through the bales may provide excellent transfer of nutrients and bacteria between the bales but is of little benefit to the internal mass within the bales. Therefore, gas production in a baled landfill is expected to be slower and occur over a longer period of time than in a conventional landfill.

Table 4.5 summarizes a number of the factors affecting gas production rates in landfills.

As discussed in the preceding section, landfill decomposition frequently occurs at a very slow rate for reasons that include low temperature and pH, inadequate moisture and nutrient deficiencies.

Table 4.5 CONTROL OF GAS PRODUCTION IN LANDFILLS

Factor	Change	Effect on CH$_4$ Rate
Cover	Increased depth	
	• reduces percolation	-
	• insulates	+
	• reduces O$_2$ (ORP)	+
		overall -
Depth	Increase	

Table 4.5 CONTROL OF GAS PRODUCTION IN LANDFILLS

Factor	Change	Effect on CH$_4$ Rate
	• insulates	+
	• reduces O$_2$ (ORP)	+
	• increases contact of biomass and organics	+
		overall +
Moisture increase by recycling	• increases contact of biomass organics, etc.	+
	• reduces O$_2$ (ORP)	+
		overall +
Additives	• lime or other buffer	+
	• phosphate	+
	• digestors - MSW, sludge	+
		overall +

References

Chian, E., and deWalle, F. 1977. Evaluation of Leachate Treatment. Volume 2: Biological and Physical-Chemical Processes. *U.S. Environmental Protection Agency* EPA-600/2/-77-186b, November.

Chian, E. S. K., and deWalle, F. B. 1979. Effects of Moisture Regimes and Temperature on MSW Stabilization. *U.S. Environmental Protection Agency* EPA/2-79-023a, Cincinnati, Ohio.

Emcon Associates. 1975. Sonoma County Waste Stabilization Study. EPA-530-SW-65d.1, U.S. EPA *Office of Solid Waste Management Programs,* Washington, D.C.

Emcon Associates. 1980. *Methane Generation and Recovery from Landfills.* Ann Arbor Science, Ann Arbor, Michigan.

Emcon Associates. 1981. State of the Art of Methane Gas Enhancement in Landfills. *Report to Argonne National Laboratory,* ANL/CNSV-23, June.

Farquhar, G., and Rovers, F. A. 1973. *Gas Production During Refuse Decomposition.* Water Air and Soil Pollution 2 (10):483–499.

Fungaroli, A., and Steiner, R. 1979. Investigation of Sanitary Landfill Behavior Volume 2, Final Report. *U.S. Environmental Protection Agency* EPA-600/2-79-053a.

Ham, R., et al. 1979. Recovery, Processing, and Utilization of Gas from Sanitary Landfills. *Report to U.S. Environmental Protection Agency,* EPA-600/2-79-001, February.

Hartz, K., Klink, R. E., and Ham, R. 1982. *Temperature Effects: Methane Generation from Landfill Surfaces.* ASCE Journal of the Environmental Engineering Division 108 (EE4): 629–638.

Lu, J. C. S., Eichenberger, B., Stearns, R. J., and Melnyk, I. 1984. Production and Management of Leachate from Municipal Landfills: Summary and Assessments. *U.S. Environmental Protection Agency* EPA-600/2-84-092, May.

Ludwig, H. 1967. Final In-Situ Investigation of Gases Produced from Decomposing Refuse. *California State Water Quality Control Board,* Oakland, CA.

McCarty, P. L. 1963. "Principles and Application in Aquatic Microbiology.' In *Rudolf Research Conference Proceedings,* Rutgers University. Ed. H. Heikeleluan and W. Danders. John Wiley, New York.

Toerien, D., and Hattingh, W. 1969. *The Microbiology of Anaerobic Digestion.* Vol. 3. Pergamon Press, Great Britain.

Zehnder, A. J. B. 1978. *Ecology of Methane Formation.* Water Pollution Microbiology 2: 349–376.

Zehnder, A. J. B. et al. 1982. Microbiology of Methane Bacteria in Anaerobic Digestion. *Elsevier Biomedical Press,* London England.

4.6 PROBLEMS

4.1 Describe the chemical and physical hazards associated with solid wastes as they may affect surrounding land uses. Indicate the degree of concern associated with each of these potential hazards.

4.2 Describe the general characteristics of leachate as a function of time. Indicate the important influences on these characteristics.

MASS BALANCE COMPUTATIONAL PROCEDURES IN LANDFILL ASSESSMENT

5.1 PRINCIPLE OF CONTINUITY OF MASS

This chapter builds on the anaerobic decomposition principles described in Chapter 4; the emphasis here is on the quantification aspects that will ultimately be useful in design. Specifically, the constituents within a landfill follow the principle of continuity of mass: There are only finite quantities of various types of materials within the refuse. Thus, as various constituents leach out and/or biodegrade over time, the remaining refuse mass that continues to produce gas, and/or leachate, is depleted over time. It is important to be able to estimate how long the depletion may take.

A great deal can be learned about the existing status of a landfill from field monitoring results. The magnitudes of specific indicator parameters are key to assessing the extent to which various reactions have progressed toward completion. Knowledge of certain principles in relation to these assessments will be useful in subsequent chapters.

Although it is easy to discuss mass balance procedures in principle and in isolation for a finite quantity of material, in reality, a landfill contains very heterogeneous material of variable composition. The uncertainty of typical historical record-keeping practices at a landfill translates into considerable uncertainty about the contents. Thus, the concepts developed here are useful for understanding the progression of landfill changes, but there are limitations to the accuracy of any predictions.

REFUSE = MASS LEAVING + MASS LEAVING +   WASTE   +   MASS TRANSFORMED
PLACED   IN LEACHATE   IN GAS   REMAINING   TO OTHER PRODUCTS

Fig. 5.1   Continuity of mass principle.

Much can be learned by combining the principles and using simple models to examine individual portions or cells within a landfill. These cells can then be used as building blocks to represent components of a complete landfill. Figure 5.1 indicates the principle of continuity in a simple equation form that can be written for total weight, specific components, or individual elements. This approach can then be used to predict the decomposition. Such approaches can be employed in studies of input and output fluxes, making it possible to assess and, in some cases, to develop bounds on the short- and long-term behavior of landfills and thus its influence on the environment.

5.2   SIMPLE MODELING APPROACHES

Simple mass balance procedures are useful in obtaining a basic understanding of gas and leachate production characteristics. Often the mathematical procedures are simple curve-fitting procedures to observed data that can be very helpful in developing a reasonable understanding. The following is a simple example of the computational steps implicit in a mass balance assessment:

E X A M P L E

Estimate the remaining site life for disposal of refuse.

Solution

Step 1 Develop a general indication of the solid waste density. Landfill densities have been generally reported in the range of 450 to 500 kg/m^3, although with good compaction, 600 kg/m^3, can be achieved. Examples of physical properties of refuse are listed in Table 5.1.

Step 2 Estimate the volume of soil required to cover each day's waste from the appropriate solid waste-to-cover ratio (such as 4 to 1).

Step 3 Estimate the tonnages of refuse that may still be landfilled, given site volume remaining.

Table 5.1 EXAMPLES OF PHYSICAL PROPERTIES OF REFUSE AS REPORTED IN TECHNICAL LITERATURE

Properties	Fungaroli (1971)		Hughes et al. (1971)			Qasim & Burchinal 1970	1974	Dept. Public Works Los Angeles (1969)			Rovers and Farquhar (1973)		Walsh & Kinman (1979)	
	Laboratory Lysimeter	Field Site	Dupage Field Site	Winnetka Field Site	Elgin Field Site			1st Report Refuse	2nd Report Refuse	Paper	Lab. Cell C3	Field Cell C1	Cell 1	Cell 2
Dry density (kg/m³)	224.3	348	—	—	—	215	429	—	580	430	157	130	312	309
Wet density (kg/m³)	—	439	—	—	—	407	592	—	—	—	314	338	480	475
Initial water content (% dry weight)	18.2	26.6	—	—	—	88.0	38.0	—	85	201	100	162	54	54
Initial moisture content (cm³/cm³)	0.04	0.09	—	—	—	0.2	0.162	—	—	—	0.157	0.21	0.162	0.164
Field capacity (cm³/cm³)	0.29	0.29	—	—	—	0.338	0.333	—	—	—	0.337	0.31	0.318	0.41
Porosity (cm³/cm³)	—	—	—	—	—	0.48	—	—	—	—	—	—	—	—
Specific yield (cm³/cm³)	—	—	0.33–0.39	0.34–0.46	0.2–0.3	0.14	—	—	0.49	0.86	—	—	—	—
Saturated conductivity (cm/h)	—	—	2.6, 26.6, 26.6	21.6, 30.6, 0.792	—	—	—	95.0	0.59	0.477	—	—	—	—

5.3 PRODUCTION RATES OF LANDFILL GAS

5.3.1 Quantities of Gas Produced

The generation of methane gas depends on decomposable organic matter such as is derived from yard and food wastes (as sources of carbon) and an abundant supply of water (for the metabolism of the methanogenic organisms). The total quantity of landfill gas to be generated from a unit mass of refuse thus depends on both the organic content of the refuse and on the environmental conditions. Knowledge of the total quantities by itself is insufficient.

It is the third phase of decomposition, methanogenic decomposition (see Section 4.2 for further details), that is a concern for landfill gas migration control; however, because the first two phases of decomposition also act to decrease the available organics, there is competition for the available substrate.

As a general characterization of methanogenic decomposition, Ham and Barlaz (1987) presented the equation:

$$C_a H_b O_c N_d + \frac{4a - b - 2c + 3d}{4} H_2O \tag{5.1}$$
$$\rightarrow \frac{4a - b - 2c - 3d}{8} CH_4 + \left(\frac{4a - b + 2c + 3d}{8} \right) CO_2 + dNH_3$$

This equation assumes that substrate material of chemical formula $C_a H_b O_c N_d$ is decomposed anaerobically to methane carbon dioxide and ammonia to the extent that the organic wastes are stabilized completely according to the relationship. Biomass production has been excluded. This equation can be used to estimate an upper bound on the amount of gas produced relative to the quantity of substrate utilized.

Table 5.2 gives values of gas generation per unit of substrate for several different substrates of importance in municipal refuse. For example, Table 5.2 indicates that the gas composition for cellulose decomposition would be 50 percent CO_2 and 50 percent CH_4. For protein decomposition, it would be 51.5 percent CH_4 and 48.5 percent CO_2, and for decomposition of fats, the gas composition would be 71.4 percent CH_4 and 28.6 percent CO_2. These results indicate that gas composition is a function of the material that is being decomposed. To the extent that it is possible to characterize materials being decomposed at a specific time in a mixed substrate such as municipal refuse, the gas composition can thus be predicted

Nevertheless, many conditions will affect the gas composition. Due to the heterogeneous nature of the landfill environment, some acid-phase anaerobic decomposition and some aerobic decomposition will occur along with the methanogenic decomposition. Because both aerobic and acid-phase anaerobic decomposition give rise to CO_2 and not to methane, there may be a higher CO_2 content in the gas generated than would be predicted from Equation 5.1 alone. Furthermore, depending on the amount of moisture present and the movement of moisture within the landfill, there may be some CO_2 dissolution.

Table 5.2   THEORETICAL CO_2 AND CH_4 GENERATION

	CO$_2$ + CH$_4$ Produced		Gas Composition (% CH$_4$)
	(L/kg decomposed)	(ft^3/lb)	
Cellulose[a]	829	13.3	50.0
Protein[b]	988	15.8	51.5
Fat[c]	1430	22.9	71.4

[a]$(C_6H_{10}O_5)$

[b]Assumed 53% C, 6.9% H, 22% O, 16.5% N, 1.25% S (S not included in calculation).

[c]Assumed $C_{55}H_{106}O_6$ as typical fat (Wertheim, 1956)

Source: Ham and Barlaz, 1987

Dry conditions reduce the activity of most organisms, but because some organisms are more susceptible to moisture content than others, the microbial balance necessary for methane formation may be disrupted. Dry conditions can also lead to increased air access to the interior of the landfill, reducing methane formation (Ham and Barlaz, 1987). The methane concentration is affected more strongly by particular landfill characteristics than is the carbon dioxide concentration, reflecting the narrow range of conditions suitable for methane producers.

Given the quantity of refuse ultimately forming methane and CO_2, and the elemental composition of the refuse, Equation 5.1 can be used to develop an estimate of the maximum amount of gas to be generated. Theoretical gas yields based on stoichiometric and biodegradability relationships generally have indicated that approximately 300 to 500 L of landfill gas are produced from 1 kg of typical municipal refuse (5 to 8 ft^3/lb). As apparent from Table 5.3, large variations in gas yields have been estimated; the variations are due in part to waste compositional differences.

Table 5.4 gives examples of the theoretical volume of gas produced per pound of decomposed matter. As evident from the data, it appears that (theoretically) equal volumes of methane and carbon dioxide are generated from anaerobic decomposition of carbohydrates and proteins, whereas the corresponding volume ratio is more than three to one for fats. This phenomenon (after Esmaili, 1974) can be explained by the preponderance of hydrogen over oxygen in the molecular structure of fatty materials

Experimental data on the nature of gases generated from sanitary landfills showed that in a refuse cell 4.6 m square (50 ft^2) and 6.1 m (20 ft^2) deep that was uniformly filled with refuse material without the addition of any intermediate soil cover and where the moisture was brought up to 80 percent, approximately equal concentrations of methane and carbon dioxide of about 40 percent by volume were observed during a period of four years (Merz and Stone, 1964).

Table 5.5 from Ham and Barlaz (1987) summarizes typical quantities of gas generated from refuse. The first row in Table 5.5 indicates the theoretical

Table 5.3 ESTIMATED TOTAL LANDFILL GAS GENERATION LEVELS FROM DIFFERENT LITERATURE SOURCES

Title	Author(s)/Date	Values Presented ft³/lb (L/kg)	Conditions
Controlled Landfill Methane Project	Emcon (1981) 6th International Landfill Gas Symposium March 1983	Average 0.30–0.54ft³/lb (19–34 L/kg) (ft³ of landfill gas per pound of refuse over 20 months)	Six test cells with various water content and sludge additions. (All values normalized on the basis of a unit weight of dry refuse.)
Recovery, Processing and Utilization of Gas from Sanitary Landfills	Ham et al. (1979)	Results from Several Authors Presented Total Landfill Gas Anderson & Callinan (1970) Theoretical	
		Boyle (1976) (Theoretical) 7.21(450)	Typical Municipal Refuse
		Golueke (1971) Theoretical 4.80(300)	Determined Volatile Component of Refuse
		Pacey (1976) (Theoretical) 1.92(120)	Determined Degradability of Organic Components
		Klein (1972) 3.84(240)	Lab Measurement Digested with Sewage Sludge
		Hitte (1976) 3.36(210)	Lab Measurement
		Pfeffer (1974) 4.16(260)	Lab Measurement
		Schwegler (1973) 3.04(190)	Estimated
		Hekimian et al (1976) 0.8(47)	Estimated
		Alpern (1973) 8.48(530)	
		City of Los Angeles (1976) 6.24(390)	Theoretical/Measurements in Landfills
		Bowerman et al. (1977) 6.40(400)	Theoretical
		Blanchet (1977) 2.08(130)	Measurement/Theoretical
		VTN Consolidated for L.A. landfills 1975) 0.8(50)	Estimate
		Merz (1964) 0.21(13.1)	Measurement (small lysimeters)
		Merz & Stone (1968) 0.064(4.0)	Measurement (lysimeter)
		Rovers & Farquhar (1973) 0.048(3.0)	Measurement (lysimeter)
		Streng (1977) 0.62(39)	Measurement
		Chian et al. (1977) 0.008(.5)	Measurement (small lysimeter)
State of the Art of Methane Gas Enhancement in Landfills	Emcon (1981); June 1981	Results from Several Authors Presented	
		Leckie 8.0(500)	Stoichiometric
		Dair & Schwegler 3.0(188)	Biodegradability
		Golueke 5.6(350)	Biodegradability
		Pfeffer 4.0(250)	Biodegradability/Theoretical

Table 5.4 THEORETICAL VOLUME OF LANDFILL GAS PRODUCED
FOR VARIOUS SUBSTRATE TYPES

	Aerobic Decomposition				Anaerobic Decomposition			
	O_2 Used		CO_2 Produced		CO_2 Produced		CH_4 Produced	
	(m^3/kg)	(ft^3/lb)	(m^3/kg)	(ft^3/lb)	(m^3/kg)	(ft^3/lb)	(m^3/kg)	(ft^3/lb)
Carbohydrate	.910	14.5	.908	14.5	.459	7.33	.457	7.3
Protein	1.100	17.6	1.070	17.1	.520	8.3	.550	8.8
Fat	1.578	25.2	1.552	24.8	.450	7.2	1.100	17.6

amount of methane and CO_2 that would be formed based on the use of
Equation 5.1 for "typical" municipal refuse. The values assume a perfect biological system and perfectly degradable materials that are converted to CO_2 and
methane. More realistically, the remaining entries of Table 5.5 indicate values
that may be obtained when not all the refuse is decomposed to methane and
CO_2. Normally, it may be assumed that only the food waste content of refuse and
perhaps only two-thirds of the paper content of refuse actually decompose to
form methane and CO_2. These assumptions allow for the fact that the refuse
does not completely decompose, that some organics are lost to cell synthesis, that
some of the refuse decomposes to end products other than methane and CO_2
(such as biomass), and that some materials do not decompose under methanogenic conditions. Such resistant materials would include plastics, lignin under

Table 5.5 TOTAL GAS (CO_2 + CH_4) PRODUCTION FROM MUNICIPAL REFUSE

	Landfill Gas Produced	
Conditions	(L/kg)	(ft^3/lb)
"Typical" U.S. municipal refuse; theoretical estimate (Eq. 5.1)[a]	520 (53% CH_4)	8.33
Weight organic components by degradability; theoretical estimate (Eq. 5.1)	100–300	1.6 – 4.8
Anaerobic digestion of refuse with sewage sludge; lag measurement	210–260	3.4 – 4.2
Lysimeters or closed container; varying success in obtaining CH_4; periods approximately 1–3 years	0.5–40	0.008 – 0.64
Full-size landfills projected from existing short-term data	50–400	0.80 – 6.4

[a]28% C, 3.5% H, 22.4% O, 0.33% N, 24.9% noncombustibles, 20.7% H_2O, (Corey, 1969)
Rounded to two digits and adjusted to 21% H_2O (wet weight) and 53% CH_4.
Source: Ham and Barlaz, 1987

anaerobic conditions, and other materials that may be protected physically or not decomposable biologically under conditions found in a landfill. These assumptions lead to total gas generation levels ranging from one-third to two-thirds of the total amount of gas that could feasibly be generated.

Excluding yard waste, the total carbon available for methane production is about 20 to 33 percent, based on the assumption that paper consists of approximately 40 percent of the total organic material available within MSW and as much as 40 to 65 percent of the available moisture (Ham et al., 1979). This may result in an equivalent decrease in the rate of methane generation (El-Fadel et al., 1989; Halvadakis et al., 1988) if currently observed rates of gas production are to be maintained.

Moisture is required for the activity of most microorganisms, including the methane bacteria. Up to a limit, it also appears that microbial activity increases with moisture content. Methane bacteria perform well in digesters where the moisture content is generally in excess of 90 percent of the wet sludge weight.

Optimum moisture conditions for microbial activity were reported by Ham et al., (1979) to be at least 50 percent and ideally 100 percent of the refuse dry weight. Chian and deWalle (1979) found that to maximize methane production, the refuse moisture content had to be above 75 percent but below 100 percent of the refuse dry weight. Gas production can be directly related to microbial activity and can therefore provide a measure of waste stabilization.

5.3.2 The Rate and Duration of Gas Generation

The total quantity of landfill gas generated depends on the organic content of the refuse as placed and to some extent on various parameters that control landfill gas generation rates. The rate at which landfill gas is generated is important to gaining an understanding of landfill gas migration potential or the viability of gas recovery as a resource. The rate of gas production depends on many factors, including refuse composition (e.g., refuse high in organic matter such as food wastes and garden trimmings will decompose rapidly), age of refuse, moisture content, pH, microbial population present, temperature, and quantity and quality of nutrients. The wide range in decomposability of matter present in refuse suggests considerable difficulty in adequately describing the rate of methane generation for a landfill. Readily decomposable substances like sugars and starches will not decompose over the same period as cellulose. Thus, it is frequent practice to consider each component individually. The total landfill decomposition is then described by summing the decomposition over all the components of interest.

Theoretical approaches to characterizing the rate of methane gas generation in a landfill generally involve developing models based on first-order kinetics. The first-order expression is :

$$\frac{dC}{dt} = -kC \tag{5.2}$$

t = time
k = an assumed first-order rate coefficient.

Because Equation 5.2 integrates to the form $C = C_0e^{-kt}$, where C_0 is the potential methane gas volume to be produced at the outset (i.e., at $t = 0$), then the half-life of the production of methane gas is:

$$\frac{C_0}{2} = C_0e^{-kt} \tag{5.3}$$

or

$$t_{1/2} = \frac{0.69}{k} \tag{5.4}$$

Equation 5.2 states that the rate of loss of decomposable matter resulting from gas production is proportional to the amount of decomposable matter remaining that will produce gas. Alternatively, the cumulative gas produced to time t is characterized by:

$$C = C_0\left(1 - e^{-kt}\right) \tag{5.5}$$

The forms of Equation 5.2 and 5.3 are identical to those used in many biological systems (e.g., BOD decomposition). If the decomposable matter can be translated directly into gas generation, as Equation 5.1 indicates, Equation 5.2 can be rewritten in terms of the rate of methane formation and the total amount of methane to be formed.

This first-order expression is the most widely used model of landfill gas generation rates for a landfill. It assumes that the factor limiting the rate of methane generation at a landfill is the quantity of material remaining in the landfill that will ultimately form methane. It assumes that other variables and factors affecting the decomposition process are not limiting the rate of methane generation, except as they can be reflected in changing the rate coefficient. Such factors include the availability of moisture, the biological availability of measured substrate, and the absence of inhibitory concentrations of toxic substances. These equations accurately model the laboratory experiments in which the growth conditions are such that the availability of substrate does, in fact, control the rate of decomposition. Examples of half-life values assuming the first-order model of Equations 5.2 to 5.5 are listed in Table 5.6.

Table 5.6 VALUES OF HALF-LIVES FOR FIRST-ORDER MODEL OF METHANE GENERATION

Waste Category	Minimum Half-Life	Maximum Half-Life
Rapidly decomposable (food, garden wastes)	1/2 year	1 1/2 years
Moderately decomposable (paper, wood)	5 years	25 years
Refractory	Infinite	Infinite

Table 5.7 LANDFILL GAS GENERATION RATES

Title	Author(s)/Date	Values Presented (ft³/lb/yr)		Conditions
Recovery, Processing and Utilization of Gas from Sanitary Landfills	Ham et al (1979) 1979	Results from several authors presented		
		Ham (1979)	0.256	Theoretical (after 5 years)
		Bowerman, et al (1977)	0.22	Estimated (literature)
		SCS Engineers (1975)	0.064–0.224	Estimated (literature)
		Boyle (1976)	0.064–0.224	Estimated (literature)
		Merz & Stone (1968)	0.08	Measurement (lysimeter)
		Streng (1977)	0.08–0.096	Measurement (lysimeter)
		Chian et al (1977)	0.0032	Measurement (lysimeter)
		Rovers & Farquhar (1973)	0.128	Measurement (lysimeter)
		Merz (1964)	0.51	Measurement (lysimeter)
		Ramaswamy (1970)	6.4	Measurement (lysimeter with high food waste)
		Beluche (1968)	0.048	Measurement
		Bishop et al (1967) Low H_2O	0.51–1.00	Pilot Scale Landfill
		Engineering Science (1967)	0.256–0.656	Test Landfill Measurement
		Carlson (1977)	0.06–0.62	Measurement
		City of Los Angeles (1976)	0.176	Measurement/Theoretical
		City of Glendale (1976)	0.032	Measurement
Palm Beach County Landfill Gas Program	SCS Engineers	February 1987	0.5	
Results of the Mountain View Controlled Landfill Project	Emcon Associates	February 1987	0.25–0.6	Six test cells

Table 5.7 (Continued) LANDFILL GAS GENERATION RATES

Title	Author(s)/Date	Values Presented (ft³/lb/yr)	Conditions
State of the Art of Methane Gas Enhancement in Landfills	Emcon Associates (1980) June 1981	Results from Several Authors Presented	
		Augenstein et al. (1976) 4.2	Lysimeter
		Buivid (1980) 0–13	Lysimeter
		Chian et al 0.00096–0.069	Lysimeter
		Moell 0.22–0.056	Simulated Landfill
		Pohland 0.000013	Simulated Landfill
		Walsh and Kinman 0.0019	Simulated Landfill
		Bradley Landfill 0.045	Field Test
		Inland Cement Landfill 0.045	Field Test
		Scholl Canyon 0.046	Field Test
		Ascon Landfill 0.070	Recovery
		Azuza—Western 0.043	Recovery
		Mountain View Landfill 0.077	Recovery
		Sheldon Arleta Landfill 0.120	Recovery
		Hewitt Landfill 0.092	Gas control
		Penrose Landfill 0.079	Gas control
Controlled Landfill Methane Project	Emcon Associates (1981)	6th International Landfill Gas Symposium 0.1–0.48 over the test period	Six test cells with various water content and sludge additions
WMI Florida Pompano Beach	WMI February 1987	0.37	"Wet" landfill

Table 5.8 LANDFILL GAS GENERATION RATES (Average to Maximum Rates from Field and Laboratory Studies)

Authors	Gas Generation Rate (m^3/dry kg/ yr)	Comments
Ramaswamy (1970)	2.8	Lab study, high percentage of readily biodegradable material
Rovers and Farquhar (1973)	0.0073	Lab study, little methane produced
Augenstein et al. (1976)	0.44	Lab study, optimized conditions
DeWalle, Chian, and Hammerberg (1978)	0.055	Lab study, little methane produced
Buivid (1980)	3.2	Lab study, optimized conditions
Pohland (1980)	0.032	Lab study, optimized conditions
Emcon Associates (1981)	0.0075	Maximum from field pumping tests—moisture state unknown
Klink and Ham (1982)	0.26	Lab study, high moisture percentage through leachate recycle
Jenkins and Pettus (1985)	0.043	Maximum from incubation of 10 field samples from one site (0.13 short-term maximum)
Pacey and Dietz (1986)	0.032	Control cell, Mountain View Project (field)
Emberton (1986)	0.037	Incubation of 73 field samples; 57% of values ≤0.037 (maximum value 0.9)
Barlaz, Milke, and Ham (1987)	0.20	Lab study, used 50% shredded fresh refuse with 50% old refuse

Alternative mathematical models include:

$$a\ zero\text{-}order\ model: \qquad \frac{dC}{dt} = -k \qquad (5.6)$$

$$a\ second\text{-}order\ model: \qquad \frac{dC}{dt} = -kC^2 \qquad (5.7)$$

Tables 5.7 and 5.8 present gas generation rates cited in the technical literature for different landfill situations.

Ham et al. (1979) divided the organic portion of municipal wastes into three gross categories described as readily decomposable, moderately decomposable, and, for all practical purposes, nondecomposable.

Food wastes are readily decomposable. Paper, wood, grass, brush, greens, leaves, oils, and paint may be considered moderately decomposable. Other organic materials such as plastics, leather, rubber, and rags are poorly biode-gradable.

The half-life of readily decomposable material such as food waste is approx-imately 0.5 to 1.5 years and that of slowly decomposable material such as paper is 5 to 25 years. Alternatively, in a moist aerobic environment or inside an anaerobic digester, these same half-lives would be on the order of half a week and 2 weeks, respectively, or two orders of magnitude lower.

Knowing the total amount of gas to be produced and the rate of gas produc-tion over some period of time, it is possible to utilize a first-order model to predict the future gas production rates for a landfill. It is common practice to make such a prediction for design and feasibility studies for landfill gas utilization systems. A typical approach is to assume a lag phase or start-up period over which little methane is produced, followed by a period of active methane generation modeled by a first-order equation using assumed half-lives of refuse components of inter-est, such as paper, to compute the constant k. A simple accounting model is then used to estimate each year's production of refuse accumulated over each of the portions of the landfill. This provides an estimate of annual gas generation rates over the disposal history at the site.

The life of methane gas generation for economic recovery by such modeling appears to be 5 to 20 years, as seen from Figure 5.2, where typical values have been utilized. The magnitudes of the curves in Figure 5.2 are dependent on the rate coefficient, k, employed.

E X A M P L E

Calculation of water consumed in the formation of landfill gas

Water is consumed during the anaerobic decomposition of the organic constituents in MSW. The amount of water consumed by the decomposition reaction can be esti-mated using the formula for rapidly decomposable material.

The mass of water taken up per kilogram of dry organic waste consumed can be esti-mated from the relation

$$C_{68}H_{111}O_{50}N + 16H_2O \rightarrow 35CH_4 + 33CO_2 + NH_3$$
$$(1741) \qquad (288) \qquad (560) \quad (1452) \quad (17)$$

The mass of water consumed per kilogram of dry, rapidly biodegradable volatile sol-ids destroyed is

$$\text{Water consumed} = \frac{288}{1741}(1.0) = 0.165\frac{kgH_2O}{kg \text{ solids destroyed}}$$

Using a gas production value of 100 L/kg solids destroyed gives the corresponding value for the amount of water consumed per liter of gas product:

$$\text{Water consumed} = \frac{0.165kg(H_2O)}{100L/kg \text{ destroyed}} = 0.00165\frac{kg \ H_2O}{L \text{ gas}}$$

Note that if a landfill continues to receive refuse, portions of the site older than 5 to 20 years will be of marginal interest for gas recovery, but this will be offset by gas production from new refuse, so that the total landfill production may remain constant.

Although many gases are produced as landfill gas, we focus here on only CO_2 and CH_4 because they represent such a large percentage (95 to 99 percent) of landfill gas. Table 5.9 indicates typical landfill gas composition. Comcor (1993) reported values for hydrogen sulfide of 15 to 45 ppm; for ethyl mercaptan/dimethyl sulfide, of 0.2 to 0.6 ppm; and for methyl mercaptan, 0.2 ppm. Table 5.10 lists values reported by Rettenberger (1987).

The trace constituents in the landfill gas (1) may be co-disposed of and volatilized or (2) they may be produced within the landfill (e.g., vinyl chloride is a byproduct of the degradation of di- and trichloroethene). The quantity and quality of nutrients available to the methanogenic bacteria is also a significant factor affecting gas generation rates. The principal nutrients required by the methanogenic bacteria include carbon, hydrogen, oxygen, nitrogen, and phosphorus. In

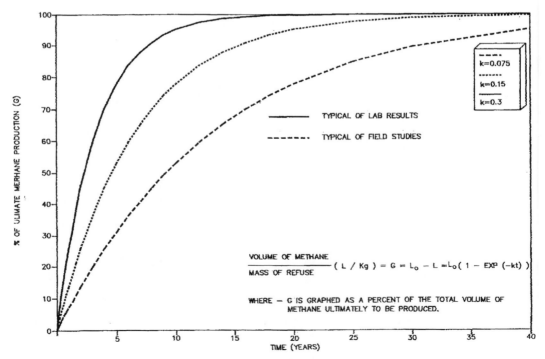

Fig. 5.2 Volume of methane produced vs time (total ultimate methane (L_0) in percent).

Table 5.9   TYPICAL LANDFILL GAS COMPOSITION

Component	Source	Typical Concentration (% by volume)	Concern
Methane (CH_4)	B[a]	50–70	Explosive
Carbon dioxide (CO_2)	B	30–50	Acidic in groundwater
Hydrogen (H_2)	B	<5	Explosive
Mercaptans (CHS)	B	.1–1	Odor
Hydrogen sulfide (H_2S)	B	<2	Odor
Solvents			
Toluene	C[b]	.1–1	Hazardous
Benzene	C	.1–1	Hazardous
Disulfates	C	.1–2	Hazardous
Others	B and C	traces	Hazardous

[a]B=Product of biodegradation
[b]C=A contaminant in the MSW

addition to these essential nutrients, the bacterial cells require limited concentrations of trace elements such as sodium, potassium, calcium, and magnesium. The application of sewage sludge to the refuse provides these needed nutrients.

5.4   PRODUCTION RATES OF LEACHATE

The generation of leachate can also be examined from the viewpoint of mass balance principles, in a manner somewhat similar to that of landfill gas production. In the situation of leachate, however, two aspects of leachate are relevant, namely, total quantity and leachate quality. Each of these will be the focus of later chapters, so we limit discussion to introductory comments to demonstrate some principles. The principle depicted in Figure 5.1 of conservation of mass applies to both water volume and chemical levels.

5.4.1 Water Quantity

There are two sources of water in the landfill—water present in the waste when landfilled and water added to the landfill. Water produced as a product of

Table 5.10   HYDROCARBONS IN LANDFILL GAS IN mg/m^3 BASED ON AIRLESS LANDFILL GAS

		(mg/m^3)			(mg/m^3)
Ethane	C_2H_6	0.8–48	Undecane	$C_{11}H_{24}$	7–48
Ethene (ethylene)	C_2H_4	0.7–31	Dodecene	$C_{12}H_{24}$	2–4
Propane	C_3H_8	0.04–10	Tridecane	$C_{13}H_{28}$	0.2–1
Butane	C_4H_{10}	0.3–23	Benzene	C_6H_6	0.03–7
Butene	C_4H_8	1–21	Ethylbenzene	C_8H_{10}	0.5–238
Pentane	C_5H_{12}	0–12	1,3,5-Methylbenzol	C_7H_8	10–25
2-Methylpentane	C_6H_{14}	0.02–1.5	Toluene	C_7H_8	0.2–615
3-Methylpentane	C_6H_{14}	0.02–1.5	m/p-xylol	C_8H_{10}	0–378
Hexane	C_6H_{14}	3–18	o-Xylol	C_6H_{10}	0.2–7
Cyclohexane	C_6H_{12}	0.03–11	Trichlorofluoromethane	CCl_3F	1–84
2-Methylhexane	C_6H_{16}	0.04–16	Dichlorofluoromethane	$CHCl_2F$	4–119
3-Methylhexane	C_6H_{20}	0.04–13	Chlorotrifluoromethane	$CClF_3$	0–10
Cyclohexane	C_6H_{12}	2–6	Dichloromethane	CH_2Cl_2	0–6
Heptane	C_7H_{16}	3–8	Trichloroemethane		
2-Methylheptane	C_8H_{18}	0.05–2.5	(chloroform)	$CHCl_3$	0–2
3-Methylheptane	C_8H_{18}	0.05–2.5	Tetrachloromethane		
Octane	C_8H_{18}	0.05–75	(carbon tetrachloride)	CCl_4	0–0.8
Nonane	C_9H_{20}	0.05–400	1,1,1-Trichloroethane	$C_2H_3Cl_3$	0.5–4
Cumole	C_9H_{12}	0–32	Chloroethane	C_2H_5Cl	0–284
Bicyclo(3,2,1)-octane-2,3-methyl-4-methylethylene	$C_{10}H_{16}$	15–350	Dichloroethene	$C_2H_4Cl_2$	0–294
			Trichloroethene	C_2HCl_3	0–182
Decane	$C_{10}H_{32}$	0.2–137	Tetrachloroethene	$C_2H_2Cl_4$	0.1–142
Bicyclo(3,1,0)hexane-2,2-methyl-5-methylethylene	$C_{10}H_{13}$	12–153	Chlorobenzene	C_6H_5Cl	0–0.2

Source: Rettenberger, 1987

decomposition is neglected in this analysis. Water added to the landfill arises from a number of possible sources:

1. from percolation of water through the landfill surface,
2. from horizontal flow through the sides,
3. from upward flow of water through the bottom.

In wet climates, the water buried with the refuse is soon overshadowed by the water added; thus, the water added controls the long-term generation. As a direct consequence, estimates of leachate quantity are determined by a mass balance reflecting initial levels, moisture additions, and outflow quantities.

The quantities of water that percolate into a landfill are a function of the water balance at the surface of the landfill (see Chapter 7 for specific details). The fate of incident precipitation in the hydrologic water balance may consist of the following: formation of surface water runoff, evaporation directly to the atmosphere, transpiration to the atmosphere through vegetation surfaces, or infiltration into the cover soils and refuse at the surface of the landfill. Water that infiltrates into the subsurface may be held in surficial soils for ultimate evaporation or transpiration. Depending on capillary action, the water held in the surficial soils can flow through plant roots to the surface during dry periods or, escaping surficial effects, can infiltrate through the refuse to become percolate and, ultimately, leachate.

Once percolation occurs, it affects leachate quantities, but the time of travel is difficult to estimate. In an idealized system, each layer must reach field capacity before moisture is passed on downward to the next layer. In reality, the landfill is not homogeneous, because there are objects that absorb moisture readily, such as paper, as well as objects that do not, such as metal, glass, and plastics. Furthermore, there are voids or channels through which water can flow ahead of the wetted front. The result is that water flows downward unevenly, reaching the bottom of the landfill as leachate long before the entire volume of refuse is at field capacity. Situations promoting channeling include the disposal of construction wastes and tree limbs, which prevent uniform compaction and result in large voids within the refuse.

In summary, the amount of leachate over the long term is a result of 1 through 3 above, with most situations being controlled by the water balance at the surface of the landfill, which, in turn, is a function of surface soils, vegetation, slope, and climate. However, once percolation occurs, it is only a question of timing as to when the leachate is formed—the volume is predictable.

5.4.2 Leachate Quality

Decomposition proceeds by physical, chemical, and biological processes. Biological processes are the most significant and in fact control, to a major extent, the chemical and physical processes. Physical decomposition involves the rinsing, physical movement, or structural change that occurs due to physical contact of water with the refuse. Chemical decomposition is chemical change arising from oxidation, reduction, or a change in pH, plus dissolution, precipita-tion, complexation, and other chemical reactions with materials in the waste. Physical and chemical decomposition are therefore heavily dependent on the amount and rate of flow of water. Chemical decomposition also depends on the chemical constituents and properties of the percolating water and the refuse it contacts. The quality of the leachate is a function of all these processes.

The depletion of oxygen results in the termination of the aerobic phase and the initiation of anaerobic processes, the first of which is the anaerobic acid phase. In this phase, degradable solid waste is converted anaerobically to CO_2, water, microorganisms and partially-degraded organics, which are mostly organic acids (especially acetic acid). Large quantities of organic materials dis-solve in the percolating water, resulting in high organic concentrations in the leachate as measured by high COD and BOD values. The combination of organic acids and dissolved CO_2 typically reduces pH levels to 5.5 and 6.5, resulting in a chemically aggressive leachate that can dissolve inorganic materials from the waste. Thus, the leachate has high specific conductance (high dissolved inor-ganic content) and high concentrations of most inorganic constituents present in the leachate.

Cellulose is a major carbohydrate found in domestic refuse. It consists of materials of different biodegradabilities, depending on the extent to which it has been processed. The biodegradability of domestic refuse cellulose is likely to be much greater than that of natural, untreated cellulose. The ratio of cellulose to hemicellulose to lignin in landfilled paper is likely to be close to that of newsprint (i.e., 70:15:15) (Rees, 1979).

Cellulose probably gives rise to glucose and cellobiose as the major products in landfill. These sugars are rapidly fermented to give CO_2, H_2, ethanol, and acetic, propionic, butyric, valeric, and caproic acids (Rees, 1979).

In the final phases of decomposition, the methanogenic phases, methano-gens convert acetic acid and hydrogen to methane and carbon dioxide. These organisms work slowly but efficiently, leaving little decomposable matter as end products (given the correct growth conditions). Because these organisms are effi-cient, they generally reduce the organic content of the leachate to low values and, by utilizing the acetic acid, raise the pH to values ranging from 7 to 8. The leachate then becomes less chemically aggressive, and the concentration of inor-ganic constituents, whose solubility is pH dependent, is reduced.

The interrelationships between the different phases of decomposition can be summarized as follows: Initially there is aerobic decomposition resulting in high CO_2 concentrations, a rapid increase in temperature, a lowering of pH, and high COD, BOD, and specific conductance levels in the leachate in contact with the waste. The landfill is unlikely to be at field capacity, so any leachate accumu-lating at the bottom of the landfill is primarily due to channeling. As the landfill reaches the anaerobic acid phase, leachate is at its highest concentration levels for COD, BOD, and specific conductance. The pH is at a minimum value—fre-quently below 6—high concentrations of CO_2 exist, and toward the end of this phase, methane begins to be generated. When the landfill reaches the methano-

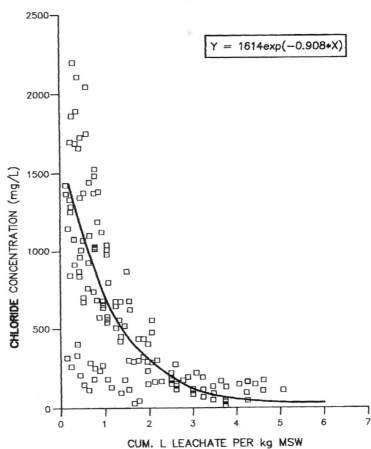

Fig. 5.3 Chloride prediction curve.

genic phase of decomposition, the pH returns to neutrality, COD and BOD are reduced because much of the COD/BOD materials are being converted to gas, the specific conductance is reduced due to the rise in pH, and methane is generated.

Consequently, COD and BOD behave in a characteristic fashion. The mobility of constituents such as heavy metals is affected by the fluctuations in pH, and there is some temporal variability in leachate quality. It must also be remembered that there are only finite quantities of the various types of constituents within the refuse mass. Thus, the most successful models of leachate quality concentration for inorganic constituents have used the cumulative volume of leachate per mass of solid waste. For example, the primary influence on chloride concentrations is the degree of leaching to which the waste has already been exposed. The lysimeter data summarized in Figure 5.3 (from Reitzel et al., 1992) indicate the leaching effect from finite quantities of chloride within the refuse mass. These results reaffirm the general concept of mass balance.

5.5   SUMMARY COMMENTS

The water balance and associated processes govern gas and leachate quantities and qualities produced over time in a landfill. Much can be learned about the ongoing activity levels within a landfill by observing the quantities and qualities of the gas and leachate generated. Different decomposition processes occur within different parts of the landfill, making interpretation sometimes rather difficult. However, the monitoring of decomposition processes in a landfill may enable prediction of future decomposition processes and rates. Further, relating waste composition to products of decomposition may enable prediction of leachate quality and gas quantity (and may improve the understanding of the processes).

References

Alpern, R. 1973. "Decomposition Rates of Garbage in Existing Los Angeles Landfills." Master's thesis, California State University at Long Beach.

Anderson, D. R. and Callinan, J. P., 1970. "Gas Generation and Movement in Landfills, Industrial Solid Waste Management." In *Proceedings of National Industrial Solid Wastes Management Conference*, University of Houston, Houston, Texas.

Augenstein, D. C., Cooney, C. L., Wentworth, R. L., and Wise, D.L. 1976. Fuel Gas Recovery from Controlled Landfilling of Municipal Waste. *Resource Recovery and Conservation* 2: 103–117.

Barlaz, M. A., Milke, M. W., and Ham, R. K. 1987. Gas Production Parameters in Sanitary Landfill Simulators. *Waste Management & Research* 5: 27–39.

Beluche, R. 1968. "Degradation of Solid Substrate in a Sanitary Landfill." Ph.D. dissertation, University of Southern California, Los Angeles.

Bishop, W. et al. 1967. "Water Pollution Hazards from Refuse Produced Carbon Dioxide." In *Advanced Water Pollution Research.* Ed. O. Jaag and H. Liebman, Water Pollution Control Federation.

Blanchet, M. J. and Staff of the Pacific Gas and Electric Company. 1977. *Treatment and Utilization of Landfill Gas: Mountain View Project Feasibility Study.* U.S. Environmental Protection Agency EPA-530/SW-583.

Bowerman, F., Rohatgi, N., Chen, K., and Lockwood, R. 1977. *A Case Study of the Los Angeles County Palos Verdes Landfill Gas Development Project.* U.S. Environmental Protection Agency EPA-600/3-77-047, July.

Boyle, W. 1976. "Energy Recovery from Sanitary Landfills: A Review of Microbial Energy Conversion." In *Proceedings of Seminar,* Gottingen, West Germany, October 4–8.

Buivid, M. G., 1980. *Laboratory Simulation of Fuel Gas Production Enhancement from Municipal Solid Waste Landfills.* Dynatech R&D Company, Cambridge, Mass.

Buivid, M. G., Wise, D. L., Blanchet, M. J., Remedios, E. C., Jenkins, B. M., Boyd, W. F., and Pacey, J. G. 1980. Fuel Gas Enhancement of Controlled Landfilling of Municipal Solid Waste. *Resource Recovery and Conservation* 6: 3–20.

Carlson, E.L. 1977. "A Study of Landfill Gas Migration in Madison, Wisconsin." Master's thesis., University of Wisconsin at Madison.

Chian, E., and deWalle, F. 1979. *Effects of Moisture Regimes and Temperature on MSW Stabilization,* U.S. Environmental Protection Agency EPA-600/2-79-023a, Cincinnati, Ohio.

Chian, E., Hammerburg, E., and deWalle, F. 1977. *Effect of Moisture Regimes and Other Factors on Municipal Solid Waste Stabilization, Management of Gas and Leachate in Landfills.* U.S. Environmental Protection Agency EPA-600/9-77-026, p. 73.

City of Glendale. 1976. *A Feasibility Study of Recovery of Methane from Parcel 1 of the Scholl Canyon Sanitary Landfill.* Report by Emcon Associates and Jacobs Engineering Co., California.

City of Los Angeles, 1976. *Estimation of the Quantity and Quality of Landfill Gas from the Sheldon-Areleta Sanitary Landfill,* Bureau of Sanitation. Research and Planning Division.

Comcor Waste Systems Ltd., 1993. *Leachate Drain Manhole Carbon Cannister Odour Control System, Report on Storrington Landfill Site,* Campbellville, Ontario.

Corey, R. C. 1969. *Principles and Practices of Incineration.* Wiley-Interscience, New York.

deWalle, F., Chian, E. S., and Hammerberg, E. 1978. *Gas Production From Solid Wastes in Landfills.* ASCE Journal of the Environmental Engineering Division 104 (EE3): 415–432.

El-Fadel, M., Finidikakis, A., and Leckie, J. V. 1989. *A Numerical Model for Methane Production in Managed Sanitary Landfills.* Waste Management and Research, 7 (1): 31–42.

Emberton, J. R. 1986. "The Biological and Chemical Characterization of Landfills." *Proceedings of Conference on Energy from Landfill Gas.* Sponsored by U.S. and U.K. Dept. of Energy, Solihull, West Midlands, U.K.

Emcon Associates. 1980. *Methane Generation and Recovery from Landfill.* Ann Arbor Science, Michigan.

Emcon Associates. 1981. State of the Art of Methane Gas Enhancement in Landfills. *Report to Argonne National Laboratory* ANL/CNSV-23, June.

Engineering Science Inc. 1966. Investigation of Movement of Combustible Gases from Sanitary Refuse Landfills. *Report to Dept. of County Engineer,* Los Angeles County, Los Angeles, Calif.

Engineering Science Inc. 1967. In-Situ Investigation of Movements of Gases Produced from Decomposing Refuse. *State Water Quality Control Board.* Publication 35. California.

Esmaili, H. 1974. *Control of Gas Flow From Sanitary Landfill.* ASCE Journal of the Environmental Engineering Division 101 (EL4, August): 555–566.

Fungaroli, A., 1971. Pollution of Subsurface Water from Sanitary Landfills. *Interim report U.S. Environmental Protection Agency,* SW-12, rg.

Golueke, C. J. 1971. *Comprehensive Studies of Solid Waste Management.* Third annual report. U.S. Environmental Protection Agency EPA-SW-10, rg.

Halvadakis, C., Finidikakis, A., Papelis, C., and Leckie, J. O. 1988. *The Mountainview Controlled Landfill Project Field Experiment.* Waste Management and Research, 6(2): 103–114.

Ham, R. et al. 1979. *Recovery, Processing and Utilization of Gas from Sanitary Landfills.* U.S. Environmental Protection Agency EPA EPA-600/2-79-00, February.

Ham, R. K. 1979. *Predicting Gas Generation from Landfills.* Waste Age 10 (November): 52–56.

Ham, R. K., and Barlaz, M. A. 1987. Measurement and Prediction of Landfill Gas Quality and Quantity. *Symposium presented at ISWA International Symposium on Process, Technology, and Environmental Impact of Sanitary Landfill,* Cagliari, Sardinia, Italy, October 20–23.

Hekimian, K. K., Lockman, W. J., and Hirt, J. H. 1976. *Methane Gas Recovery for Sanitary Landfills.* Waste Age 7(December).

Hitte, S. J. 1976. *Anaerobic Digestion of Solid Waste and Sewage Sludge into Methane.* Compost Science 17(1): 26–30.

Hughes, A. et al, 1971. *Hydrogeology of Solid Waste Disposal Sites in NE Illinois,* Final Report, U.S. Environmental Protection Agency SW-12d.

Jenkins, R. L. and Pettus, J. A. 1985. "The Use of In-Vitro Anaerobic Landfill Samples for Estimating Landfill Gas Generation Rates." *Proceedings of the First Symposium on Biotechnicological Assurances in Process and Municipal Wastes for Fuels and Chemicals,* Minneapolis, Minn. Ed. A. Antonopoulos. Argonne National Laboratory Report ANL/CNSV-TM-167.

Klein, S. A. 1972. *Anaerobic Digestion of Solid Wastes.* Compost Science 13(1): 6–10.

Klink, R., and Ham., R. 1982. *Effects of Moisture Movement on Methane Production in Solid Waste Landfill Samples.* Resources and Conservation 8: 29.

Lockman Associates, and Ham, R. 1979. *Recovery, Processing and Utilization of Gas from Sanitary Landfills.* U.S. Environmental Protection Agency EPA-600/2-79-001, Cincinnati, Ohio.

Los Angeles County and Engineering Science, Inc. 1969. *Development of Construction and Use Criteria for Sanitary Landfills.* Interim report. U.S. Dept. of Health, Education, and Welfare.

Merz, R. C. 1964. Investigation to Determine the Quantity and Quality of Gases Produced during Refuse Decomposition. *Final report to State Water Quality Control Board.* Agreement No. 12–13. U.S. CEC Report 89-10. University of Southern California.

Merz, R., and Stone, R. 1964. *Gas Production in a Sanitary Landfill.* Public Works 95(2) 84–87.

Merz, R., and Stone, R. 1968. *Quantitative Study of Gas Produced by Decomposing Refuse.* Public Works 99(11) 86fl87.

Pacey, J. 1976. "Methane Gas in Landfills: Liability or Asset?" In *Proceedings of Congress on Waste Management Technology and Resource and Energy Recovery.* U.S. Environmental Protection Agency EPA-SW-8p.

Pacey, J. G., and Dietz, A. M., 1986. "Gas Production Enhancement Techniques." In *Proceedings of Conference on Energy from Landfill Gas.* Ed. J. R. Emberton and R. F. Emberton. U.S. and U.K. Depts. of Energy, Solihull, West Midlands, U.K.

Pfeffer, J. T. 1974. *Reclamation of Energy from Organic Waste.* U.S. Environmental Protection Agency EPA-679/2-74-016 (NTIS Report PB-231 176).

Pohland, F. 1980. "Leachate Recycle as a Management Option." Prepared for presentation at Leachate Management Seminar, University of Toronto, Canada, November.

Qasim, S. R. and Burchinal, J. C., 1970. *Leaching from Simulated Landfills.* Journal of the Water Pollution Control Federation 42: 371–379.

Ramaswamy, J. 1970. "Nutritional Effects on Acid and Gas Production in Sanitary Landfills." Ph.D. diss., West Virginia University at Morgantown.

Rees, J. F., 1979. "The Fate of Carbon Compounds in the Landfill Disposal of Organic Matter." Presented at *Reclamation of Solid Wastes, 30th IUPAC General Assembly,* Davos, Switzerland, Sept. 2–10.

Reitzel, S., Farquhar, G., and McBean, E. 1992. *Temporal Characterization of Municipal Solid Waste Leachate.* Canadian Journal of Civil Engineering (August) 668–679.

Rettenberger, G. 1987. "Trace Composition of Landfill Gas." *Proceedings of International Symposium on Process, Technology and Environment Impact of Sanitary Landfill,* Cagliari, Italy, October 19–23.

Rovers, F. A., and Farquhar, G. J. 1973. Infiltration and Landfill Behavior, ASCE Journal of Environmental Engineering Division 99 (EE-5,) 671–690.

Schwegler, R. E. 1973. "Energy Recovery at the Landfill." Presented at *11th Annual Seminar and Equipment Show, Governmental Refuse Collection and Disposal Association,* Santa Cruz, Calif. November 7–9.

SCS Engineers. 1975. Environmental Impact Report on "Industry Hills Civic-Recreation-Conservation Project." Interim report. Prepared for the Industry Urban-Development Agency, Long Beach, Calif. April 16.

Streng, D. R. 1977. *The Effects of Industrial Sludges on Landfill Leachates and Gas, Management of Gas and Leachate in Landfills.* U.S. Environmental Protection Agency EPA-600/9-77-026, September, p. 41.

VTN Consolidated Inc. 1975. *Environmental Impact Reports on NRG NU Fuel Company's Landfill Gas Processing System.* Report to City of Rolling Hills Estates, Calif. January.

Walsh, J., and Kinman, R. 1979. "Leachate and Gas Production Under Controlled Moisture Conditions." *Symposium Municipal Solid Waste; Land Disposal - Proceedings of the 5th Annual Research Symposium,* EPA-600/79-023a, March 26–28, 41–61.

Wertheim, E. 1956. Introducing Organic Chemistry, 3rd ed., McGraw-Hill, New York.

Wigh, R. J. 1984. "Comparison of Leachate Characteristics." From *Municipal Solid Waste Test Cells.* U.S. Environmental Protection Agency EPA-600/2-84-124, July.

5.6   PROBLEMS

5.1 Municipal solid waste composition has changed considerably in the last decade due to societal changes in wastes, and recycling. Not only does the gas production vary with time but also with the time at which the waste was generated.

A decision was made to add an additional 5 m of waste to an older landfill site that had a refuse depth of approximately 20 m. A year later, gas samples were collected, and the concentration of CO_2 was over 70%, and the methane concentration was less than 30%. The engineers then became alarmed that there was low methane production, so they checked the quantity of nutrients, pH, and temperature and observed no abnormalities. No toxic substances were detected that would destroy the methanogenic bacteria.

Explain in detail the two main variables that may be responsible for the low methane production.

5.2 Explain why the COD concentration in landfill leachate decreases over the life of a landfill (from time $t = 0$ to t greater than 10 years).

5.3 What would be the impact of landfilling shredded, biodegradable waste on the quantity of leachate and gas generated?

5.4 What are the options for managing landfill gas?

5.5 Refuse in a landfill with an in situ density of 800 lb/yd^3 of dry solids is projected to attain field capacity when the moisture content reaches 50%. Determine how much water (in in./ft.) can be added to the refuse, beyond the initial 25% moisture content of the refuse.

5.6 A city with a population of 100,000 is expected to have a 3% increase in population per year. A new landfill site is opened to handle all waste from the city for the next 25 years. Present waste production rates are as follows:

Present Waste Production	Waste Type (tonnes/yr)
Food waste	9,500
Yard waste	6,000
Paper	11,940
Plastic	3,260
Glass	2,590

Assume that 35.4% of organics are readily decomposable, 61.0% are moderately decomposable, and 3.6% are refractory, with respective half-lives of 1, 5 and 20 years.

(a) Calculate the theoretical gas yield rates over the life of the landfill. Assume first-order kinetics.

(b) The city expects to divert 30% of food waste, 60% of yard waste, and 50% of paper from the landfill to recycling and compost facilities. Calculate the theoretical gas yield rates over the life of the landfill.

HYDROGEOLOGIC
PRINCIPLES

6.1 INTRODUCTION

The theory of porous media flow has been described in a number of references, and the intent here is not to duplicate such efforts. Instead, we focus strictly on the development of an understanding of the basic principles needed to evaluate the migration of leachate and gas constituents within and immediately adjacent to landfills. Thus, the principles outlined are intended only to develop a sufficient understanding to allow the calculation of gas and leachate migration in a simple hydrogeologic system. In addition, we assume that only relatively simple hydrogeologic systems must be dealt with. To address more sophisticated systems, we refer to sources of more advanced information.

6.2 DARCY'S LAW

Darcy's law is an empirical law developed in 1856 for flow through porous media. It states that the flow of a fluid in a porous medium is equal to the product of a constant multiplied by the gradient of the force driving the fluid through the system divided by the porosity of the medium. Darcy's law provides an accurate description of the flow of groundwater in almost all hydrogeologic environments (the limitations in this regard relate to the assumption of laminar flow) and holds for saturated flow and for unsaturated flow. Specifically, Darcy's Law states that q, the specific discharge, is directly proportional to the hydraulic head gradient:

$$q = (-K)\frac{\Delta h}{\Delta l} = (-K)\frac{dh}{dl}$$ (6.1)

where: K = a proportionality constant, referred to as the fluid (hydraulic) conductivity (L/T), that is used to describe the flow in a given porous medium
dh/dl = the fluid gradient, or loss of head (or energy) in the direction of flow (L/L) (the negative sign is introduced because dh/dl is negative in the direction of flow)

Specific discharge q has dimensions of velocity (L/T). Ranges for hydraulic conductivity K for various geologic materials are given in Figure 6.1 (Freeze and Cherry, 1979). Figure 6.1 shows that for a given hydraulic gradient dh/dl the specific discharge will be much greater for materials such as sand than for a glacial till or clay.

The specific discharge, q, is also referred to as the macroscopic velocity and represents the bulk movement of groundwater through a porous medium. In a porous medium the actual pathway followed by an individual particle of water consists of a tortuous route through the interconnected pore spaces of the medium. Thus, specific discharge is not the same as the microscopic velocity associated with the actual paths of individual particles of water. The microscopic velocities are real, but difficult to measure. An estimate of the microscopic velocities can be obtained from the average linear velocity $\bar{v}$:

$$\bar{v} = \frac{q}{n}$$ (6.2)

where: n = porosity.

The true microscopic velocities are generally larger than $\bar{v}$, because the water particles must travel along irregular paths that are longer than the linearized path represented by $\bar{v}$ (Freeze and Cherry, 1979); see Figure 6.2. Thus, $\bar{v}$ does not represent the average velocity of the water particles traveling through the pore spaces.

Despite its simplicity, the "advection" model in Equation 6.2 gives a useful order-of-magnitude estimate of contaminant flow rate.

An alternative form of Darcy's law is written as:

$$Q = -KiA$$ (6.3)

where: i = fluid gradient (dh/dl),
A = the cross-sectional area.

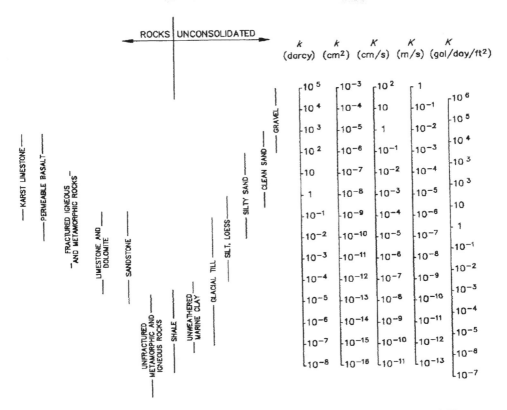

Fig. 6.1 Range of values of hydraulic conductivity and permeability. (source: Freeze and Cherry, 1979)

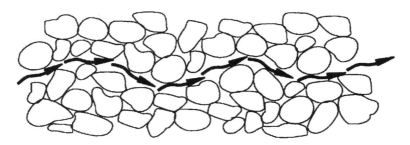

Fig. 6.2 Microscopic concept of groundwater flow.

The fluid conductivity, K, can be characterized by :

$$K = \frac{k\rho g}{\mu} \qquad (6.4)$$

where: k = specific or intrinsic permeability (L^2, e.g., cm^2),
ρ = mass density of the fluid (M/L^3, e.g., g/cm^3),
μ = dynamic viscosity (M/LT, e.g., g/cm·◊s),
g = acceleration due to gravity (L/T^2, e.g., cm/s^2).

As the name implies, the fluid conductivity differs for different fluids in response to the density and viscosity of the fluid, whereas the intrinsic permeability is a function of the soil type.

Intrinsic permeability has dimensions of L^2 and can vary by more than 12 orders of magnitude from highly permeable gravels to the relatively impermeable tills and clays. The intrinsic permeability k is proportional to the square of the particle diameter and is a function of the medium only. The fluid conductivity, K, as noted in Equation 6.4, is affected by the fluid properties as well. The fluid properties can reflect conditions for both gases and liquids. The adjustment to different fluids is an important capability (Schmidtke et al. 1992). When fluids other than water are being considered, the differences in viscosity and/or density are reflected in Equation 6.4 in combination with the intrinsic permeability of the soil in estimating the fluid conductivity.

Note that due to the nonisotropic compaction ongoing in a landfill, directional aspects are relevant. The intrinsic permeability for compacted solid waste in a landfill is frequently 10^{-11} to 10^{-12} m^2 in the vertical direction and 10^{-10} m^2 in the horizontal direction.

The application of Darcy's law has both upper and lower limits. For the head loss to vary linearly with velocity, the flow must be governed by the viscous forces within the fluid; therefore, the flow regime must be laminar. At very high rates of flow, the flow regime changes from laminar to turbulent conditions, and Darcy's law becomes invalid. The upper limit for which Darcy's law is valid is usually defined by the ratio of inertial forces to viscous forces (Reynold's number, Re) which is represented by:

$$Re = \frac{\text{inertial forces}}{\text{viscous forces}} = \frac{\rho q d}{\mu} \qquad (6.5)$$

where: ρ and μ = the fluid density and viscosity, respectively,
q = specific discharge,
d = length.

The value of d has been defined in different sources as a mean pore dimension or a mean grain diameter. Darcy's law is valid as long as the Reynolds number, based on average grain diameter, does not exceed some value between 1 and 10 (Bear, 1972).

The lower limit to Darcy's law may be reached in flow through fine-grained materials of low fluid conductivity (e.g., clays). In this case the pores of the material may be so small that the liquid molecules are attracted by the charge of the material particles. This attraction effectively increases the liquid viscosity. As a result, small fluid gradients may not be sufficient to overcome these forces (unaccounted for by Darcy's law). However, Freeze and Cherry (1979) suggest that this limit is usually of very little practical importance because the flow rates are already exceedingly small.

6.3   CHARACTERIZATION OF FLOW-FIELD PARAMETERS

6.3.1   Fluid Conductivity

Fluid conductivity, K, can be estimated using either laboratory or field techniques. Laboratory techniques typically consist of collecting and placing a sample of the porous medium in either a constant-head or a falling-head permeameter.

The technical apparatus for laboratory evaluation of Darcy's equation under constant-head conditions is schematically depicted in Figure 6.3. Analytical techniques for variable head conditions (i.e., falling head) also exist (e.g., see Peck et al., 1974).

Distilled water is generally used in soil mechanics for fluid conductivity/permeability determinations. However, tests have shown a decrease in permeability with distilled water when compared with those where pore fluid is used as the permeant. Therefore, to determine the effect of leachate or other waste liquids, the baseline fluid conductivity/permeability is often obtained using 0.1 N $CaSO_4$ (calcium sulfate).

An alternative method for estimating hydraulic conductivity for coarse-grained materials is the Hazen method, by which the hydraulic conductivity may be estimated from:

$$K = 0.01 \, (d_{10})^2 \tag{6.6}$$

where: K = hydraulic conductivity (m/s) (defined again because the equation is unit specific),
    0.01 = a constant (m/s(mm)^2),
    d_{10} = effective soil particle size (mm), where ten percent of the particles in a sample are of smaller size.

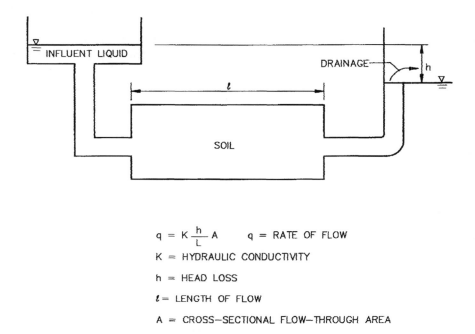

$$q = K \frac{h}{\ell} A \qquad q = \text{RATE OF FLOW}$$

K = HYDRAULIC CONDUCTIVITY

h = HEAD LOSS

ℓ = LENGTH OF FLOW

A = CROSS–SECTIONAL FLOW–THROUGH AREA

Fig. 6.3 Constant-head laboratory apparatus for determining hydraulic conductivity.

For example, for $d_{10} = 0.015$ mm,

$$K = 0.01\,(0.015)^2 \frac{\text{m}}{\text{s} \cdot \text{mm}^2}\,\text{mm}^2$$

$$= 2.3 \times 10^{-6}\,\text{m/s}$$

Typical field methods for estimating K include (1) slug tests, (2) injection tests, and (3) pumping tests.

Slug tests involve instantaneous displacement of a known volume of water and recording the response with respect to time at the tested well. Two types of slug tests are (1) rising head (volume extracted) and (2) falling head (volume added).

Injection tests and pumping tests involve the addition or removal of water, respectively, for an extended period of time (1 hour to several days) and recording of response (water level measurement) with respect to time at the tested well and at other wells (i.e., monitoring wells) in the vicinity of the tested well.

In general, hydraulic conductivity estimates obtained by laboratory and slug tests are representative of the hydrogeologic conditions local to the tested well, whereas the results of injection and pumping tests are more representative of a larger portion of the formation being tested. For additional information on these tests, see Hvorslev (1951), Bouwer and Rice (1976), Ziegler (1976),and Driscoll (1986).

6.3.2 Hydraulic Gradient

Hydraulic gradients are required for calculating flows according to Darcy's law and determining flow velocities. Hydraulic gradients are determined through measurements of hydraulic head at different locations in the subsurface. In the saturated zone, measurements are made using single and nested standpipe piezometers. In the unsaturated zone, measurements are obtained with hydraulic tensiometers (Kirkham, 1964; Richards, 1965).

Basic devices used to measure hydraulic head within the saturated zone are, either: a tube or pipe that is screened (i.e., open) within the flow zone in which the elevation of water level is measured; or, a pressure transducer. The water level in the piezometer in an unconfined aquifer gives the elevation of the water table (or free surface), that is, the point where the pressure is zero (not counting atmospheric pressure).

To estimate the hydraulic gradient in an aquifer (confined or unconfined) water level measurements are made using single standpipe piezometers (see Figure 6.4). The intake to the piezometer is usually a section of slotted pipe referred to as the screen. For a confined saturated zone, a seal between confining layers, such as depicted in Figure 6.4(b), must be installed. The seal is installed to ensure that the monitored levels are not being influenced by the overlying unconfined unit. When hydraulic head measurements are required for a system consisting of multiple aquifers, nested piezometers are used. For the situation shown in Figure 6.4(c), the confined aquifer has a larger hydraulic head than the unconfined aquifer. Thus, the vertical flow direction is upward from the confined aquifer to the unconfined aquifer.

By placement of two piezometers separated by a known distance, l, the hydraulic gradient, i, in the field can be easily measured, as depicted in Figure 6.5.

6.3.3 Capillarity

Between two fluids in contact with each other, or a fluid in contact with a solid, there is a free interfacial energy created by the difference between the forces that attract the molecules toward the interior of each phase and those that attract them to the contact surface. As these forces increase, fluids such as pore water are retained in the porous medium above the elevation of the water table.

In the laboratory, the interfacial tension in a capillary tube causes water to rise and form a meniscus. The height of this rise is a function of the radius of the tube. In straight columns, the principles are easy to depict, as indicated in Figure 6.6.

In porous media, the shape of the interface is very complex due to the tortuosity of the interstices between the soil grains. Nevertheless, the capillary rise

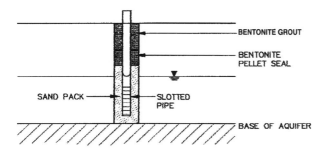

a) UNCONFINED AQUIFER INSTALLATION

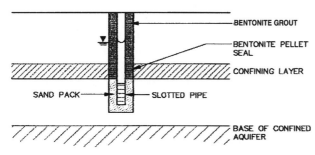

b) CONFINED AQUIFER INSTALLATION

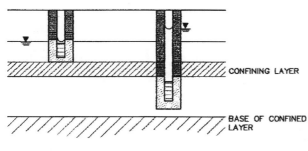

C) NESTED INSTALLATION

Fig. 6.4 Typical piezometer arrangements to measure hydraulic head.

potential is real. This capillary rise occurs in the subsurface between the unsaturated zone and the saturated zone. Table 6.1 gives values of both hydraulic conductivity and capillary rise for different soil types and refuse types.

The relationship to quantify the capillary rise is written as:

$$P_c = \frac{2\sigma_r}{r}\cos a \qquad (6.7)$$

where: σ_r = surface tension of the fluid in gm/cm^2,
r = radius of the tube in cm,
a = contact angle between the surface of the fluid and the
wall of the tube.

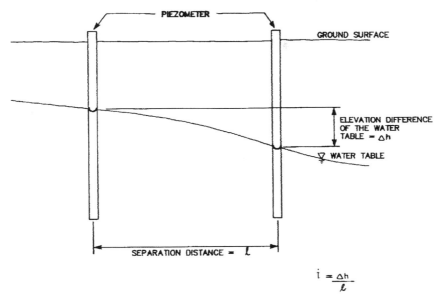

Fig. 6.5 Field determination of hydraulic gradient using two piezometers.

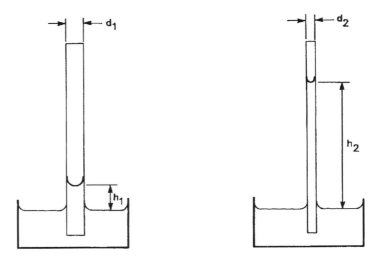

Fig. 6.6 Schematic of
capillary pressure con-
cepts.

$$d_1 > d_2$$

$$h_1 < h_2$$

6.3.4 Porosity

Refuse and soils naturally contain a percentage of space that may be occupied by fluids. This space is called the porosity. The porosity, n, is defined as:

Table 6.1   APPROXIMATE AVERAGE HYDRAULIC CONDUCTIVITY AND CAPILLARY HEAD OF SOILS

Soil Type	USCS Soil Type	Hydraulic Conductivity (cm/s)	Capillary Head (cm)	Reference
Gravel	GP	10^{-1}	—	Lutton et al. (1979)
	GW	10^{-2}	6	
	GM	5×10^{-4}	68	
	GC	10^{-4}	—	
Sand	SP	5×10^{-2}	—	Lutton et al. (1979)
	SW	10^{-3}	60	
	SM	10^{-3}	112	
	SC	2×10^{-4}	—	
Silt	ML	10^{-5}	180	Lutton et al. (1979)
	MH	10^{-7}	—	
Clay	CL	3×10^{-8}	180	Lutton et al. (1979)
	CH	10^{-9}	200-400	
Refuse Type				
Refuse as placed		10^{-3}	—	Oweis et al. (1990)
Shredded refuse		$10^{-2} - 10^{-4}$	—	Fungaroli and Steiner (1979)
Baled refuse	—	7×10^{-4}	for dense specimens (unit weight of 71 lb/ft^3)	Fang (1983)
		1.5×10^{-3}	for loose specimens (unit weight of 35.8 lb/ft^3)	

$$n = \frac{V_v}{V_T} \qquad (6.8)$$

where: V_v = volume of the voids,
V_T = total unit volume, and
V_s = the volume of the solid portion, is $V_T - V_v$.

Ranges of typical porosity values for various materials are provided in Table 6.2. Both the hydraulic conductivity and porosity of municipal refuse are

Table 6.2 POROSITY VALUES FOR VARIOUS MATERIALS

Material Type	Porosity (%)
Clay	45–55
Sand	35–40
Gravel	30–40
Sand and gravel	20–35
Peat	85–90
Municipal waste	30–40

highly variable, depending on the waste composition, density, and stage of decomposition. Hydraulic conductivity has been reported as ranging from 1×10^{-2} to 1×10^{-5} cm/s (Hughes, 1971; Fungaroli and Steiner, 1979), with a majority of the measurements within an order of magnitude of 1×10^{-3} cm/s (as indicated in Table 6.1).

In general, a portion of the porosity consists of dead-end, nonconnected pores through which advective migration cannot occur. Thus, in terms of migration of fluids, although porosity is necessary and important, the effective porosity is of greater concern. The effective porosity, n_e, of a material is a measure of the interconnected pore space through which a fluid will flow. It is usually represented as a percentage of the material's total volume. Values of effective porosity are extremely important, as they allow the advective flow velocity of the fluid to be calculated according to the equation:

$$v_a = \frac{-Ki}{n_e} \qquad (6.9)$$

where: v_a = advective flow velocity.

The other variables are as defined previously. Thus, the advective flow velocity of a medium is a function of its effective porosity.

Materials with extremely low values of n_e result in high groundwater flow velocities and, in the case of chemicals passing through landfill liners, potentially short times to breakthrough.

Although clays can exhibit relatively high total porosity values (40 to 70 percent), only a small percentage of the pore spaces may be interconnected and thus contributing to the effective porosity. Thus, in unsaturated flow through low-permeability soil, it is likely that values of n_e approach zero and the interstitial groundwater flow velocities become high (Cartwright, 1984). However, the volume of fluid migrating through the few small interconnected pores is small.

6.3.5 Moisture Content

Darcy's law and the concepts of hydraulic head and fluid conductivity have been developed with respect to a saturated porous medium. In the unsaturated zone, the pore spaces are partially filled with water and partially filled with air. The volumetric moisture content, Θ, is defined as:

$$\Theta = \frac{V_w}{V_v} \tag{6.10}$$

where: V_w = volume of water,
V_v = total volume of pore space,
V_a = volume of air ($V_v = V_w + V_a$).

For saturated flow (i.e., $V_a = 0$), $\Theta = n$, and for unsaturated flow, $\Theta < n$.

In general, the lower the degree of saturation in the unsaturated zone, the lower the hydraulic conductivity of the material. The influence of varying moisture conditions on groundwater flow is presented in greater detail in Freeze and Cherry (1979).

6.4   CHEMICAL MIGRATION

The subject of chemical migration consistent with leachate and/or groundwater movement is the result of a series of processes including advection, dispersion, and retardation. Although not intended to be a detailed treatise, brief comments on each of these processes are essential in the discussion of the fundamentals of chemical migration. Thus, the focus here is specific to the development of the fundamental principles. Additional discussion regarding the magnitudes and the application of these processes to individual problems will be introduced at appropriate locations within the text, in particular, in Chapter 12.

6.4.1 Advection

The process by which solutes move due to the bulk motion of the flowing groundwater is termed *advection* (the same process is also referred to as *convection* or *plug flow* in various texts).

Table 6.3 SUMMARY OF FACTORS CAUSING HYDRODYNAMIC DISPERSION

Molecular diffusion	Due to chemical gradients
Mechanical mixing	On a microscopic scale results from
	• velocity differences across individual pores
	• velocity differences in different pore channels
	• fluctuations in the streamline with respect to mean flow direction

A simple model describing advection assumes that the solute moves at the average linear groundwater velocity, $\bar{v}$, as determined by Darcy's Law (Equation 6.2), as:

$$\bar{v} = \frac{q}{n} = \frac{-K}{n} \frac{\varsigma h}{\varsigma l} \qquad (6.11)$$

This model can be used to calculate the rate of movement of a leachate through a landfill liner, thus determining the "time to breakthrough." It also provides an estimate of the rate at which a chemical plume will move horizontally within an aquifer.

6.4.2 Dispersion

Although the concept of advection is useful as an indicator of the rate of chemical migration, in reality there is also a tendency for the solute to spread out from the path it would be expected to follow according to simple advection. The process creating this spreading is called *hydrodynamic dispersion*. It occurs as a result of molecular diffusion and mechanical mixing, and results in the dilution of the solute.

Molecular diffusion is the phenomenon whereby molecular constituents of a fluid move under the influence of their thermal-kinetic energy from a zone of high species concentration to one of low concentration. Mechanical mixing, on the other hand, is caused by velocity variations within the pore network. Table 6.3 summarizes these two processes.

Of the two, mechanical mixing dominates at moderate to high groundwater velocity; chemical or molecular diffusion dominates at low velocities.

Mathematically, hydrodynamic dispersion is represented by the dispersion coefficient D_l. Because D_l is a function of both mechanical mixing and chemical diffusion, D_l is not a constant but depends on the flow velocity. Therefore,

$$D_l = \alpha_l \bar{v} + D^* \qquad (6.12)$$

where: α_l =   so-called dispersivity (L),
 D^* =   apparent or effective diffusion coefficient of the solute species in porous material.

In turn,

$$D^* = wD'$$ (6.13)

where: D' = diffusion coefficient in 'free' solution
         w = empirical coefficient that takes into account the solid phase on rates of diffusion.

Example values are $D' = 10^{-9}$ m^2/s, and $w = 0.01$ to 0.5 (see Freeze and Cherry, 1979). The spreading action of dispersion is illustrated schematically in Figure 6.7.

6.4.3 Chemical Retardation

There is often a significant difference between the observed velocity of groundwater and the velocity of some of the dissolved species that it transports. This difference arises from the retardation of the chemical species by attenuation processes (see Chapter 12). It occurs either within the groundwater or through interaction with the solid phase of the porous medium, both unsaturated and saturated, through which the groundwater is traveling. To represent this behavior, the term relative ionic velocity is defined as the ratio of the average velocity of the ionic species, v_c, to the average velocity of the advecting groundwater v:

$$\frac{\bar{v}}{\bar{v}_c} = 1 + \frac{\rho_b}{n}K_d$$ (6.14)

where: ρ_b = bulk density of the porous medium,
         K_d = retardation coefficient.

Values of K_d vary from zero (nonreactive) to approximately 10^3 ml/g or more. If K_d is greater than 10 ml/g, the solute is effectively immobile. Additional discussion on natural attenuation processes is presented in Chapter 12.

Consequently, ions that are considered to be conservative (i.e., nonretardable) have a relative ionic velocity of unity. At the other extreme, ionic species that are irreversibly removed from solution have a relative ionic velocity of zero. Thus, the retardation factor is defined as the inverse of the relative ionic velocity.

E X A M P L E

The specific discharge at a site was determined to be 1.6×10^{-8} m/s at a site with a bulk density of waste of 1.6 g/cm^3. Assume a saturated medium with a constant water content of 0.4 (i.e., $\theta = 0.4$). Determine the average velocity of strontium (Sr^{2+}) if $K_d^{Sr^{2+}} = 10$ ml/g.

Solution

From saturated media assumption, $\theta = n$. Then

$$\bar{v} = \frac{q}{n} = \frac{1.6 \times 10^{-8}}{0.4} = 4.0 \times 10^{-8} \text{m/s} \qquad (6.15)$$

From Equation 6.14

$$\frac{4.0 \times 10^{-8}}{\bar{v}_c} = 1 + \frac{1.6 \text{g/cm}^3}{0.4} \left(10 \frac{\text{ml}}{\text{g}} \right) \qquad (6.16)$$

$$\bar{v}_c = 9.76 \times 10^{-10} \text{m/s}$$

6.5   ANALYTICAL SOLUTIONS

Analytical solutions can assume many different forms and degrees of complexity. However, there is one analytical solution that represents the simplest and most well-known equation describing dispersion through a porous medium. This equation is described here. Interested readers are referred to Van Genuchten and Alves (1982) for further reading.

For a continuous step function with inflow of concentration C_0 introduced at the origin $l = 0$ (after Ogata, 1970),

$$\frac{C}{C_0} = \frac{1}{2} \left[\text{erfc} \left(\frac{l - \bar{v}t}{2\sqrt{D_l t}} \right) + \exp \left(\frac{\bar{v}l}{D_l} \right) \text{erfc} \left(\frac{l + \bar{v}t}{2\sqrt{D_{lt}}} \right) \right] \qquad (6.17)$$

where: l = distance along the flow path,
D_l = coefficient of hydrodynamic dispersion in the direction of the flow path,
$\bar{v}$ = rate of advective transport,
C = concentration of the fluid in the column or column output (mass of solute per unit volume of solution).
C_0 = influent concentration.

In situations where flows are significant and mechanical mixing dominates chemical diffusion, Equation 6.15 simplifies to give

$$\frac{C}{C_0} = \frac{1}{2} \text{erfc} \left(\frac{l - \bar{v}^t}{\sqrt{D_{lt}}} \right) \qquad (6.18)$$

Additional discussion on chemical migration is presented in Chapter 12.

References

Bear, J. 1972. *Dynamics of Fluids in Porous Media.* American Elsevier, New York.

Bouwer, H., and Rice, R. C. 1976. *A Slug Test for Determining Hydraulic Conductivity of Unconfined Aquifers with Completely or Partially Penetrating Wells.* Water Resources Research 12 no. 3 (June): 423–428.

Cartwright, K. 1984. Shallow Land Burial of Municipal Wastes in Groundwater Contamination (Studies in Geophysics). National Research Council (U.S.), *Geophysics Study Committee.* National Academy Press, pp. 67–77.

Driscoll, F. G. 1986. Groundwater and Wells, 2d ed., Johnson Division, St. Paul, Minn.

Fang, H. Y. 1983. Physical Properties of Compacted Disposal Materials. Unpublished report.

Freeze, R. A., and Cherry, J. A. 1979. *Groundwater.* Prentice-Hall, Englewood Cliffs, N. J.

Fungaroli, A., and Steiner, R. 1979. Investigation of Sanitary Landfill Behavior, Volume 1, *Final Report. U. S. Environmental Protection Agency* EPA-600/2-79-053a.

Hughes, A., Schleicher, J. A. and Cartwright, K. 1971. Hydrology of Solid Waste Disposal Sites in N.E. Illinois, Final Report. *U. S. Environmental Protection Agency* EPA-SW12d.

Hvorslev, M. J. 1951. Time Lag and Soil Permeability in Groundwater Observations, *Waterways Experiment Station,* Vicksburg, Miss., April.

Kirkham, D. 1964. *"Soil Physics."* In Handbook of Applied Hydrology, ed. V. T. Chow. McGraw-Hill, New York.

Lutton, R. J., Regan, G. L., and Jones, L.W. 1979. Design and Construction of Covers for Solid Waste Landfills, *U. S. Environmental Protection Agency* EPA-600/2-79-165.

Ogata, A. 1970. Theory of Dispersion in a Granular Medium, *U.S.G.S. Professional Paper* 411-I.

Oweis, I. S., Khera, R. P. 1990. *Geotectonically of Waste Management.* Butterworths. Cambridge, England.

Peck, R. B., Hanson, W. E., and Thornburn, T. H. 1974. *Foundation Engineering.* John Wiley, Toronto, Ontario.

Richards, S. J. 1965. "Soil Suction Measurements with Tensiometers." In Methods in Soil Analysis, part 1, Ed. C. A. Black. *American Society of Agronomy,* Madison, Wisconsin, pp. 153–163.

Schmidtke, K., McBean, E., and Rovers, F. 1992. Collection Well Design for DNAPL Recovery. *ASCE Journal of the Environmental Engineering,* Division 118 No. 2 (Mar/Apr): 183–195.

Van Genuchten, M. Th., and Alves, W. J. 1982. Analytical Solutions of the One-Dimensional Correctetive - Dispersive Solute Transport Equation, *U.S. Salinity Lab,* Riverside, Calif.

Ziegler, T. W. 1976. *Determination of Rock Mass Permeability.* NTIS Report 5-76-2, January.

6.6 PROBLEMS

6.1 A sample of soil was tested and found to have the following properties: $D_{10} = 0.0018$ cm, $n = 0.25 = (V_v / V_T)$, and $n_e = 0.2$.

Calculate (a) the hydraulic conductivity and (b) the volume percentage of the solid portion of the sample.

List all assumptions.

6.2 A constant-head permeameter was set up as shown in Figure 6.3. The length and area of the soil column were 20 cm and 25 cm^2, respectively. At a head differential of 10 cm, the measured flow rate was 55 cm^3/min. Calculate the hydraulic conductivity in (a) centimeters per second (b) feet per day.

Based on Figure 6.1, what type of material is this?

6.3 Three monitoring wells were installed as shown in Figure 6.8.

Water levels were measured, and the following water elevations were determined:

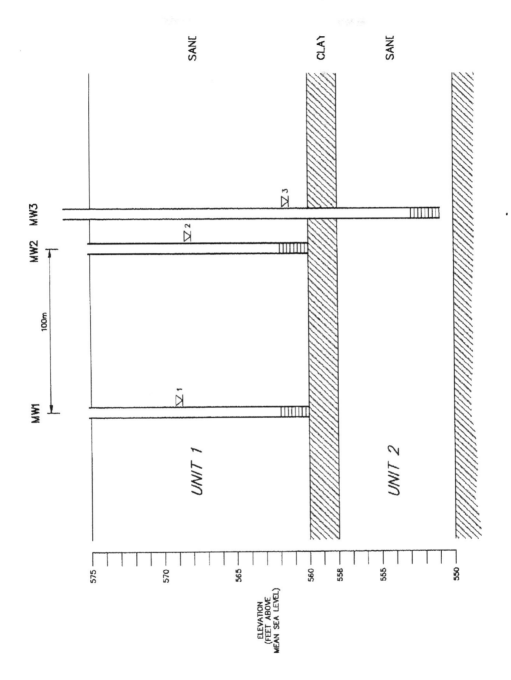

Fig. 6.8

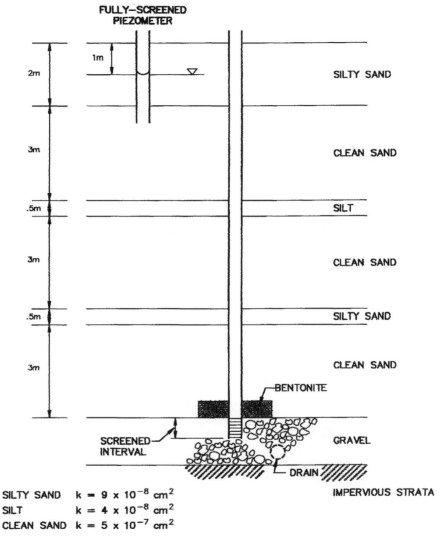

SILTY SAND $k = 9 \times 10^{-8}$ cm^2
SILT $k = 4 \times 10^{-8}$ cm^2
CLEAN SAND $k = 5 \times 10^{-7}$ cm^2

ASSUME SPATIAL DIMENSIONS OF THE AREA ARE 100 x 300m.

ANSWER   (a)   300m^3
         (b)   4.73m

Fig. 6.9

$MW_1 = 573.21$ masl

$MW_2 = 569.89$ masl

$MW_3 = 565.37$ masl

Calculate (a) the horizontal gradient between MW_1 and MW_2 and (b) the vertical gradient through the confining layer.

What is the direction of groundwater flow between units 1 and 2? Using the soil properties in Problem 6.1, calculate (a) the Darcy flow, (b) the average linear flow, and (c) the advective flow velocities in unit 1.

6.4 For groundwater, ρ is typically 1 g/cm^3, and μ μ = 1 centipoise. For the materials described in Problems 6.1 and 6.2, calculate the intrinsic permeability. If the fluid properties are $\rho = 0.8$ g/cm^3 and $\mu = 1.26$ centipoise, what is the hydraulic conductivity of the materials?

6.5 Assume a layered soil system in which there is negligible inflow or outflow of water through the lower layer or sides (confined as indicated in Figure 6.9). The average influx at the water table is estimated as 1.16×10^{-7} m/s/yr. The water table is 3 ft. below the ground surface. Find (a) the expected volume of water generated per day and (b) the rise in water level in piezometer #3.

6.6 Four different soil layers exist, each with a thickness of 10 m. The hydraulic conductivities of the individual layers are 5×10^{-4} cm/s, 4×10^{-5} cm/s, and 5×10^{-4} cm/s, progressively, from top to bottom layers. If the differential head is 110 m at the top and 82 m at the bottom, calculate the hydraulic head at the three changes in soil strata.

WATER BALANCE MODELING FOR A LANDFILL

7.1 INTRODUCTION

As demonstrated in Chapters 4 and 5, moisture is of critical importance in the biological stabilization rate of refuse. Moisture is essential because it affects the availability of nutrients and bacteria necessary for biological decomposition. Knowledge of the water content and water movement within the landfill is also of great importance in planning for leachate collection systems, for leachate treatment and for liner design. As a result, the estimation of moisture fluxes within landfills is extremely important in gaining an understanding of the ongoing decomposition processes within a landfill.

In this chapter we examine in considerable detail the various processes within the overall water balance. The long-term water balance is controlled in large part at the surface of the landfill and thus is a function of surface soils, vegetation, slope and climate. As a result, we focus on these surface processes, but we also examine the range of water balance components active within the refuse.

An understanding of the moisture flow within the landfill is developed from water balance principles. The water balance principle is one of mass conservation. There is an initial water content at the time of placement of the refuse, and inputs and outputs of water over time to and from the refuse. Thus, the water balance is a mathematical technique for keeping track of the moisture inputs to storage within, moisture outputs from a landfill and for characterizing the active features at the landfill surface that influence the percolation into the refuse (e.g., surface runoff, evapotranspiration, and infiltration). Percolation recharges the water content within the refuse, ultimately resulting in leachate collection by the tile- and barrier-system at the base of the landfill, some migration through the barrier, and discharge to the environment.

7.2 COMPONENTS OF WATER BALANCE MODELING

7.2.1 Elements of the Water Budget within the Landfill

Water balance modeling is a general term for characterizing, over time, the changes in water content within the refuse. There are two sources of water in the landfill—the water present in the waste when landfilled (primary leachate) and the water added to the landfill from precipitation and groundwater inputs (secondary leachate). In wet climates, the primary leachate is soon overshadowed by secondary leachate, which controls the long-term generation. Secondary leachate arises from percolation of water through the landfill surface, from horizontal flow through the sides, or from upward flow of water through the bottom.

It is necessary to characterize the water contributions and losses from the landfill over time to be able to predict the leachate generation rates. To accomplish these many and varied tasks, the methodology must include numerous features associated with the water balance, as schematically depicted in Figure 7.1. Although the concepts exist for the mathematical characterization of the numerous processes depicted in Figure 7.1, varying degrees of sophistication can be employed. For example, the amount of leachate that will be generated can be estimated simply as a percentage of precipitation. Kmet (1982) indicated that the following percolation rates could be expected for a municipal solid waste landfill in Wisconsin: percolation during the active operating period of the site, 25 to 75 cm/yr (30 to 100 percent of incident precipitation); and percolation after final cover placement, 5 to 15 cm/yr (7 to 20 percent of incident precipitation). Although simple, these values are highly uncertain. Actual rates will vary depending on the type and thickness of cover soil, slope of the surface cover, and operational practices. The resulting need for adjustments in the percentages to apply them to a different site are highly questionable.

More sophisticated methodologies (than the percentage assignment) include (1) the Water Balance Method (WBM) (Fenn et al., 1975) and (2) the Hydrologic Evaluation of Landfill Performance (HELP) Model (Schroeder et al., 1983) and its updates.)

Both these methodologies address the same problem. Differences between the two approaches arise in the technical bases employed and the computational sophistication. Only the HELP model provides the overall water balance of a landfill, because the WBM restricts its focus to the surface layer of the site, in terms of percolation. Once infiltrating water gets below the evapotranspiration depth, it becomes percolation and eventually leachate. Thus, both approaches

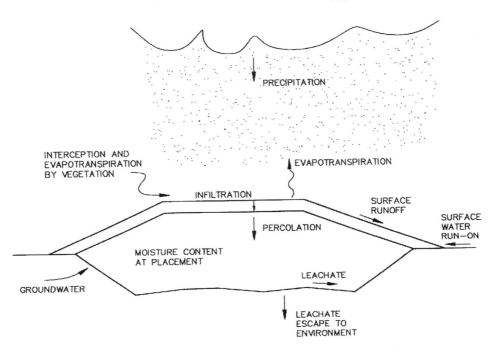

Fig. 7.1 Schematic of components of water balance within a landfill.

examine the surface water balance, but only the HELP model continues the examination once percolation quantities are calculated.

7.2.2 The Water Balance Method (WBM) Principle

The water balance method uses conservation of mass, by first determining the major segments of precipitation that detract from percolation (e.g., interception by vegetation). The method accounts for precipitation after it strikes the ground. The incident precipitation can form surface water runoff, evaporate directly to the atmosphere, transpire to the atmosphere through vegetation surfaces, or infiltrate into the cover soils and refuse at the surface of the landfill. This accounting takes into consideration nine computational elements, as listed in Table 7.1 and schematically depicted in Figure 7.2.

Water movement by percolation through the surface layer is caused by gravity. A further reduction of moisture content (beyond gravitational effects) can be achieved by evapotranspiration and biological degradation within the refuse. The difference between moisture content and the water-holding capacity of the refuse is the potential for moisture uptake. This potential depends on the waste density, which changes over time.

Table 7.1: COMPUTATIONAL ELEMENTS OF WATER BALANCE AT THE LANDFILL SURFACE

Abbreviation	Component
PET	Potential evapotranspiration
R/O	Surface runoff
I	Infiltration
ST	Soil moisture storage
ΔST	Change in soil moisture storage over time
I-PET	Infiltration minus potential evapotranspiration
ΣNeg (I-PET)	The sum of the negative values of (I-PET)
AET	Actual evapotranspiration
PERC	Percolation

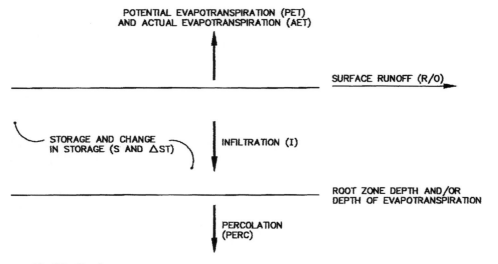

Fig. 7.2 Surface water components of water balance.

Table 7.2: DEFINITION OF SOME WATER BALANCE TERMS

Soil water content	The ratio of the weight of water in a soil or refuse to unit total weight.
Field capacity	The soil water content after drainage by gravity. A field capacity of 30 percent by volume corresponds to 30 cm/100 cm.
Wilting point	The lowest soil water content that can be achieved by plant transpiration.
Available moisture (available to plants)	The difference between the soil water content at field capacity and the wilting point.

Percolation *(PERC$_t$)* into the refuse from the surface layer may be estimated as

$$PERC_t = P_t - R/O_t - \Delta ST_t - AET_t \qquad (7.1)$$

where t denotes the time associated with each of the surface water
components.

Note that actual evapotranspiration is less than or equal to potential evaporation $(AET_t \le PET_t)$, subject to the availability of water. Percolation is that component that ultimately becomes leachate. A number of methods are available for computing R/O_t, ΔST_t, and AET_t.

Of the computational elements of the water balance listed in Table 7.1 used in calculating percolation rates, three of the four are speculative (R/O, PET, ST, and possibly I) (after Gee, 1981) and are based on empirical or experimental methods. All others are derived from the first three and precipitation.

To assist in the subsequent discussion, it is useful to establish some terminology, as listed in Table 7.2, for soil water content, field capacity, wilting point, and available moisture.

Moisture content of solid wastes is usually expressed as a percentage of the weight of moisture per unit weight of wet or dry material. In the wet-weight method of measurement, the moisture in a sample is expressed as a percentage of the wet weight of the material; in the dry-weight method, it is expressed as a percentage of the dry weight of the material.

Values of field capacity and available moisture capacity of refuse as reported in the technical literature are listed in Table 7.3.

Reference	Wet Density kg/m³ (lb/yd³)	Dry Density kg/m³ (lb/yd³)	Initial Moisture Content cm/cm (in./ft)	Field Capacity cm/cm (in./ft)	Available Moisture Capacity cm/cm (in./ft)
Ham (1980)				.333(4.0)	
				.333(4.0)	
Hughes et al. (1971)	Uncompacted			.100(1.2)	
	Compacted			.350(4.2)	
Fungaroli (1971)	384(647)		.039(0.47)	.286(3.43)	.247(2.96)
	408(687)		.164(1.97)	.294(3.53)	.130(1.56)
	410(690)		.192(2.30)	.325(3.90)	.133(1.60)
			.204(2.45)	.346(4.15)	.142(1.70)
California			.163(1.95)	.297(3.56)	.133(1.60)
Rovers & Farquhar (1973)	315(530)		.160(1.92)	.302(3.62)	.142(1.7)
	339(570)		.210(2.52)	.310(3.72)	.100(1.2)
Stone (1974)		475(800)		.283–.850(3.4–10.2)	
Sonoma Co.	625(1052)		.167(2.0)	.375(4.5)	.208(2.5)
	622(1047)		.150(1.8)	.283(3.4)	.133(1.6)
Walsh & Kinman (1982)	480(808)	312(526)	.167(2.0)	.318(3.82)	.152(1.82)
	474(798)	309(520)	.165(1.98)	.404(4.85)	.238(2.85)
	476(802)	310(522)	.166(1.99)	.368(4.42)	.202(2.42)
Wigh (1979)	391(658)	303(510)	.083(1.0)	.367(4.4)	.283(3.4)
	430(724)	314(528)	.117(1.4)	.325(3.9)	.208(2.5)
	596(1004)	406(683)	.192(2.3)	.375(4.5)	.183(2.2)
Fungaroli & Steiner (1979)	394(563)	283(476)	.052(0.62)	.342(4.1)	.290(3.48)

Table 7.3: FIELD CAPACITY AND AVAILABLE MOISTURE CAPACITY OF MUNICIPAL REFUSE AS REPORTED BY VARIOUS RESEARCHERS

7.2.3 Formative Processes

A series of individual processes warrant discussion.

Runoff   Runoff from the landfill surface is typically computed using the Soil Conservation Service (SCS) runoff curve number (U.S. Dept. of Agriculture, 1972). Factors such as surface slope and roughness are not considered directly in estimating runoff and, hence, infiltration. However, they are reflected by the selection of the runoff curve number. The runoff coefficient is the fraction of precipitation that forms surface runoff.

Evapotranspiration   Evapotranspiration, representing the combination of evaporation plus transpiration (consumptive use by vegetation), represents an important component of the water budget at a landfill. Evapotranspiration comes from water that has infiltrated into the surficial soils or refuse. The fluxes of evapotranspired water depend on capillary flow and the flow through the plant roots to the surface during dry periods. A key element in the determination of available soil moisture is the root penetration depth. How the root systems will develop is uncertain because the gases generated during waste decomposition likely result in oxygen displacement, precluding full root development.

In summary, evapotranspiration occurs from the root zone depth of the vegetation and/or the evaporative zone depth, whichever is the larger. In the event of water in excess of field capacity, infiltrate will continue to flow downward, escaping surficial effects, to become percolation.

The quantity of evapotranspiration is approximately equal to the frequently reported "lake evapotranspiration," which is about 0.7 times the pan evaporation level (Lutton, 1979).

E X A M P L E

Determine the maximum potential evapotranspiration of available moisture from a 1.5 m deep clay loam cover, vegetated with alfalfa. The root zone depth is 1.0 m.

	CLAY LOAM
FC = field capacity (m/m)	0.3
WP = wilting point (m/m)	0.13

Solution

Choosing the smaller of the depth of soil or root zone gives the available moisture as $(0.3 - 0.13)\,1.0 = 0.17$ m water, the maximum possible evapotranspiration.

Percolation   Once water has percolated below the evaporative zone depth, it becomes percolation. Theoretically, percolation adds to the moisture content of layers of the soil and refuse in the landfill beginning at the uppermost layer, until no more moisture can be held against the force of gravity in that layer.

The moisture content, or field capacity, for municipal refuse is approximately 55 percent moisture on a wet-weight basis with variations from this level depending on the refuse characteristics, density and voids. Therefore, during the first years after placement, part of the infiltrated precipitation is adsorbed or stored in the voids and pores of the refuse. In addition, some water is consumed during decomposition processes. As a consequence, higher leachate production rates can be expected with "age" of the refuse. The resulting time lag between applied moisture and leachate collected can be seen in the results illustrated in Figure 7.3 (after Fungaroli and Steiner, 1979). For the first year, water added did not result in significant leachate. However, after 3 years, leachate generation rates essentially mirrored the quantities of water added with a lag due to travel time through the refuse.

Field Capacity–Wilting Point–Available Moisture The difference between the moisture content at field capacity and the wilting point is a soil's (or refuse's) available water content or moisture-holding capacity. The amount of water a given soil will hold is a function of its texture, with heavily textured soils holding more water than lightly textured soils. Figure 7.4 illustrates the variabilities of these features for different soil types. Alternatively, for refuse, the available moisture is the difference between the field capacity 0.55 and wilting point of 0.17. The field capacity of noncompacted commingled wastes from residential and commercial sources is in the range of 50 to 60 percent.

EXAMPLE

Assume the existence of a vegetative topsoil 0.65 m deep. Estimate the available water, assuming the existence of 0.5 m of root penetration.

Solution

Determine the lesser of the soil thickness versus the root depth: The root zone is the lesser, at 0.5 m and therefore governs. Assuming a moisture content of 200 mm/m topsoil, the available moisture is simply 200 mm/m $\times$ 0.5 m, or 100 mm.

Note that leachate collection at the base of a landfill will occur before the field capacities of the overlying soil and refuse column are reached. Leachate migrates, or short-circuits, by specific routes through the refuse in voids or channels through which water can flow readily without moving the wetted front downward. The result is that water flows downward unevenly, reaching the bottom of the landfill as leachate long before the landfill is at field capacity. Examples of situations promoting channeling include landfilling of baled refuse or refuse containing demolition wastes and tree limbs. These phenomena are, in general, not included in the water balance models (to be described later), where the existence of a main wetting front is anticipated. Instead, the water balance

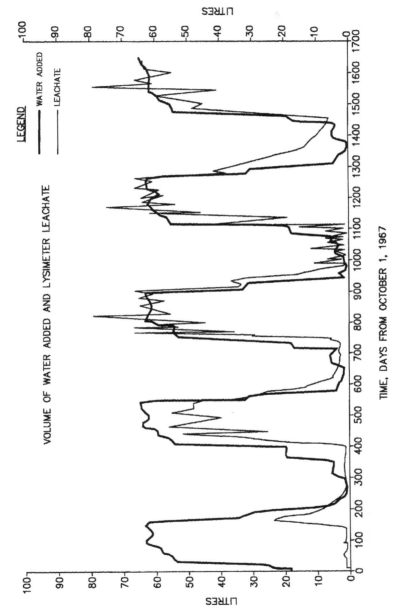

Fig. 7.3 Time lag between applied moisture and leachate collected.

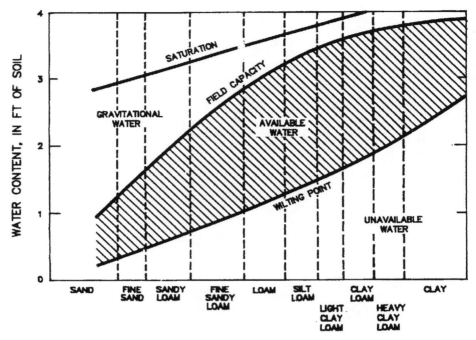

Fig. 7.4 Water storage capacities for various soil types. (Lutton et al., 1979)

models assume plug flow or idealized conditions, with each layer reaching field capacity before moisture is passed on downward to the next layer.

In terms of magnitude of soil or waste moisture-holding capacity, several findings are noted:

- Ham (1980) found that for waste received at the Madison Shredding Plant the moisture content was generally 15 to 20 percent in the winter months and 35 to 45 percent in the summer months. The higher moisture content in the summer was attributed to increased rainfall and increased quantities of grass clippings and leaves associated with summer refuse.
- Kmet (1982) suggested a value for field capacity of refuse of 4 in./ft. Due to the presence of initial moisture levels in the waste, the available moisture capacity for new water would be approximately 2 in./ft. An initial 35 percent moisture content (dry-weight basis) converts to 2 in./ft for refuse at a density of 800 lb/yd^3 (1080 lb/yd^3 wet density).

- Fungaroli and Steiner (1979) determined from experiments that the field capacity of refuse decreases with decreasing particle size. They reported the equation

$$\text{Field capacity} = 2.6 \ \ln D - 14.0 \tag{7.2}$$

where D = wet density (lb/yd^3).

The time required for the refuse and liner to reach field capacity can be theoretically determined (e.g., Fungaroli, 1971; Moore, 1980). Although theoretical methods hypothesize that it takes several years for refuse to reach field capacity, experience with clay-lined sites in Wisconsin has indicated that a significant quantity of leachate is generated within 1 to 2 years of beginning site operation (Kmet et al., 1981).

EXAMPLE

Estimate the time needed to reach field capacity through 10 m of MSW refuse. Assume that one-third of precipitation infiltrates during three years of active site operation (30 cm/yr) and that the infiltration rate is 6 cm/yr following final cover placement. Assume a field capacity of 0.3 cm/cm for the refuse (Table 7.3) and an initial moisture content of 0.16 cm/cm.

Solution

Advancement of wetted front during active site operation:

$$\frac{30 \text{ cm}}{\text{yr}} \times 3 \text{ yr} \times \frac{0.3 \text{ cm}}{\text{cm}} = 27 \text{ cm}$$

Advancement of wetted front following placement of final cover:

$$\frac{6 \text{ cm}}{\text{yr}} \frac{(0.3 - 0.16) \text{ cm}}{\text{cm}} = \frac{42 \text{ cm}}{\text{yr}}$$

Time to reach field capacity after final cover placement:

$$\frac{10 \text{ m} - 0.27 \text{ m}}{42 \text{ cm}/\text{yr}} \frac{(100 \text{ cm})}{\text{m}} = 23.2 \text{ yr}$$

As apparent from these calculations, in theory it takes a very long time for a landfill to reach field capacity. However, short-circuiting will occur so that leachate will be encountered long before the 3 + 23.2 = 26.2 years indicated.

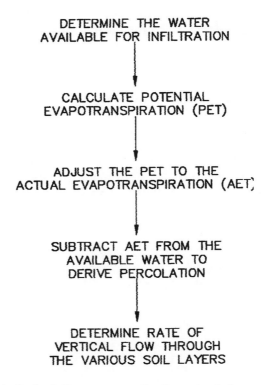

Fig. 7.5 Flowchart of calculation sequence for the water balance method.

7.2.4 Multilayer Conditions

There are several noteworthy points regarding the derivation of the water balance models.

- The results indicate a strong seasonality in leachate quantities. The degree to which this variation occurs depends on the temporal pattern of percolation into the drainage layer and the dampening effect of the refuse.
- The applications of the various equations to landfill concerns have involved adjustment of models that were originally developed for other purposes (e.g., hydrologic budget for field crops). Thus, most of the existing methods were developed for use in areas such as watersheds having naturally placed soils rather than the mixed subsoil layers commonly found at landfills. Below the vegetative layer lies the hydraulic barrier, which is the first element of the cover designed specifically to minimize the passage of infiltration.
- The surface integrity of the landfill cover is presumed.

Water Balance Components	January	February	March	April	May	June	July	August	September	October	November	December	Year	Activity
T (°F)	19.3	22.6	32.1	46.1	56.3	66.5	71.3	70.3	61.8	51.6	36.7	24.1		Determine potential evapotranspiration
i	0.0	0.0	0.0	1.97	4.50	7.65	9.31	8.96	6.13	3.25	0.37	0.0	I = 42.1	
UPET	0.0	0.0	0.0	0.04	0.08	0.12	0.14	0.14	0.10	0.06	0.01	0.0		
r	24.3	24.6	30.6	33.6	37.9	38.4	38.7	36.0	31.2	28.5	24.3	23.1		
PET	0.0	0.0	0.0	1.34	3.03	4.61	5.42	5.04	3.12	1.71	0.24	0.0	24.51	Determine water available for infiltration
›(in.)	1.57	1.04	2.23	2.90	3.37	3.75	3.66	2.99	3.20	2.13	2.16	1.70	30.7	
C r/o	0.18	0.18	0.18	0.18	0.18	0.18	0.18	0.18	0.18	0.18	0.18	0.18		
R/O	0.28	0.19	0.40	0.52	0.61	0.68	0.66	0.54	0.58	0.38	0.39	0.31		
I	1.29	0.85	1.83	2.38	2.76	3.07	3.00	2.45	2.62	1.75	1.77	1.39		
I-PET	1.29	0.85	1.83	1.04	-0.27	-1.54	-2.42	-2.59	-0.50	0.04	1.53	1.39		Determine actual evapotranspiration
ACCWL				0.0	-0.27	-1.81	-4.23	-6.82	-7.32					
ΔST	4.0	4.0	4.0	4.0	3.73	2.51	1.35	0.70	0.61	0.65	2.18	3.57		
ÆST	0.0	0.0	0.0	0.0	-0.27	-1.22	-1.16	-.65	-0.09	0.04	1.53	1.39		
AET	0.0	0.0	0.0	1.34	3.03	4.29	4.16	3.10	2.71	1.71	0.24	0.0		
PERC	0.86	0.85	1.83	1.04	0.0	0.0	0.0	0.0	0.0	0.0	0.0	0.0	3.17	Resulting percolation

Source: Modified from Kmet, 1982.
Notes:
C r/o = 0.18 (coefficient recommended by Fenn et al. (1975) for heavy soil @ 2–7%)
No special provisions for snowmelt runoff
Final cover has 4 in. available water.

Table 7.4: WATER BALANCE MODEL APPLICATION USING SPREADSHEET APPROACH FOR DELAFIELD SITE

7.3   WATER BALANCE METHOD (WBM)

The quantification of evapotranspiration is a key to the water balance for a land-fill. The evapotranspiration is a function of temperature, humidity, wind, solar radiation, available water, plant type, and growth stage. Obviously, it is a complicated function and has been the focus of considerable research.

The so-called Water Balance Method (WBM) was developed in the 1940s and 1950s for quantifying evapotranspiration by Thornthwaite and Mather (1957) and was adapted by Mather and Rodriguez (1978) and Fenn et al. (1975) for landfill conditions. The method computes evapotranspiration (ET) from an empirical equation that calculates the potential evapotranspiration as a power function of mean monthly air temperature. Evapotranspiration is then assumed to be equal to the potential ET multiplied by the ratio of the actual soil moisture content to the field capacity soil moisture content, which is defined generally as the moisture content when drainage from the soil begins. Soil moisture in excess of the field capacity is drainage/percolation not affected by ET. The WBM has been developed to obtain monthly estimates and its use for shorter periods is not recommended (Mather, 1974). The empirical method involves three steps:

1. A heat index is obtained for each of the 12 months of the year and summed to create an annual index.
2. The daily potential evapotranspiration is obtained from the heat index by use of tables.
3. The potential evapotranspiration is adjusted for month and day lengths with correction factors.

Table 7.5: SEQUENCE OF CALCULATIONS AS A PART OF THE WBM PROCEDURE

1.	T	Enter the average monthly temperature (°F).
2.	i	Using the monthly temperature, determine the monthly heat index for each month. For months with $T < 32$ °F, i = 0. Sum the i values to obtain I (the yearly heat index) (see Table C.1, in Appendix C).
3.	$UPET$	Using the monthly temperature and the yearly heat index, find the Unadjusted Potential Evapotranspiration (see Table C.2).
4.	r	Using the site latitude find the monthly correction factor for sunlight duration (see Table C.3, in Appendix C).
5.	PET	Multiply the monthly UPET by the monthly r to obtain the Adjusted Potential Evaporation for each month (inches of water).
6.	P	Enter the average monthly precipitation (inches of water).
7.	$C\,r/o$	Enter the appropriate runoff coefficient to calculate the runoff for each month (see Table 7.6, this Chapter)..
8.	r/o	Multiply the monthly precipitation by the monthly runoff coefficient to calculate the runoff for each month (inches of water).

Source: After Kmet, 1982.

Table 7.5: SEQUENCE OF CALCULATIONS AS A PART OF THE WBM PROCEDURE

9.	*I*	Subtract the monthly runoff from the monthly precipitation to obtain the monthly infiltration (inches of water).
10.	*I- PET*	Subtract the monthly adjusted potential evapotranspiration from the monthly infiltration to obtain the water available for storage (inches of water).
11.	*ACC WL*	Add the negative *I-PET* values on a cumulative basis to obtain the cumulate water loss . *Note:* Start the summation with zero accumulated water loss for the last month having *I-PET* > 0 (inches of water).
12.	*ST*	Determine the monthly soil moisture storage (inches of water) as follows: a) Determine the initial soil moisture storage for the soil depth and type (see Table 7.7). b) Assign this value to the last month having *I-PET* > 0. c) Determine *ST* for each subsequent month having *I-PET* <0 (see Table C.4, Appendix C). d) For months having *I-PET* ≥ 0, add the *I-PET* value to the preceding month's storage. Do not exceed the field capacity. Enter the field capacity if the sum exceeds this maximum.
13.	Δ*ST*	Calculate the change in soil moisture for each month by subtracting the *ST* for each month from the preceding month (inches of water).
14.	*AET*	Calculate the actual evapotranspiration as follows (inches of water): a) Wet months *I-PET* ≥ 0: *AET = PET* b) Dry months *I-PET* < 0: *AET = PET + (I-PET − ΔST)* *Note:* For months when I-PET is negative, the evapotranspired amount is the amount potentially evapotranspired plus that available from excess infiltration that would otherwise add to soil moisture storage plus that available from previously stored soil moisture.
15.	*PERC*	Calculate the percolation as follows (inches of water): a) Dry months *I-PET* < 0: *PERC* = 0 b) Wet months *I-PET* > = 0: *PERC = (I-PET − ΔST)* Sum the percolation values for the year to obtain total annual leachate production per unit area.
16.	*P*	*P = PERC + AET + ΔST + r/o*

Source: After Kmet, 1982.

Thornthwaite and Mather (1957) found that the actual ET is usually less than the potential ET because moisture availability is seldom ideal. Although brief, the preceding outlines the computational steps of the Water Balance Method. Figure 7.5 illustrates, in flowchart form, the sequence of calculations within the Water Balance Method. In essence, the WBM is a relatively simple set of calculations that can be performed by hand and/or by computer spreadsheet. As an example, the details are outlined in a spreadsheet type of format in Table 7.4. The sequence of calculations for each of the rows/columns in the spreadsheet is summarized in Table 7.5.

Table 7.6: RUNOFF COEFFICIENTS

Surface Conditions: Grass cover (slope)	Runoff Coefficient
Sandy soil, flat, 2%	0.05 – 0.10
Sandy soil, average, 2–7%	0.10 – 0.15
Sandy soil, steep, 7%	0.15 – 0.20
Heavy soil, flat, 2%	0.13 – 0.17
Heavy soil, average, 2–7%	0.18 – 0.22
Heavy soil, steep, 7%	0.25 – 0.35

Source: Fenn et al., 1975

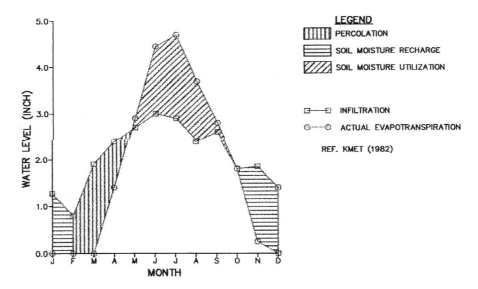

Fig. 7.6 Example of water balance model outputs.

The results of the calculation sequence are graphed in Figure 7.6(Appendix C). To discuss the graph, it is easiest to start with the month of October

1. From October to February, the infiltration through the cover is larger than the actual evapotranspiration—thus, the difference goes into storage in the surface layer. However, at the end of February, the storage capacity in the surface layer is reached.
2. From February until May, the infiltration continues to be larger than the actual evapotranspiration, yet the surface layer is at field capacity. Therefore, additional water contributions create percolation (and ultimately leachate).

3. Commencing in May, the evapotranspiration exceeds infiltration. To meet this demand (for evapotranspiration), water comes out of surficial storage. Because the cumulative excess demand (evapotranspiration minus infiltration) does not exceed the available moisture-holding capacity of the surface layer, potential evapotranspiration equals actual evatranspiration. The increasing depletion of the soil moisture continues until October, when once again, infiltration rates exceed evapotranspiration rates and the calculation procedure returns to (1) above.

Table 7.7: PROVISIONAL WATER HOLDING CAPACITIES FOR DIFFERENT COMBINATIONS OF SOIL AND VEGETATION

Soil Type	Available Water		Root Zone		Applicable Soil Moisture Retention Table	
	mm/m	in./ft	in.	ft	mm	in.
Shallow-Rooted Crops (spinach, peas, beans, beets, carrots, etc.)						
Fine sand	100	1.2	.50	1.67	50	2.0
Fine sandy loam	150	1.8	.50	1.67	75	3.0
Silt loam	200	2.4	.62	2.08	125	5.0
Clay loam	250	3.0	.40	1.33	100	4.0
Clay	300	3.6	.25	.83	75	3.0
Moderately Deep-Rooted Crops (corn, cotton, tobacco, cereal grains)						
Fine sand	100	1.2	.75	2.50	75	3.0
Fine sandy loam	150	1.8	1.00	3.33	150	6.0
Silt loam	200	2.4	1.00	3.33	200	8.0
Clay loam	250	3.0	.80	2.67	200	8.0
Clay	300	3.6	.50	1.67	50	6.0
Deep-Rooted Crops (alfalfa, pastures, shrubs)						
Fine sand	100	1.2	1.00	3.33	100	4.0
Fine sandy loam	150	1.8	1.00	3.33	150	6.0
Silt loam	200	2.4	1.25	4.17	250	10.0
Clay loam	250	3.0	1.00	3.33	250	10.0
Clay	300	3.6	.67	2.22	200	8.0
Orchards						
Fine sand	100	1.2	1.50	5.00	150	6.0
Fine sandy loam	150	1.8	1.67	5.55	250	10.0
Silt loam	200	2.4	1.50	5.00	300	12.0

Table 7.7: PROVISIONAL WATER HOLDING CAPACITIES FOR DIFFERENT COMBINATIONS OF SOIL AND VEGETATION

Soil Type	Available Water		Root Zone		Applicable Soil Moisture Retention Table	
	mm/m	in./ft	in.	ft	mm	in.
Clay loam	250	3.0	1.00	3.33	250	10.0
Clay	300	3.6	.67	2.22	200	8.0
Closed Mature Forest						
Fine sand	100	1.2	2.50	8.33	250	10.0
Fine sandy loam	150	1.8	2.00	6.66	300	12.0
Silt loam	200	2.4	2.00	6.66	400	16.0
Clay loam	250	3.0	1.60	5.33	400	16.0
Clay	300	3.6	1.17	3.90	350	14.0

Notes:

These figures are for mature vegetation. Young cultivated crops, seedlings, and other imma-
ture. vegetation will have shallower root zones and, hence, have less water available for the
use of. the vegetation. As the plant develops from a seed or a young sprout to the mature
form, the root. zone will increase progressively from only a few inches to the values listed
above. Use of a series of soil moisture retention tables with successively increasing values of
available moisture permits the soil moisture to be determined throughout the growing season.

The thickness of soil from which evatranspiration can occur is normally
determined by the depth of the root zone. For example, if a landfill surface was
covered with deep-rooted crops (pastures, grasses), from Table 7.7, the 4 in. of
moisture storage capacity could result from any of the following:

(a)   3.33 ft. of fine sand   1.2 in./ft $\times$ 3.3 ft = 4 in.

(b)   1.67 ft of silt loam   2.4 in./ft $\times$ 1.67 ft = 4 in.

(c)   0.5 ft of clay overlain by 0.92 ft of   3.6 in./ft $\times$ 0.5 ft + 2.4
       silt loam   in./ft $\times$ 0.92 ft =
                                                          1.8 + 2.2 = 4.0 in.

Thus, the Water Balance Method is a relatively simple model that proceeds
through a calculation sequence on a monthly time basis. Implicit within the pro-

cedure are a series of assumptions as listed in Table 7.8 that must be appropriate for the situation at hand.

Table 7.8: BASIC ASSUMPTIONS IN WATER BALANCE MODEL (*Source:* After Gee, 1981)

1.	The sole source of infiltration/percolation is precipitation falling directly on the landfill's surface. All surface runoff from adjacent areas is assumed to be diverted around the landfill surface.
2.	Groundwater does not enter the landfill.
3.	All water movement through the landfill is vertically downward.
4.	The landfill is at field capacity at the start of calculations.
5.	No recycle of leachate or co-disposal of liquid occurs.

Accuracy of the WBM Procedure   The calculation sequence for the WBM procedure is relatively inaccurate. Lu et al. (1981) used the WBM to determine leachate generation and compared the results with leachate measurements from five different landfill sites representing different geographical areas and site conditions. The WBM error ranged from 1.32 percent to .5389 percent. Gee (1981) reported the error of the WBM as 94 percent.

7.4 HYDROLOGIC EVALUATION OF LANDFILL PERFORMANCE

The Hydrologic Evaluation of Landfill Performance (HELP) model is an alternative water balance model (e.g., Schroeder et al., 1984a and b). The HELP model is based on the same hydrologic principles as the WBM but utilizes a much more detailed sequence of calculations. The model also has the ability to examine water fluxes throughout the complete vertical profile of a landfill, such as depicted in Figure 7.7.

Specifically, the HELP model is a quasi-two-dimensional model that performs daily calculations for the water budget over a one-dimensional column of the landfill. Surface runoff, evapotranspiration, vertical percolation, and lateral drainage are calculated using daily time increments.

The HELP model represents a landfill by using three types of layers, namely, vertical percolation layers, lateral drainage layers, and barrier soil liners. In the vertical percolation layer, percolation is modeled as independent of the depth of water. Neither saturated soil above the layer nor lateral drainage is

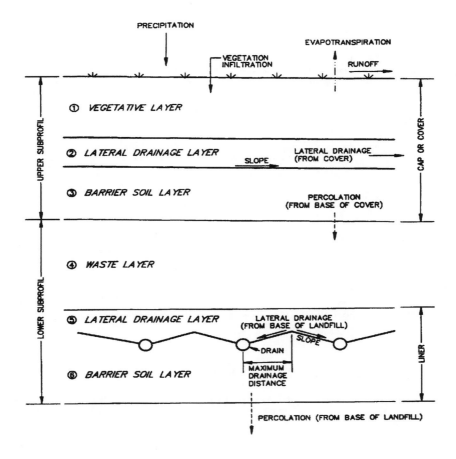

Fig. 7.7 Landfill profile assumed in HELP model.

permitted. The second type of layer, for lateral drainage, is a more permeable medium in which vertical percolation is calculated in the same manner as in the vertical percolation layer, and lateral drainage is permitted. The lateral drainage is evaluated based on the magnitude of the leachate mound that forms. Finally, the barrier soil is a low-conductivity soil used to restrict vertical water flow. Only vertical percolation is permitted within this layer, and flow is modeled as a function of the depth of water saturated above the base of this layer. These three types of layers are then combined to represent a landfill profile such as the one depicted in Figure 7.7.

Due to its nonproprietary nature and the flexibility of the model to represent a wide array of landfill configurations, the HELP model is now essentially the "model of requirement" used by the EPA. A number of reports and papers have been prepared in support of the model (e.g., Schroeder et al., 1983, 1984, 1988; Peyton and Schroeder, 1988), and the interested reader is directed to those information sources for specific details.

We indicate some of the more pertinent features when the model is used as a tool for water balance modeling:

1. The model carries out the calculation sequence on a daily basis. As a direct result, the data requirements are substantial. However, the model provides default parameters that may be used, making the model very simple to employ, and requires input of data only when the default parameters are inadequate. Table 7.9 summarizes the data inputs.

TABLE 7:9: SUMMARY OF HELP MODEL INPUT DATA

Category	Details
Climate data	Daily precipitation—three options exist: 1. Use a default precipitation option (utilizes 5 years of historical meteorologic data; 2. Input precipitation data; 3. Generate a sequence of precipitation events using a rainfall generator.
	Mean monthly solar radiation—these data are generated by the model
	Mean monthly temperature—these data are generated by the model
Soil Data	Saturated hydraulic conductivity
	Soil porosity
	Evaporation coefficient
	Field capacity
	Wilting point
	Minimum infiltration rate
	SCS runoff curve number
	Initial soil water content (vol/vol)
Vegetation data	Crop type
	Crop cover
	Leaf area indices
	Winter cover factor
	Evaporative zone depth
Design data	Number of layers

TABLE 7:9: SUMMARY OF HELP MODEL INPUT DATA

Category	Details
Design data (continued)	Layer thickness
	Layer slope
	Lateral flow distance
	Surface layer of landfill
	Leakage fraction (for synthetic membrane liners, the value is to be between 0 and 1)
	Runoff fraction from waste

2. Moisture that enters the topsoil layer is determined by a daily infiltration procedure that considers the amount of antecedent moisture, the density of vegetation on the surface, and the evaporative and runoff potential.

3. The rate of vertical moisture flow in the soil varies with the soil moisture content. When the soil is fully saturated, the flow rate is equal to the saturated hydraulic conductivity in accordance with Darcy's law. The unsaturated hydraulic conductivity in the layer is calculated based on the average soil moisture within the layer and is defined by a linear function of the soil moisture as follows:

$$ K_u = K_s \left[\frac{\Theta - \Theta_r}{\Theta_s - \Theta_r} \right]^{3 + 2/\lambda} $$

where: K_u = unsaturated hydraulic conductivity,
  K_s = saturated hydraulic conductivity,
  θ = soil moisture (vol/vol),
  θ_r = residual soil moisture (vol/vol)(i.e., field capacity),
  θ_s = saturated soil moisture (vol/vol),
  λ = pore-size distribution index.

When the moisture content is reduced to field capacity, the flow rate is zero. The parameters θ_r and λ are constants in the Brooks-Corey equation relating soil water content to capillary pressure (Brooks and Corey, 1964). The HELP model program contains default values of all required soil characteristics based on soil texture class.

The soil moisture is transported from one layer to another by a storage procedure computed each day and proceeds sequentially from the top layer to the bottom layer, assuming free drainage at the bottom of each layer. Because free drainage is assumed for each layer, the hydraulic gradient for all layers except barrier layers is equal to unity. Hence, the flow is equal to the unsaturated hydraulic conductivity.

4. Lateral drainage in the drainage layers is computed analytically from a lin-
    earized Boussinesq equation. (This is necessary to obtain a
    one-dimensional approximation to a two-dimensional problem. The
    two-dimensional character of leachate mounding will be developed in
    Chapter 10).

5. Snow is assumed to remain as snowpack when the daily mean temperature
    is less than 32°F.

6. As an indication of the array of default information built into the model,
    Table 7.14 lists the default soil characterizations for the 21 soil types from
    which a user may select.

7. The HELP model does not include special leachate migration routes; the
    existence of a main wetting front is assumed. Thus, leachate will impinge
    on a barrier liner before the model's results indicate, as discussed in
    Section 7.2.3.

The HELP model is valuable in that it assists in the development of an
understanding of the ramifications of different landfill designs. Calibration and
verification studies of the HELP model have been reported in several locations
(e.g., Peyton and Schroeder, 1988; Schroeder and Peyton, 1987). Generally, positive findings have been reported.

The accuracy of the model in terms of predicting leachate quantities is
always a concern (see Table 7.11 for examples of comparison with measured
results). The standard deviations indicate that the estimates must be treated
only as useful guidelines. The designer is not in a position (normally) to check
the accuracy of the model, since the site is not yet constructed. The program is
under continuing development, so a model user should ensure that the most
recent version of the model is employed. Improvements to the HELP model are
being developed for snowmelt, winter runoff, unsaturated hydraulic conductivities, and selection of evaporative zone depths.

Several aspects of the HELP model are noteworthy:

• The model does not model the aging of a liner.
• The model does not develop a water balance over the history of development
  of a landfill site. For example, the model can be used to examine the impact
  of sequential placement of cells, one on top of another, and then placement
  of a final cover only by successive model uses. Input of the initial moisture
  conditions, given the ending conditions for a previous phase of the landfill
  sequence of operation, is very cumbersome.
• The HELP model does not model leachate quality.
• There is a tendency to underpredict the surface runoff coefficient, because a
  daily time increment is used. This underprediction occurs because the rain-

Table 7.10: DEFAULT SOIL CHARACTERISTICS

SOIL TEXTURE CLASS			MIR[d] (in./h)	Porosity (vol/vol)	Field Capac- ity (vol/vol)	Wilting Point (vol/vol)	Hydraulic Conductiv- ity (in./h)	CON[e] (mm/day) 0.5
HELP[a]	USDA[b]	USCS[c]						
1	CoS	GS	0.500	0.351	0.174	0.107	11.95	3.3
2	CoSL	GP	0.450	0.376	0.218	0.131	7.090	3.3
3	S	SW	0.400	0.389	0.199	0.066	6.620	3.3
4	FS	SM	0.390	0.371	0.172	0.500	5.400	3.3
5	LS	SM	0.380	0.430	0.16	0.060	2.780	3.4
6	LFS	SM	0.340	0.401	0.129	0.075	1.000	3.3
7	LVFS	SM	0.320	0.421	0.176	0.090	0.910	3.4
8	SL	SM	0.300	0.442	0.256	0.133	0.670	3.8
9	FSL	SM	0.250	0.458	0.223	0.092	0.550	4.5
10	VFSL	MH	0.250	0.511	0.301	0.184	0.330	5.0
11	L	ML	0.200	0.521	0.377	0.221	0.210	4.5
12	SIL	ML	0.170	0.535	0.421	0.222	0.110	5.0
13	SCL	SC	0.110	0.453	0.319	0.200	0.084	4.7
14	CL	CL	0.090	0.582	0.452	0.325	0.065	3.9
15	SICL	CL	0.070	0.588	0.504	0.355	0.041	4.2
16	SC	CH	0.060	0.572	0.456	0.378	0.065	3.6
17	SIC	Ch	0.020	0.592	0.501	0.378	0.033	3.8
18	C	CH	0.010	0.680	0.607	0.492	0.022	3.5
19	Waste		0.230	0.520	0.320	0.190	0.283	3.3
20	Barrier soil		0.002	0.520	0.450	0.360	0.000142	3.1
21	Barrier soil		0.001	0.520	0.480	0.400	0.0000142	3.1

[a]Soil classification system used in the HELP Model
[b]Soil classification system used by the U.S. Department of Agriculture
[c]Unified Soil Classification System
[d]MIR = minimum infiltration rate
[e]CON = evaporation coefficient (transmissivity).

Table 7.11: PERCENTAGE DIFFERENCES BETWEEN HELP MODEL PREDICTIONS AND FIELD MEASUREMENTS AS REPORTED FOR DIFFERENT SITES

COMPONENTS OF WATER BALANCE

	Runoff		Evapotranspiration Plus Change in Moisture Storage		Lateral Leachate Drainage	
	Mean	Std. Dev	Mean	Std. Dev	Mean	Std. Dev
Test Cells						
Covered cells, Wisconsin (Ham, 1980)	19.3	31.8	-1.3	4.7	-3.0	21.2
Uncovered cells, Wisconsin (Ham, 1980)	-16.3	16.8	-3.2	11.8	170.	167.
Sonoma County test cells						
(Sonoma County, 1975)	14.8	6.4	-19.0	15.4	-52.4	56.3
Boone County test cell (Wigh, 1984)	NR	NR	NR	NR	-14.6	NR
Landfills						
Brown County						
Daily cover					65	
Final cover					-29	
Eau Claire County						
Daily sand cover					138	
Interim sludge cover (uncompacted clay)					-96	
Interim sludge cover (clayey loam)					-31	
Final cover					127	
Marathon County						
Daily cover					29	
Final cover					-33	

Notes:
As abstracted from Peyton and Schroeder, 1988
All values in percentages
NR—not reported

fall rate from a short, intensive rainfall is averaged out over the daily time increment, making the intensity much lower.

* The synthetic-liner leakage fraction is a function of hole size, depth of leachate ponding, and saturated hydraulic conductivity of the underlying soil. There is considerable uncertainty in the assignment of the leakage fraction in the HELP model.

7.5 SUMMARY COMMENTS

Leachate generation is not likely to result in a constant flow throughout the year or from year to year but will follow a pattern somewhat similar to that of precipitation.

The water balance models described in this chapter provide good insight into the patterns of leachate production and are definitely useful as design tools. It is important, however, to keep in mind that the assignments of the coefficients in the models are difficult, and many processes are ongoing. There is considerable uncertainty in the computed answers.

References

Brooks, R. H., and Corey, A. T. 1964. "Hydraulic Properties of Porous Media." *Hydrology Paper no. 3.*, Colorado State University, 27 pp.

Fenn, D., Hanley, K. J., and DeGeare, T.V. 1975. *Use of the Water Balance Method for Predicting Leachate Generation from Solid Waste Disposal Sites.*   U. S. Environmental Protection Agency EPA-530-SW-169, Cincinnati, Ohio.

Fungaroli, A. A. 1971. *Pollution of Subsurface Water from Sanitary Landfills*, U. S. Environmental Protection Agency. Interim report SW-12, rg.

Fungaroli, A., and Steiner, R. 1979. *Investigation of Sanitary Landfill Behavior - Volume I, Final Report*, U. S. Environmental protection Agency EPA-600-2-79-053a.

Gee, J. R. 1981. "Prediction of Leachate Accumulation in Sanitary Landfills." In *Proceedings of the Fourth Annual Madison Conference of Applied Research and Practice on Municipal and Industrial Waste.* Dept. of Engineering and Applied Science, University of Wisconsin Extension at Madison, pp. 170–190, September.

Ham, R. K. 1980. *Decomposition of Residential and Light Commercial Solid Waste in Test Lysimeters*, U. S. Environmental Protection Agency SW-190c.

Hughes, G. M., Landon, R. M. and Farvolden, R. N. 1971. *Hydrogeology of Solid Waste Disposal Sites in Northeastern Illinois*, U. S. Environmental Protection Agency SW-12d.

Kmet, P. 1982. *EPA's 1975 Water Balance Method: Its Use and Limitations.* A Wisconsin Dept. of Natural Resources Guidance Report, Madison, October.

Kmet, P., Quinn, K., and Slavik, C. 1981. "Analysis of Design Parameters Affecting the Collection Efficiency of Clay-Lined Landfills." Paper presented at 4th Annual Madison Conference of Applied Research and Practice on Municipal and Industrial Waste, Madison, Wis., September 28–30.

Lu, J. C. S., Morrison, R. D., and Stearns, R. J. 1981. "Leachate Production and Management from Municipal Landfills: Summary and Assessment." In *Proceedings of the Seventh Annual Research Symposium, Land Disposal, Municipal Solid Waste,*U.S. Environmental Protection Agency EPA-600/9-81-002a, Cincinnati, Ohio.

Lutton, R. J. 1979. *Soil Cover for Controlling Leachate from Solid Waste, Municipal Solid Waste: Land Disposal.* U.S. Environmental Protection Agency EPA-600-9-79-023a, August.

Lutton, R. J. 1980. *Evaluating Cover Systems for Solid and Hazardous Waste.* Report to Municipal Environmental Research Lab. U.S. Environmental Protection Agency SW-867, Cincinnati, Ohio.

Lutton, R. J, Regan, G. l., and Jones, L. W. 1979. *Design and Construction of Covers for Solid Waste Landfills.* U.S. Environmental Protection Agency EPA-600-2-79-165, August.

Mather, J. R. 1974. *Climatology: Fundamentals and Applications.* McGraw-Hill, New York.

Mather, J. R., and Rodriguez, P.A. 1978. *The Use of the Water Budget in Evaluating Leaching Through Solid Waste Landfills,* PB80-180888. U.S. Dept. of the Interior, Washington, D.C.

Moore, C. A. 1980. *Landfill and Surface Impoundment Performance Evaluation Manual,* U. S. Environmental Protection Agency SW-869. Cincinnati, Ohio.

Perrier, E. R., and Gibson, A. C. 1980. *Hydrologic Simulation on Solid Waste Disposal Sites.* U.S. Environmental Protection Agency EPA-SW-868.

Peyton, R., and Schroeder, P. R.1988. Field Verification of HELP Model for Landfills. *ASCE Journal of the Environmental Engineering Division* (April): 247–268.

Rovers, F. A., and Farquhar, G. J. 1973. Infiltration and Landfill Behavior. *ASCE Journal of the Environmental Engineering Division* 99 (October): 671–690.

Schroeder, P. R., and Peyton, R. L. 1987a. *Verification of the Hydrologic Evaluation of Landfill Performance Model Using Field Data.* U.S. Environmental Protection Agency EPA-600-2-37/050.

Schroeder, P. R., and Peyton, R. L. 1987b. *Verification of the Lateral Drainage Component of the HELP Model Using Physical Models.* U.S. Environmental Protection Agency EPA-600-2-87-049, Cincinnati, Ohio, June, p. 117(b).

Schroeder, P. R., Morgan, J. M., Wolski, T. M., and Gibson, A. C. 1983. T*he Hydrologic Evaluation of Landfill Performance Model* Volume 1, *User's Guide for Version 1*, United States Environmental Agency, EPA-DF-85/001a.

Schroeder, P. R., Gibson, A. C., and Smolen, M. D. 1984a. *Hydrologic Evaluation of Landfill Performance (HELP) Model*: Volume 2. *Documentation for Version 1.* U.S. Environmental Protection Agency, EPA-530-SW-84-010, Cincinnati, Ohio, June, p. 256.

Schroeder, P.R., Morgan, J. M., Wolski, T. M., and Gibson, A. C. 1984b. *Hydrologic Evaluation of Landfill Performance (HELP) Model.* Volume 1. *User's Guide for Version 1.* U.S. Environmental Protection Agency EPA-530-SW-85-009, Cincinnati, Ohio, June, p. 120.

Schroeder, P. R., Gibson, A. C., and Smolen, M. D. 1988. *The Hydrologic Evaluation of Landfill Performance Model.* Volume 2, *Documentation for Version 2.* U. S. Environmental Protection Agency.

Stone, R. 1974. *Disposal of Sewage Sludge into a Sanitary Landfill.* U.S. Environmental Protection Agency EPA-SW-71d.

Thornthwaite, C. W., and Mather, J. R. 1957. Instructions and Tables for Computing Potential Evapotranspiration and the Water Balance. Drexel Institute of Technology, Laboratory of Climatology, Centerton, N.J. *Publications in Climatology*10(3).

U.S. Dept. of Agriculture, Soil Conservation Service. 1972. *National Engineering Handbook.* Section 4, *Hydrology.* U.S. Government Printing Office.

Walsh, J. J., and Kinman, R. N. 1982. "Leachate and Gas Production Under Controlled Moisture Conditions." In *Proceedings of the Eighth Annual Research Symposium on Land Disposal of Solid Wastes.* U.S. Environmental Protection Agency EPA-600-9-82-002.

Wigh, R. J. 1979. *Boone County Field Study Interim Report.* U.S. Environmental Protection Agency EPA-600-2-79-058.

Wigh, R. 1984. *Landfill Research at the Boone County Field Site. Final Report.* Municipal Environmental Research Laboratory, Office of Research and Development, U.S. Environmental Protection Agency EPA-600-2-84-124, Cincinnati, Ohio, July.

7.6 PROBLEMS

7.1. Landfill sites are filled as a series of successive cells. Sketch the progressive cell completion and the consequent leachate production rates that you might expect if the site will be completed in 20 years.

7.2 Design a landfill cover for a landfill in your community, using the HELP model. You want to keep the leachate quantities to 2 in. per year or less to avoid encountering large treatment costs.

   (a) Design a minimal cover system that just meets the stipulation.

   (b) Examine the implications on the water balance of an elaborate system to reflect considerable protection of the groundwater regime from contamination by the leachate. Briefly support your basis for the utilization of various default parameters used in the HELP model.

   (c) Discuss what you think are the most difficult parameters to assign in the utilization of the HELP model for this application.

7.3 Determine the leachate quantities generated for a landfill cap consisting of 2 ft of vegetative soil with vegetation having an 18 in. deep root zone. Compare the results using the Water Balance Method and the HELP model, assuming meteorological conditions in your community.

7.4 Determine the leachate quantities generated per unit surface area for a landfill located at 25°N latitude, using the Water Balance Method.

CHAPTER **8**

LANDFILL COVER DESIGN

8.1 THE ROLE OF THE LANDFILL COVER

Landfill leachate is generated as a combination of water (1) infiltrating through the landfill cover, (2) being co-disposed with the refuse, or (3) being recharged from groundwater.

Water infiltrating through the landfill cover is of particular interest in this chapter. Changing the design features of the landfill cover can affect changes in the overall water balance within the landfill. The design adopted must take into account numerous considerations, including costs and long-term maintenance implications. In this chapter we focus on developing and describing the principles on which design decisions can be made.

The water infiltrating through the landfill cover picks up soluble materials during passage through the refuse. For many landfills, then, a major design consideration is on minimizing the leachate production during the operational sequence of filling the landfill, and minimizing postclosure leachate production. (*Note*: For some landfills, the cover is not designed with a view toward minimizing infiltration but instead is designed to capture water to promote rapid stabilization of the refuse. However, except where specifically noted, the focus in this chapter on cover design is toward minimizing leachate production).

Lowering leachate quantities via surface cover design reduces expenses involved in subsequent leachate handling and treatment systems, systems that must remain operable for a long time. However, it is necessary to evaluate the

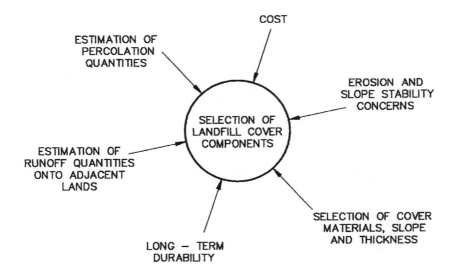

Fig. 8.1 Considerations to be reflected in selection of landfill cover components.

cost-effectiveness of the landfill cover relative to the costs of acceptable disposal of the leachate—there are trade-offs involved. In this chapter we examine the characteristics and design features that have proven effective in ensuring the integrity of the cover and its ability to perform as designed. In Chapter 11 we examine the leachate treatment factors that are needed to complete any trade-off assessment.

In an effort to minimize leachate production, certain design elements can be used, including (1) increasing the surface slopes to enhance surface runoff; (2) placing a relatively impermeable cover that will minimize life-cycle costs; and (3) containing the landfill gas to enable proper management.

Material and construction costs of these design initiatives must be considered relative to the potential costs of leachate treatment, given the changes in waste disposal volumes. Therefore, the evaluation procedure must include examination of the impacts of alternative design initiatives at the landfill surface.

The generally large areas that landfill covers protect, and the thickness and numbers of individual layers within them, make covers a cost-intensive component of landfill facility design. The surface cover is not a single element but a series of components functioning together. Landfill cover design must reflect performance objectives, material selection, and cost, as schematically illustrated in Figure 8.1.

8.1.1 The Cover as a System of Components

The designer has no control over factors such as weather, that affect the water balance at a landfill. Other features will function adequately only if they are designed as part of the overall cover system. For example, a clay soil cover can be placed to minimize percolation. However, if it is left exposed to wind, rain, freezing, or desiccation, the clay cover will crack and not retain its integrity and will be subject to erosion.

Consequently, in normal practice the design of a cover employs a series of components. The role of individual components varies, but their combined effect is intended to achieve the desired objectives. Note that there is not one combination of design components that is used consistently from one site to another—the design of landfill covers is far from routine. Different designers have different preferences. Thus, we focus on the individual principles and features associated with the different elements.

Frequently, two or three distinct materials are combined in layers within the cover to take advantage of the favorable attributes of each. The primary component in the layered system is usually the hydraulic barrier, which is frequently clay. However, because soil barriers are susceptible to deterioration by cracking when exposed at the surface, a buffer layer above is required to protect the clayey soil. Thus, the basic cover contains two primary layers: (1) the surface layer and (2) the hydraulic barrier layer.

Additional or contributing layers assist the primary function of minimizing the downward passage of surface water into the refuse. Table 8.1 lists these lay-

Table 8.1   PRIMARY ROLE OF VARIOUS COMPONENTS WITHIN A LANDFILL COVER

Cover Component	Primary Role
Vegetative soil cover	Reduces infiltration and wind erosion, and provides rootzone and temporary moisture retention
Filter layer	Prevents sifting of overlying cover soil into the drainage layer
Drainage layer	Provides a lateral path for water to exit rapidly
Clay layer	Minimizes infiltration through the cover

ers and their primary roles. Possible sequences of the individual layers are illustrated in Figures 8.2(a) through (c). These sequences of layers are only examples of the many different configurations utilized by designers.

The sand drainage layer noted in Figure 8.2(c) was added below the surface or vegetative layer but above the hydraulic barrier layer to facilitate lateral drainage of the infiltrating water. In addition, a filter layer is frequently required. This may be a horizon of cohesionless soil carefully selected for its particle-size gradation or it may be a geotextile fiber. The intent is to allow passage of infiltrating water but to stop downward piping (movement) of fine soil particles from the vegetative layer into the sand drainage layer, which otherwise

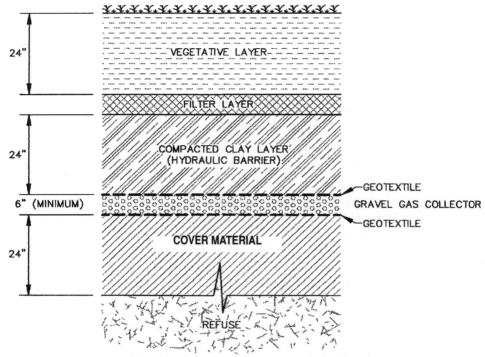

Fig. 8.2 (a) Layered sequence within a landfill cover—Type I.

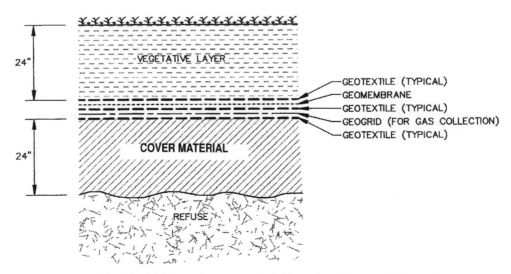

Fig. 8.2 (b) Layered sequence within a landfill cover—Type II.

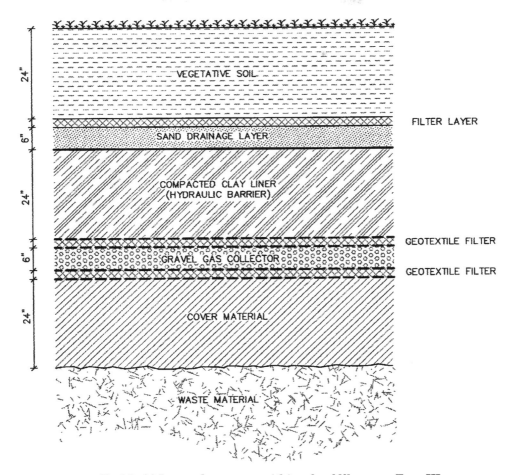

Fig. 8.2 (c) Layered sequence within a landfill cover—Type III.

may lead to plugging of the drainage layer and/or the gas collection via the gravel layer.

8.1.2 Objectives of Cover Design

As noted previously, the primary purpose of the landfill cover is to minimize postclosure leachate generation arising from percolation of rainfall and melting snow, and to convert the percolation into surface runoff and/or evapotranspiration without eroding the cover. Additional performance objectives of landfill cover design include the following:

- To operate with a minimum of postclosure maintenance
- To allow the site to be returned to some beneficial use as quickly as possible
- To make the site aesthetically acceptable to nearby residents
- To accommodate settlement
- To prevent the blowing of litter or dust onto adjacent properties
- To suppress fire dangers
- To contain gases and vapors
- To allow placement as each cell is completed.

Cover design must be governed by a need to conform with stated performance objectives.

Additional cover design attributes have been suggested by Hatheway and McAneny (1987), including that the cover be water- and erosion-resistant, that it provide stability against slumping, cracking, and slope failure, and that it be resistant to cold weather distress and to disruption by animals or plants.

In examining features of design attributes and performance objectives, we will focus on (a) the development of the characteristics and use of a number of techniques for surface covers and (b) the examination of various mathematical models in predicting surface runoff quantities that drain to surrounding lands during major storm events.

It is noteworthy that a sophisticated cover system may delay stabilization of the refuse. As was described in Chapters 4 and 5, moisture content and moisture movement affect the rate of solid waste decomposition and thus the rate of methane generation. The rate of methane generation has been shown to increase proportionally with moisture content (Emcon, 1975). Hartz and Ham (1983) showed, for example, a linear correlation between moisture content and methane production both before and during leachate recirculation. Thus, the implementation of a tight cover has the following implications:

1. Methane gas production rates will be significantly lower at landfill sites with a tight cover because the amount of infiltration into the landfill is reduced.
2. Waste material will decompose at a much slower rate. This will result in a longer period before this site can be used for some purposes.

These consequences must be considered during design of the landfill cover. The response may include leachate recycling and gas collection and venting systems.

8.2   COMPONENTS OF THE HYDROLOGIC CYCLE

The water balance modeling in Chapter 7 required mathematical analyses of the hydrologic budget in the landfill cover as a part of the overall calculation sequence. However, the surface hydrologic budget calculations in Chapter 7

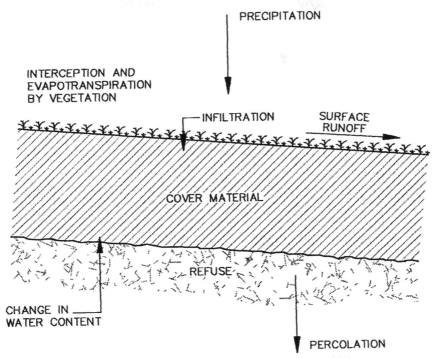

Fig. 8.3 Components of water balance at landfill cover.

were handled in a relatively simplistic manner; for specific design features such as the control of off-site flows, it is necessary to consider the components of the hydrologic cycle in more detail.

The water balance models in Chapter 7 were one-dimensional (allowing variations in the vertical direction only). However, design features must reflect the spatial variations that exist in runoff processes (e.g., different slopes result in different surface runoff). Of particular concern, then, are design considerations that influence the integrity of the surface cover, such as the thickness of the cover, the slope, and soil characteristics. These considerations will involve a more detailed assessment of the hydrologic budget processes at the surface.

Many of the models that are useful in making these hydrologic budget assessments have been "borrowed" from watershed hydrology, the technical aspects of which have been studied for many years. Many of the techniques are highly empirical (e.g., regression - based models). The criterion for selecting a particular model to apply to landfill cover design must be its ability to examine different design initiatives. Selection of a hydrologic model from the many that exist (see Viessman et al., 1989, for further reading) must also reflect the availability of the data.

The elements of the hydrologic cycle pertinent to a landfill cover are schematically depicted in Figure 8.3. These components include interception, evapotranspiration, infiltration, surface runoff, and subsurface lateral flow.

Interception   A portion of the precipitation is intercepted by vegetation. Interception is that segment of the gross precipitation input that wets and adheres to above-ground vegetation until it is returned to the atmosphere through evaporation. The amount of water intercepted is a function of the season of the year, the vegetative species, the age and density of the vegetation, and the character of the storm.

Table 8.2 METHODS FOR ESTIMATING EVAPOTRANSPIRATION QUANTITIES

Method	Comment
Lysimeter	Consists of an upright vegetation soil cell so designed that input water and water percolating out the bottom can be accurately measured. The difference is evapotranspiration.
Adjusted pan-evaporation	Since evaporation from a typical pan 4 ft in diameter and 10 in. deep differs from free water evaporation, a pan coefficient is applied to the pan-evaporation value (normally 0.7).
Thornthwaite method (Thornthwaite and Mather, 1957; Mather and Rodriguez, 1978)	This empirical method involves three steps. A heat index is obtained for each of the 12 months and summed for the annual index. Daily potential evapotranspiration is obtained from the heat index by use of tables. Finally, the potential evapotranspiration is adjusted for month and day lengths with correction factors (see Chapter 7).

Evapotranspiration   Root systems of plants absorb water in varying quantities. This water is transmitted through the plant and escapes through pores in the leaf system via transpiration. The determination of evapotranspiration is generally based on empirical equations (see Table 8.2) or pan-evaporation data. Evapotranspiration is a function of solar radiation, differences in vapor pressure between a water surface and the overlying air, temperature, wind, atmospheric pressure, and type of vegetation.

Within a storm, vapor pressure gradients are reduced and evaporation usually becomes insignificant. As a result, it is common practice to make no deduction from gross precipitation for evapotranspiration during a storm. However, once the storm is over, evapotranspiration is usually the main process by which a cover soil loses water, and thus, it is a major factor in determining leachate generation.

Infiltration   Infiltration is the process by which water penetrates the ground surface and clarify into the soil. Infiltrating water is driven by a combination of gravitational and capillary forces. As capillary pores become filled, the gravitationally driven water descends to greater depths and encounters increased resistance due to reduction in the large pore spaces and perhaps a barrier such as clay (Lutton, 1979).

Table 8.3   EXAMPLES OF ALTERNATIVE MODELS FOR QUANTIFYING INFILTRATION

Source

Holtan (1961)

$$I = GI \bullet a \bullet S_a^{1.4} + I_c$$

where: I = infiltration
GI = growth index of vegetation in percent of maturity
S_a = available storage in equivalent inches of water
a = constant
I_c = constant rate

Huggins and Monke (1966)

$$I = I_c + A \left(\frac{S - F}{T_p} \right)^p$$

where:
A and P = coefficients
S = the storage potential of a soil overlying the less permeable layer
F = total volume of water that infiltrates
T_p = total porosity of soil lying over the less permeable layer

Horton (1933, 1939)

$$I = I_c + (I_0 - I_c) \, e^{-ct}$$

The equation approaches a stable minimum, where: I_0 = an initial infiltration rate
I_c = a decay constant having dimensions of (T^{-1})
c = a constant
t = time

Note: See, for example, Viessman et al., (1989) for further examination of alternative infiltration methods.

Many factors influence the infiltration rate, including the condition of the soil surface and its vegetative cover, the properties of the soil such as its porosity and hydraulic conductivity, and the current moisture content of the soil.

Mathematical procedures for calculating infiltration vary in sophistication from the application of reported average rates for specific soil types and vegetative covers, to the use of conceptually rigorous differential equations governing the flow of water in unsaturated porous media.

Table 8.3 gives the more commonly utilized equations for infiltration. An indication of the difficulty in estimating infiltration is apparent from the list of factors that affect it (Table 8.4).

If a soil system already exists, it is possible to measure infiltration in the field with an infiltrometer, which consists of a metal tube forced into the soil. The quantities of water added are monitored. The preferred infiltrometer system has an outer concentric ring that produces a buffer region as indicated in Figure 8.4.

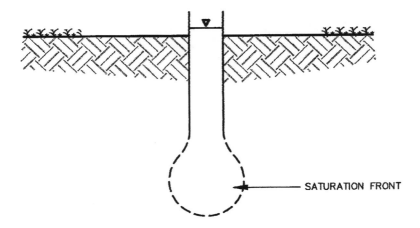

SINGLE—RING INFILTROMETER

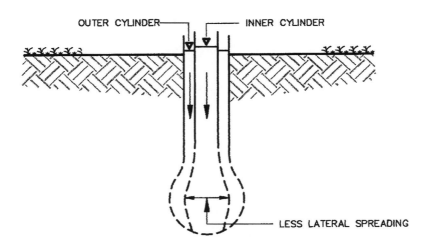

DOUBLE—RING INFILTROMETER

Fig. 8.4 Infiltration as Measured by Field Infiltrometer.

Surface Runoff Surface runoff develops after the initial demands of interception, infiltration, and surface storage have been satisfied. Surface runoff quantities are a function of numerous phenomena including surface slope, antecedent moisture conditions, and vegetative cover. Again, a variety of mathematical representations have been developed; however, frequent use is made of the SCS methodology for carrying out the calculations (Soil Conservation Service, 1972).

The empirically derived curves that summarize the SCS methodology are indicated in Figure 8.5. In fact, utilization of the curves bypasses many of the hydrologic processes just discussed. Given an estimate of the curve number that

Table 8.4 FACTORS AFFECTING INFILTRATION INTO SOIL

1. Porosity = f (soil density, grain size and shape, organic content)

2. Initial moisture content

3. Slope and degree of surface deformation

4. Raindrop size and impact velocity

5. Rainfall intensity and duration

6. Inwash of fine materials

7. Vegetation

8. Temperature

Source: After Gee, 1981.

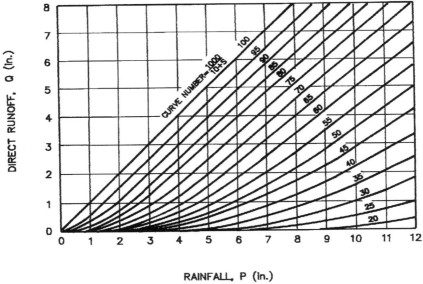

RAINFALL, P (in.)

Fig. 8.5 Estimation of direct runoff amounts from storm rainfall.

is assumed to characterize the surface cover, and given a rainfall quantity, the surface runoff can be read directly from Figure 8.5. Alternatively, runoff values can be obtained from a tabular array of runoff depths for various curve numbers and rainfall quantities, as found in Table 8.5. The key to effective hydrologic analyses is in the assignment of the curve number. The magnitude of the curve number can reflect a series of different features including surface slope, antecedent moisture condition, vegetative cover intensity, duration of storm, and cover soil type. Curve numbers frequently employed in landfill cover applications are included in Table 8.6.

Table 8.5 RUNOFF DEPTH FOR SELECTED CURVE NUMBERS AND RAINFALL AMOUNTS

Rainfall	Runoff Depth (in inches) for Curve Number of												
	40	45	50	55	60	65	70	75	80	85	90	95	98
1.0	0.00	0.00	0.00	0.00	0.00	0.00	0.00	0.03	0.08	0.17	0.32	0.56	0.79
1.2	0.00	0.00	0.00	0.00	0.00	0.00	0.03	0.07	0.15	0.27	0.46	0.74	0.99
1.4	0.00	0.00	0.00	0.00	0.00	0.02	0.06	0.13	0.24	0.39	0.61	0.92	1.18
1.6	0.00	0.00	0.00	0.00	0.01	0.05	0.11	0.20	0.34	0.52	0.76	1.11	1.38
1.8	0.00	0.00	0.00	0.00	0.03	0.09	0.17	0.29	0.44	0.65	0.93	1.29	1.58
2.0	0.00	0.00	0.00	0.02	0.06	0.14	0.24	0.38	0.56	0.80	1.09	1.68	1.77
2.5	0.00	0.00	0.02	0.08	0.17	0.30	0.46	0.65	0.89	1.18	1.53	1.96	2.27
3.0	0.00	0.02	0.09	0.19	0.33	0.51	0.71	0.96	1.25	1.59	1.98	2.45	2.77
3.5	0.02	0.08	0.20	0.35	0.53	0.75	1.01	1.30	1.64	2.02	2.45	2.94	3.27
4.0	0.06	0.18	0.33	0.53	0.76	1.03	1.33	1.67	2.04	2.46	2.92	3.43	3.77
4.5	0.14	0.30	0.50	0.74	1.02	1.33	1.67	2.05	2.46	2.91	3.40	3.92	4.26
5.0	0.24	0.44	0.69	0.98	1.30	1.65	2.04	2.45	2.89	3.37	3.88	4.42	4.76
6.0	0.50	0.80	1.14	1.52	1.92	2.35	2.81	3.28	3.78	4.30	4.85	5.41	5.76
7.0	0.84	1.24	1.68	2.12	2.60	3.10	3.62	4.15	4.69	5.25	5.82	6.41	6.76
8.0	1.25	1.74	2.25	2.78	3.33	3.89	4.46	5.04	5.63	6.21	6.81	7.40	7.76
9.0	1.71	2.29	2.88	3.49	4.10	4.72	5.33	5.95	6.57	7.18	7.79	8.40	8.76
10.0	2.23	2.89	3.56	4.23	4.90	5.56	6.22	6.88	7.52	8.16	8.78	9.40	9.76
11.0	2.78	3.52	4.26	5.00	5.72	6.43	7.13	7.81	8.48	9.13	9.77	10.39	10.76
12.0	3.38	4.19	5.00	5.79	6.56	7.32	8.05	8.76	9.45	10.11	10.76	11.39	11.76
13.0	4.00	4.89	5.76	6.61	7.42	8.21	8.98	9.71	10.42	11.10	11.76	12.39	12.76
14.0	4.65	5.62	6.55	7.44	8.30	9.12	9.91	10.67	11.39	12.08	12.75	13.39	13.76
15.0	5.33	6.36	7.35	8.29	9.19	10.04	10.85	11.63	12.37	13.07	13.74	14.39	14.76

Caution: A frequent practice is to use a *daily* time increment in the calculation sequence for the hydrologic water balance at the surface (e.g., as utilized in the HELP model). Thus, daily rainfall levels are utilized, which implies that rainfall occurs uniformly over the 24-hour period. This assumption effectively decreases the intensity of the rainfall (because the more frequent situation is a short-duration intensive storm). The result is that the calculations imply more infiltration and less surface runoff than would actually be the situation. The net result is that the approach using SCS values on a daily time increment will tend to underpredict surface runoff and overpredict leachate quantities.

Subsurface Lateral Flow   Subsurface lateral flow occurs when infiltrated rainfall meets an underground zone of low transmission (the barrier (clay) layer) and is diverted laterally. The intercepted water then travels above the barrier layer to either (a) the external surface of the landfill, at which it appears as a seep or spring, or (b) the drainage collection tiles, where it is collected.

Runoff Velocity and Time of Concentration The integrity of the surface cover is directly related to the velocity of the surface runoff, which is estimated for different slope and surface covers, using Manning's equation. Figure 8.6 characterizes (from Manning's equation, e.g., see Bedient and Huber, 1988) the velocity for different slopes and surface covers. Values of Manning's coefficients for different types of surface cover are listed in Table 8.7.

E X A M P L E

From Figure 8.6, runoff from a landfill cover at a 5 percent slope, over a grassy surface, will experience a velocity of 3.5 ft/s.

Given the velocity, it is a simple matter to calculate the time of travel for a distance of 1000 ft:

$$\text{Time of travel } = \frac{\text{Distance of travel}}{\text{Velocity}} = \frac{100 \text{ ft}}{3.5 \text{ ft/s}} = 4.8 \text{ min} \tag{8.1}$$

The time of travel can then be used to calculate the time of concentration, t_c, which is the time required for the most remote location in the drainage catchment to contribute to a point of interest, such as a culvert under a road near the landfill site. The culvert may be required to pass all flows, for example, on a 5-year basis (i.e., to fail the water only once on average, within a 5-year period).

An alternative procedure is to use the Kirpich formula, which estimates the time required for water to flow from the most remote point in the watershed to

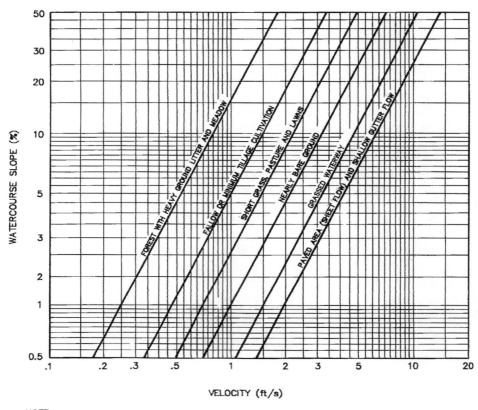

Fig. 8.6 Average velocities for estimating travel time for overland sheet flow.

NOTE:

ft/s x .3048 = m/s

the point of interest as,

$$t_c = 0.0078 L^{0.77} S^{-0.385} \qquad (8.2)$$

where:

t_c = time of concentration (min),
L = maximum length of flow (ft),
S = watershed gradient (ft/ft).

Table 8.6 RUNOFF CURVE NUMBERS FOR SELECT CONDITIONS (after SCS, 1975)"

Cover Description		Curve Numbers for Hydrologic Soil Group (2)			
Cover Type	Hydrologic Condition	A	B	C	D
Pasture, grassland or range: continuous forage for grazing (1) -	Poor	68	79	86	89
	Fair	49	69	69	84
	Good	39	61	74	80
Meadow: continuous grass, protected from grazing and generally mowed for hay	—	30	58	71	78
Brush: brush-weed-grass mixture with brush the major element (1)	Poor	48	67	77	83
	Fair	35	56	70	77
	Good	30	48	65	73
Farmsteads: buildings, lanes, driveways and surrounding lots	—	59	74	82	86

Notes:
(1) Average runoff condition, and initial abstraction of 0.2 S (see Viessman et al, 1989)
(2) Poor: <50% ground cover
    Fair: 50 to 75% ground cover
    Good: >75% ground cover
(3) Actual curve number if less than 30: Use CN = 30 for runoff computations

Table 8.6: To Assign Soil Group, Use The Following:

Soil Group	Description
A	Lowest Runoff Potential. Includes deep sands with very little silt and clay, also deep, rapidly permeable loess
B	Moderately Low Runoff Potential. Mostly sandy soils less deep than A, and loess less deep or less aggregated than A, but the group as a whole has above-average infiltration after thorough wetting.
C	Moderately High Runoff Potential. Comprises shallow soils and soils containing considerable clay and colloids, though less than those of Group D. The group has below-average infiltration after presaturation.
D	Highest Runoff Potential. Includes mostly clays of high swelling percent, but the group also includes some shallow soils with nearly impermeable subhorizons near the surface.

Table 8.7   MANNING COEFFICIENTS FOR SELECTED SURFACES

	Manning's Number Range
Open Channels, Lined (Straight Alignment)	
Gravel bottom, sides as indicated:	
1. Formed Concrete	0.017 – 0.020
2. Random stone in mortar	0.020 – 0.023
3. Dry rubble (rip rap)	0.023 – 0.033
Concrete-lined excavated rock:	
1. Good section	0.017 – 0.020
2. Irregular section	0.022 – 0.027
Open Channels, Excavated (Straight Alignment, Natural Lining)	
Earth, uniform section:	
1. Clean, recently completed	0.016 – 0.018
2. Clean, after weathering	0.018 – 0.020
3. With short grass, few weeds	0.022 – 0.027
4. In gravelly soil, uniform section, clean	0.022 – 0.028
Earth, fairly uniform section:	
1. No vegetation	0.022 – 0.025
2. Grass, some weeds	0.025 – 0.030
3. Dense weeds or aquatic plants in deep channels	0.030 – 0.035
4. Sides clean, gravel bottom	0.025 – 0.030
5. Sides clean, cobble bottom	0.030 – 0.040
Dragline excavated or dredged:	
1. No vegetation	0.028 – 0.033
2. Light brush on banks	0.035 – 0.050
Rock:	
1. Based on design section	0.035
2. Based on actual mean section:	
a) Smooth and uniform	0.035 – 0.040
b) Jagged and irregular	0.040 – 0.045
Channels not maintained, weeds and brush uncut:	
1. Dense woods, high as flow depth	0.06 – 0.12
2. Clean bottom, brush on sides	0.05 – 0.08
3. Clean bottom, brush on sides, highest stage of flow	0.07 – 0.11
4. Dense-brush high stage	0.10 – 0.14
Highway Channels and Swales With Maintained Vegetation *(Values shown are for velocities of 2 and 5 ft/s)*	
Depth of flow up to 0.7 foot:	
1. Bermuda grass, Kentucky bluegrass, buffalo grass:	
a) Mowed to 2 inches	0.07 – 0.045
b) Length 4 – 6 inches	0.09 – 0.05
2. Good stand, any grass:	
a) Length about 12 inches	0.18 – 0.09

Table 8.7 (Continued)MANNING COEFFICIENTS FOR SELECTED SURFACES

	Manning's Number Range
b) Length about 24 inches	0.30 – 0.15
3. Fair stand, any grass:	
a) Length about 12 inches	0.14 – 0.08
b) Length about 24 inches	0.25 – 0.13
Depth of flow 0.7 – 1.5 feet	
1. Bermudagrass, Kentucky bluegrass, buffalograss:	
a) Mowed to 2 inches	0.05 – 0.035
b) Length about 4–6 inches	0.06 – 0.04
2. Good stand, any grass:	
a) Length about 12 inches	0.12 – 0.07
b) Length about 24 inches	0.20 – 0.10
3. Fair stand, any grass:	
a) Length about 12 inches	0.10 – 0.06
b) Length about 24 inches	0.17 – 0.09

Street and Expressway Gutters

Concrete gutter with asphalt pavement:	
1. Smooth	0.013
2. Rough	0.015
Concrete pavement:	
1. Float finish	0.014
2. Broom finish	0.016

Natural Stream Channels

Minor streams (surface width at flood stage less than 100 feet)	
1. Fairly regular section:	
a) Some grass and weeds, little or no brush	0.030 – 0.035
b) Dense growth of weeds, depth of flow materially greater than weed height	0.035 – 0.05
c) Some weeds, light brush on banks	0.035 – 0.05
d) Some weeds, heavy brush on banks	0.05 – 0.07
e) For trees within channel, with branches submerged at high stage, increase all above values	0.01 – 0.02
2. Irregular sections, with pools, slight channel meander; increase values given in l. a) – e)	0.01 – 0.02
3. Mountain streams, no vegetation in channel, banks usually steep, trees and brush along banks submerged at high states:	
a) Bottom of gravel, cobbles and few boulders	0.04 – 0.05
b) Bottom of cobbles, with large boulders	0.05 – 0.07
Flood plains (adjacent to natural streams):	
1. Pasture, no brush	
a) Short grass	0.030 – 0.035
b) High grass	0.035 – 0.05

Table 8.7   (Continued)MANNING COEFFICIENTS FOR SELECTED SURFACES

	Manning's Number Range
2. Cultivated areas	
a) No crop	0.03 – 0.04
b) Mature row crops	0.035 – 0.045
c) Mature field crops	0.04 – 0.05
3. Heavy weeds, scattered brush	0.05 – 0.07
4. Light brush and trees	
a) Winter	0.05 – 0.06
b) Summer	0.06 – 0.08
5. Medium to dense brush	
a) Winter	0.07 – 0.011
b) Summer	0.10 – 0.16

Rational Method The rational method is a mathematical formulation widely utilized in engineering for the design of storm sewers, based on the equation

$$Q = CiA \qquad (8.3)$$

where:

C = dimensionless runoff coefficient,
i = the rainfall intensity (in./hr)
A = the contributing drainage area (in acres)

The flow, Q, is in cubic feet per second. The units described must be utilized, since the conversion is 1 ft^3/s = 1.008 acre-in./hr. As a result, the unit conversion is considered to be included in the runoff coefficient.

Table 8.8: RUNOFF COEFFICIENTS FOR THE RATIONAL FORMULA

Topography and Vegetation	Open Sand Loam	Clay and Silt Load	Tight Clay
Woodland			
Flat, 0-5% slope	0.10	0.30	0.40
Rolling, 5-10% slope	0.25	0.35	0.50
Hilly, 0-30% slope	0.30	0.50	0.60
Pasture			
Flat	0.10	0.30	0.40
Rolling	0.16	0.36	0.55
Hilly	0.22	0.42	0.60
Cultivated			
Flat	0.30	0.50	0.60

Table 8.8: RUNOFF COEFFICIENTS FOR THE RATIONAL FORMULA

	Values of C in $Q = CiA$		
Rolling	0.40	0.60	0.70
Hilly	0.52	0.72	0.82

Source: U.S. Department of Interior, 1982

The calculated flow, Q, from Equation 8.3 represents a peak flow. The flow recurrence is associated with the return frequency of the rainfall. Therefore, if a rainfall of intensity i begins instantaneously and continues indefinitely, the rate of runoff will increase until the time of concentration, t_c, when all the watershed is contributing to flow at the design point. As a result, the return frequency of the rainfall intensity translates to an equivalent return frequency of the runoff flow, Q, (at least insofar as can be represented by this simple equation).

Some key assumptions are implicit with the rational method, namely,

1. The rainfall is assumed to occur at a uniform intensity over the entire watershed. This assumption is reasonable for landfill design concerns because of their relatively small spatial extent.
2. The rainfall occurs at a uniform intensity for a duration equal to the time of concentration. (More sophisticated procedures exist when more detailed assessments need to be made; see, Chow et al. (1988) for further details.)
3. The frequency of the runoff equals that of the rainfall used in the equation.
4. The runoff coefficient is the same for all storm events.

Table 8.9   DESIGN FACTORS IN INFILTRATION/PERCOLATION CONTROL

	Comments
Material selection	permeability is a direct indication of a material's tendency to allow percolation of water
Compaction	effective in reducing infiltration and percolation
	may interfere with grass seeding
Soil layering	lower layers act to impede percolation while upper layers support vegetation, provide erosion protection and help retain capillary water in the lower (less permeable) layers
Increased thickness	increases water storage capacity and reduces detrimental effects of cracks and settlement
Discontinuities	avoid differential settlement
	creates depressions detaining runoff
	keep bulky objects away from upper part of waste

Table 8.9 DESIGN FACTORS IN INFILTRATION/PERCOLATION CONTROL

	Comments
Surface slope	a steeper slope increases surface runoff
	at less than 3 percent, surface irregularities act as traps
	5 percent is best
	at steep slopes, an assessment of slope stability is necessary (e.g. see Lambe and Whitman, 1969)
Ditching and drainage	to assist in conveying the water off the landfill

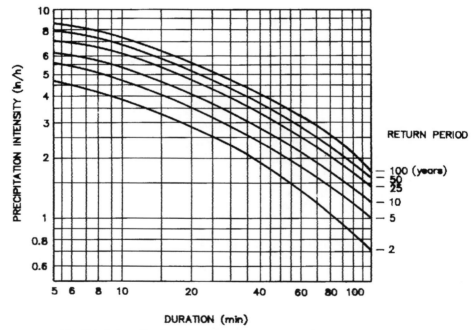

Fig. 8.7 Intensity-duration-frequency curve of maximum rainfall in
Chicago, IL, USA.

The least precise aspect of applying the rational method to landfill-related
concerns is in the assignment of the runoff coefficient. Table 8.8 gives runoff
coefficients for various combinations of terrain and soil textures.

Flow Recurrence Calculation Estimation of flows associated with specific recurrence intervals requires utilization of the time of concentration and the
intensity-duration-frequency (*idf*) curves specific to a particular region. The idf
curves are of the form depicted in Figure 8.7 and show the relationships among

the intensity, duration, and frequency of precipitation. The idf curves are site-specific and may be obtained from local agencies.

E X A M P L E

A landfill with a time of concentration of runoff of 20 minutes will have a precipitation intensity of 3.5 in./h for a storm with a recurrence interval of 5 years (see Figure 8.7). In addition, the total rainfall for this storm will be $1/3$ (3.5), or 1.17 in. of water.

8.3   DESIGN FACTORS IN INFILTRATION/PERCOLATION CONTROL

Another major concern with hydrologic budget assessment at the landfill surface is estimation of the percolation rate. A number of design elements of the landfill cover can have a large influence on leachate levels. Some of these factors are briefly detailed in Table 8.9 and discussed further below.

Material Selection Selection of materials is subject to availability and cost. The normal practice is to utilize soil as the final cover material for municipal solid waste sites, but it is also possible to use geomembranes. Design and performance aspects of geomembranes will be examined more fully in Chapter 9 in relation to liners; however, where the cost of purchase and implementation of geomembranes is warranted for surface covers, they may be incorporated into the landfill cover system.

Compaction Compaction of soils lowers infiltration/percolation rates but involves additional costs of equipment use. On highly compacted landfill surfaces, ponding of rainwater can be observed because there is inadequate slope.
    Poor compaction may result when soil pore water is frozen during cold weather conditions. Some improvements may be achieved by adding a freeze-point suppressant (e.g., calcium chloride).

Thickness A guideline for the minimum cover thickness is $T \geq 2R$, where T is cover thickness and R is relief, defined for this criterion as the vertical distance from the high point to the low point of irregularities on the top surface of the solid waste (within a 7 to 15 m spatial region).
    The thickness in excess of this minimum may be governed by many factors, including coverage, gas migration, infiltration, trafficability and support requirements, and freeze/thaw or dry/soak effects. Increased thickness involves additional expenditure, increases the water storage capacity, increases percolation time, and suppresses the formation of cracks. Thickening the cover may be a direct and effective procedure for reducing gas migration through the cover (because soil moisture helps impede gas movement), especially to the extent that

the increased thickness enhances maintenance of high moisture content within the cover soil.

The low bearing capacity of some solid waste landfills may be circumvented by increasing the soil thickness in the cover. In this way, the relatively strong soil resists punching and rotational shear. Past experience with buildings on landfills is replete with cases of structural damage from differential settlement and hazards from accumulation of methane and other gases.

Differential Settlement   Prior to the 1950s, most landfills were operated as open-burning dumps. Although this caused air pollution, the end result was that most of the material left in the landfill was nonbiodegradable. This meant that little settlement occurred over time. Given today's operational procedures, covers must be designed to accommodate long-term settlement or subsidence of the cover. Subsidence arises because of one or more of the following factors:

- A reduction in void space and compression of loose materials due to self-weight and the weight of overlying materials
- Volume changes from biological decomposition and chemical reaction
- Loss of volume due to dissolution into leachate
- Movement of smaller particles into larger voids
- Settlement of underlying soil materials beneath the landfill.

The intent in many designs is to maintain a 5 percent slope toward the edge. The design geometry of the cap is controlled by the need to move surface waters away from the landfill both in the short-term and after long-term settlements have occurred. Because the random bulk of the contained refuse tends to prevent consistent compaction, significant settlements of the cap are possible. Without an adequate initial slope, excessive settlements of the waste can produce localized depressions that allow surface water to pond.

The factors influencing the settlement include fill height, stress history, magnitude of decomposable component, moisture content, and initial refuse density or void ratio.

An average settlement is 11 percent of the overall depth, but the settlement may reach 30 percent (although with current practices of better compaction at placement, this will decrease). The majority of this settlement occurs in the first year. Projections of 13 percent of the original compacted refuse depth were expected by Headrick and Southard (1978). Yen and Scanlon (1975) found that the rate of settlement decreases logarithmically with time. Edil et al. (1990) fit mathematical curves to settlement data for four different landfills.

Discontinuities and Surface Slope   The causes of subsidence have already been noted and may involve a significant settlement level. Therefore, if the refuse depth is variable, problems may result, as schematically depicted in Figure 8.8.

With fewer putrescibles being placed in landfills in the future (as a result of the 3Rs), the integrity of future cover systems is expected to improve because there will be less settlement as a result of less decomposition of waste and more

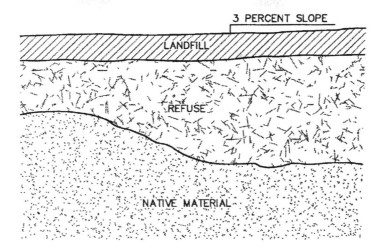

BEFORE SETTLEMENT

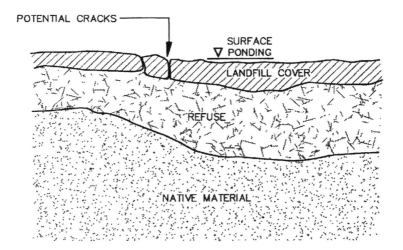

AFTER SETTLEMENT

Fig. 8.8 Disruption of surface slope by excessive settlement.

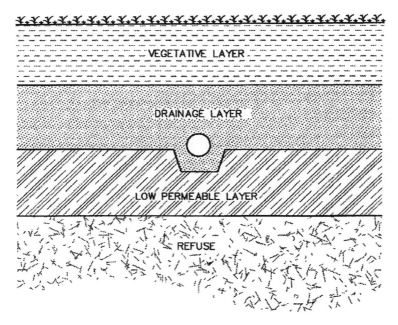

Fig. 8.9 Subsurface drainage.

intensive efforts at getting good compaction. The reduction of refuse through the reduction of yard and food wastes will also affect the settlement and structural characteristics of a landfill. With less plant material being disposed of, less gas and liquid will be formed. Thus, consolidation resulting from the decay of plant material will be reduced.

The ability to minimize consolidation through more efficient compaction will present a greater number of options for the reclamation and development of a former landfill site (Emberton and Parker, 1987). It was observed by Eliassen in 1942 (Tchobanoglous et al., 1977) that 90 percent of the settlement of landfills occurred within the first 5 years, when biodegradable material is transformed into gas and liquid. Eliassen also observed that landfills that received extensive compaction experienced only 50 percent of the settlement observed in landfills with minimal compaction.

Subsurface Drainage Subsurface drainage involves placement of drainage tile within the cover system to collect, infiltrate, and provide a ready conduit to transport the water off the cover. A schematic of the configuration is shown in Figure 8.9. Such a system can be utilized, but obviously the expenditure involved must be warranted.

Vegetation Interception and evapotranspiration by planted vegetation will help prevent percolation and help maintain the integrity of the cover in a number of additional respects. These will be examined in Section 8.5.

8.4   Cover Materials for Final and Daily Cover

A number of criteria exist for creation of a hydraulic barrier above the waste surface. Both soil materials and non-soil materials are utilized, although soils are most commonly used. The geotechnical classification and property determination of soil is a well established body of knowledge. Lutton et al. (1979) have rated various soils for effectiveness in controlling percolation.

Examples of non-soil materials used for surface cover are listed in Table 8.10. Reasons for using materials other than soils may be that the site lacks sufficient quantities of soil for daily cover or that the available soil is unbalanced in quantity between cohesive and cohesionless types (Hatheway and McAneny, 1987). As apparent from the listing in Table 8.10, a number of alternatives exist for non-soil materials. Additional possibilities are included in Table 8.11. Success at the Keele Valley Landfill site location has been obtained using mulched leaves as a daily cover. Other daily cover materials are foundry sands and foams.

Geomembranes are typically recommended at 40 to 60 mil (1 mil = 0.001 in.) thickness. A soil buffer or geotextile must be provided above and below the geomembrane for protection to avoid tearing and puncture of the geomembrane. The geotextile must be placed in a "relaxed" fashion, with slack. Additional discussion of these placement needs is presented in Chapter 9.

Table 8.10 NON-SOIL LANDFILL COVER MATERIALS

Material	Advantages	Disadvantages
Asphalt	Traditional use as hydraulic barrier Ductile and often chemically durable. Wide variety of water-based emulsions Available in panels or rolls	Low tensile strength; needs geotextile reinforcement Need careful design and installation
Industrial Wastes	Abundant and cheap where available Predictable chemistry; often inert to vapors/gases Ideal for drainage and load-bearing application	Many available only as less-desirable fine-grained materials

After Hatheway and McAneny (1987)

Table 8.10 NON-SOIL LANDFILL COVER MATERIALS

Material	Advantages	Disadvantages
Geomembranes	Thin sheets take the place of thicker soil layers Pre-formed as relatively large panels Small leaks less serious than when used as liners Very low permeability Large body of geotechnical knowledge in application	May be vapor-gas degradable Requires careful installation Cannot be exposed to elements Low tensile strength Uncertain life span under various in-place conditions
Geotextiles	Ideal for: filtration drainage separation reinforcement armoring	Limited to essentially secondary roles, in conjunction with use of soil Lack information on long-term performance
Soil-Cement	Relatively low permeability Can be formulated on site	Inflexible and brittle Difficult to utilize as major cover component

After Hatheway and McAneny (1987)

8.5 VEGETATIVE GROWTH AND SURFACE PREPARATION

Although early landfill sites were located far from residential areas, rapid urban and suburban expansions have brought many once-remote sites within developed areas. Thus, landfills frequently need to be reasonable in appearance.

In large part, because of the proximity to urban areas, completed landfill sites are being developed to include such features as parks, golf courses, nature areas, and bicycle paths. As a result, an effective vegetative growth must be established and maintained on the final cover soil to achieve any of these desired end uses.

Table 8.11 EXAMPLES OF NON-SOIL COVER ALTERNATIVES

Portland Cement Concrete or Mortar
Bituminous Concrete or Mortar
Bitumen-sulfur Concrete
Sprayed Bituminous Membranes
Sprayed Sulfur Membrane

Table 8.11   **EXAMPLES OF NON-SOIL COVER ALTERNATIVES**

Portland Cement Concrete or Mortar
Polyurethane Foam
Pre-fabricated Bituminous Membrane
Plastic and Rubber Membrane
Fly Ash
Bottom Ash and Slag
Incinerator Residue
Mill Tailings
Plant Sludges

Vegetation also serves other purposes relevant to the water balance and maintenance of the integrity of the surface cover. The vegetation helps resist erosion and promotes evapotranspiration by:

- Creating a leaf layer above the soil, reducing the kinetic energy of the rainfall and thereby decreasing erosion
- Decreasing the wind velocity, thereby decreasing soil erosion
- Decreasing the water runoff velocities
- Minimizing soil crusting. Heavy loams, characterized by high clay content and dense structure have excessive swell/shrink behavior. These types of soils can often crack, allowing infiltration through the cover. However, the cracks close soon thereafter due to the nature of these soils.
- Decreasing the freezing depths. Frozen soil landfill covers and exposed liners are subject to thermal contractions and increases in tensile stresses. Potential cracking may very well occur over the full depth of freezing (greater than 2 ft in some locations). Since frozen ground is weak, the formation of the cracks can dramatically increase infiltration.

Selection of the appropriate vegetative species is an important consideration in design of the landfill cover. The likelihood of species' survival and ability to function as desired depends on factors such as climate, site characteristics, ability for roots to withstand landfill gas in the root zone, and soil properties. The vegetation must be both persistent and yet not have roots that might penetrate beyond the vegetative layer. The vegetation should be indigenous to the area, hardy, and drought resistant. The regional Soil Conservation Service, highway department, or county agriculture extension agent can provide such information.

The topsoil forming the protective top cover must be selected and constructed to support the vegetation by allowing sufficient surface water to infiltrate into the topsoil and by retaining enough plant-available water to sustain plant growth through drought periods. Particle-size distribution, structure, and organic matter content influence the quantity of available water a given soil can retain and must be considered in selecting the topsoil material. In general, medium-textured soils, such as loam soils, have the best overall characteristics for seed germination and plant root system development.

A soil with a texture resembling loam should be chosen for areas where trees and shrubs will be planted, since these types of vegetation require looser, deeper soil for root development than do grasses and ground covers. The soil should be tested for pH, Mg, Ca, P, NO_3, NH_4, K, Cu, Fe, Zn, Mn, conductivity, particle-size distribution, bulk density, and organic matter. Soil pH should be in the range 5 to 8 to be suitable for most plants. If the pH is over 8, necessary elements for plant growth may not be soluble. A pH of less than 5 may cause some elements to become toxic. In general, soil pH should be above 6.5.

Tests for fertilizer and lime requirements are essential to provide the appropriate balance. The three major nutrients necessary are nitrogen, phosphorus, and potassium. Soil amendments have included digested municipal sewage, sludge, and manure.

In some situations, the mowing of the grass is appropriate and for others, it is not. Some advantages of mowing the grass are that it keeps out unwanted species, helps in fire control, and allows easy inspection of surface integrity. Some disadvantages of mowing are that it increases the compaction of topsoil and increases costs ($20–$45 /acre/mowing).

The thickness of cover soil needed to support the vegetation is a function of the vegetation type. A guideline frequently utilized is that the cover soil thickness must be two times the depth of the vegetative root zone. Thus, the necessary thickness for grasses is 24 to 30 in. and for shrubs, 36 to 42 in.

The soil on recently covered landfills must be stabilized soon after spreading to prevent erosion. Mulch and tack should be used to hold the seed in place in sloping areas and to conserve moisture. Mulch should be applied uniformly within 48 hours after seeding at the rate of 3000 to 4000 lb/acre. Mulching materials include cereal or grain, straw, prairie hay, and fiber mulch. Microterracing is recommended before seeding and mulching for steep, sloping ground. Generally it is best to embed the seed in the soil. Special techniques such as hydroseeding may be required for steep slopes inaccessible to conventional equipment.

Figure 8.10 illustrates the application of seed and mulch. Grasses and other ground covers selected must be landfill-tolerant species. Trees and shrubs should not be planted until one or two years after grass has been planted. If the grasses cannot grow due to gases from the landfill, other, deeper-rooted species are not likely to survive. A barrier can be used beneath each tree-planting area to help protect the root system from the landfill gases, as depicted in Figure 8.11.

Roots from plants and trees can penetrate very deeply into the subsoil. The depth is obviously dependent upon the type of plant, type of soil, geographic loca-

Fig. 8.10 Application of seed and mulch.

tion, etc. but depths of many feet are not uncommon. Thus, the plastic or metal planter depicted in Figure 8.11 serves the additional purpose of containing the root system.

Further discussions on available plants and site-selection criteria can be found in Lee et al. (1976).

8.6   GAS CONTROL

As a direct consequence of placement of a comprehensive cover to minimize leachate, there may be a potential for gas buildup under the cover. A gas collection layer may therefore be required. Such a gas collection system might take the form of a low-permeability layer of 2 ft of clay and a 20 mil flexible membrane liner, a surface water drainage layer, and a cover layer capable of supporting vegetation. The control system employs a laterally extensive porous gas control layer as close as feasible above the refuse. Figure 8.12 depicts a typical passive collection system. This gas collection layer may be an active or a passive collection system based on rate of landfill gas generation and any desire or requirement to collect the gas for flaring or stripping (cleaning) and/or reuse as a power source by industry or utility. In the case of an active collection system, the gas is pumped to one or more centralized systems.

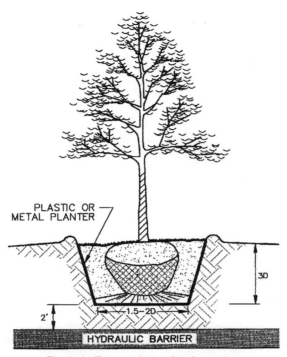

Fig. 8.11 Tree and woody plant planter.

A coarse particle-size gradation in the gas collection layer is required to minimize the clogging effect of anaerobic biologic slimes. The position of the gas drainage layer in the cover system is placed below the low permeable cover to intercept the gases rising from waste cells. This gas collector layer must allow the gases to freely flow to the vent pipe that leads to the atmosphere and provide for drainage of condensate. A minimum slope in the collector pipe of 2 percent is required to maintain the gas flow. Slopes ranging from 2 to 5 percent are common (Richardson and Koerner, 1987).

8.7 MAINTENANCE PROBLEMS

Among the many problems associated with maintenance of the surface cover are the following:

- **Difficulty of maintenance of the vegetation**—Many attempts to vegetate landfills have been unsuccessful. Flower et al. (1978) described the abnormally high incidence of vegetation die-off on landfills and examined the stress on vegetation of low moisture, and elevated CO_2 and CH_4 concentrations. As a result of landfill gas in the root zone, trees produce less shoot and stem growth (Gilman et al., 1981). Gilman et al. (1981, 1983) determined that landfill gases in the root system of trees and shrubs must

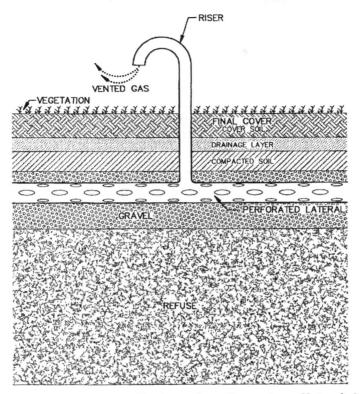

Fig. 8.12 Passive collecting and venting system of laterals in
gravel trenches above waste cell

be minimized to promote good vegetative growth. Three gas barrier systems were effective at accomplishing this (a) a soil trench underlain with plastic sheeting over gravel and vented by means of vertical PVC pipe, (b) a 0.9 m mound of soil underlain with 30 cm of clay, and (c) a 0.9 m soil mound with no clay barrier.

Maintenance of the vegetation is very important, since the vegetation serves many purposes, as indicated in the previous section. An additional concern is that during periods of dryness, roots contract, leaving passages that allow rapid percolation of incident rainfall. Thus, the root system must be limited in depth. One means of controlling root growth and also keeping landfill gas out of the root zone is to use planters, as depicted in Figure 8.11.

Environmental Soil Conditions—Examples include freezing and thawing, wetting and drying, root penetration, and burrowing animals. These problems are important in terms of runoff conditions, pore blockage, and tensile cracking.

A particular concern in the northern tier of the United States and throughout Canada is potential cracking through the full depth of freezing (up to 2 m). Soil has a thermal coefficient of contraction almost three times higher than that of steel, so a small decrease in temperature quickly generates tensile stresses. Because frozen ground is weak in tension, initial fracturing commences at the ground surface and penetrates the cover soils to the depth needed to relieve the tensile stresses.

Andersland and Al - Moussawi (1987) reported that the depth of crack formation is dependent on soil type, temperature, water content, and surface effects (vegetation and snow cover). They calculated the maximum frost depths for three sites, as listed in Table 8.12. The greatest frost depths correspond to sand with a low water content. However, frozen soil under a constant stress will deform with time in a viscous manner; the resulting creep behavior will reduce the tensile stresses over the period of time in which temperature is decreasing.

Table 8.12 MAXIMUM FROST DEPTH FOR THREE LOCATIONS FOR SAND, SILT, AND CLAY SOILS

	Frost Depth* (cm)		
	Sand γ_d = 15.0 kN m^{-3} w = 8.3% S = 3.0%	Silt γ_d = 14.0 kN m^{-3} w = 26.2% S = 80.0%	Clay γ_d = 16.0 kN m^{-3} w=24.2% S = 100%
Fargo, North Dakota 1973/74	210	155	196
Madison, Wisconsin 1978/79	162	110	133
Lansing, Michigan 1983/84	115	83	100

Notes:

* Based on modified Berggren equation (Aldrich & Paynter, 1953)

γ_d soil dry density kNm^{-3}

w water content (percent dry weight basis)

S degree of soil saturation

- The ground surface was assumed to have no snow or turf cover.

Source: After Andersland and Al-Moussawi (1987).

- **Erosion due to excessive velocities**—If the velocity of runoff becomes high, erosion of the surface cover will occur. Erosion velocities must be maintained at less than the values indicated in Table 8.13 for grass-lined channels and Table 8.14 for different soil types.
- **Off-Site flows**—It is necessary for post-development runoff levels to be equal to predevelopment runoff levels. Otherwise, the post-development runoff will affect the adjacent land uses and may precipitate damage claims.

8.8   SOIL LOSS ESTIMATION

Engineering assessment for soil conservation on the landfill cover requires knowledge of the factors that cause loss of soil and those that reduce the losses. Fortunately, there has been extensive research on agricultural soil losses. Since 1930, controlled studies on experimental plots and small watersheds have supplied much valuable information on these relationships. Analyses of these data have resulted in the development of the Universal Soil Loss Equation, which predicts annual soil loss as the product of six quantifiable factors, namely,

$$A = RKLSCP \tag{8.4}$$

where
A = soil loss (tons /acre),
R = rainfall factor,
K = soil erodibility factor,
L = slope-length factor,
S = slope-gradient factor,
C = cropping-management factor,
P = erosion control practice factor.

Table 8.13   MAXIMUM PERMISSIBLE WATER VELOCITIES IN CHANNELS LINED WITH UNIFORM STANDS OF VARIOUS GRASS COVERS

| | | Maximum Permissible Water Velocity | | | |
| | | Erosion-Resistant Soils | | Easily-Eroded Soils | |
Vegetation or Cover	Slope Range (%)	(m/s)	(ft/s)	(m/s)	(ft/s)
Bermuda grass	0 - 5	2.4	8.0	1.8	6.0
	5 - 10	2.1	7.0	1.5	5.0
	>10	1.8	6.0	1.2	4.0

Source: U.S. Department of Transportation, 1961

Table 8.13   MAXIMUM PERMISSIBLE WATER VELOCITIES IN CHANNELS LINED WITH UNIFORM STANDS OF VARIOUS GRASS COVERS

Vegetation or Cover	Slope Range (%)	Maximum Permissible Water Velocity			
		Erosion-Resistant Soils		Easily-Eroded Soils	
		(m/s)	(ft/s)	(m/s)	(ft/s)
Buffalo grass					
Kentucky bluegrass	0 - 5	2.1	7.0	1.5	5.0
Smooth brome	5 - 10	1.8	6.0	1.2	4.0
	>10	1.5	5.0	0.9	3.0
Blue grama					
Lespedeza sericea					
Weeping lovegrass					
Yellow bluestem	0 - 5	1.1	3.5	0.75	2.5
Kuzu					
Alfalfa					
Crabgrass					

Source: U.S. Department of Transportation, 1961

Table 8.14   PERMISSIBLE CHANNEL VELOCITIES FOR DIFFERENT SOIL TYPES

Soil Type or Lining (earth walls with no vegetation)	Maximum Permissible Velocity		
	Clear Water (ft/s)	Water Carrying Fine Silts (ft/s)	Water Carrying Sand and Gravel (ft/s)
Fine sand (noncolloidal)	1.5	2.5	1.5
Sandy loam (noncolloidal)	1.7	2.5	2.0
Silt loam (noncolloidal)	2.0	3.0	2.0
Ordinary firm loam	2.5	3.5	2.2
Volcanic Ash	2.5	3.5	2.0
Fine gravel	2.5	5.0	3.7
Silt clay (very colloidal)	3.7	5.0	3.0
Graded loam to cobbles (noncolloidal)	3.7	5.0	5.0

Note: ft/s x .3048 = m/s

Table 8.14 PERMISSIBLE CHANNEL VELOCITIES FOR DIFFERENT SOIL TYPES

Soil Type or Lining (earth walls with no vegetation)	Maximum Permissible Velocity		
	Clear Water (ft/s)	Water Carrying Fine Silts (ft/s)	Water Carrying Sand and Gravel (ft/s)
Graded silt to cobbles (colloidal)	4.0	5.5	5.0
Alluvial silts (noncolloidal)	2.0	3.5	2.0
Alluvial silts (colloidal)	3.7	5.0	3.0
Coarse gravel (noncolloidal)	4.0	6.0	6.5
Cobbles and shingles	5.0	5.5	6.5

Note: ft/s x .3048 = m/s

Each of these factors must be assigned for a particular site. More extensive information is available in numerous sources, including Beasley (1972).

Rainfall Factor, R Soil losses are proportional to the kinetic energy of a storm and its intensity. The rainfall erosion index at a particular location is an average annual total. The lines in Table 8.13 with the same erosion index value are termed iso-erodents.

If estimates of soil losses other than the average annual losses are desired, then values other than average annual values of the erosion index must be substituted for R. Estimates of soil loss from a single storm that will be exceeded once in 1, 5 and 20 years are indicated in Table 8.15.

E X A M P L E

The average annual value of R for Minnesota is 100. Alternatively, $R = 25$ for the 1-year storm. In other words, 25 percent of the annual erosion is associated with the typical maximum storm within that year.

Soil Erodibility, K The soil erodibility depends on the physical and chemical properties of the soil. Some soils erode more readily than others even when rainfall and slope are the same. The difference is due to the properties of the soil itself, although the degree to which erosion occurs must be found via experiments. The values of K for different soil types are indicated in Table 8.16.

Table 8.15 EXPECTED MAGNITUDES OF SINGLE STORM EROSION
INDEX VALUES (TO BE SUBSTITUTED FOR 'R' IN THE EROSION EQUATION)

Location	1 Year	10 Years	20 Years
Alabama—Montgomery	62	145	172
Arkansas—Little Rock	41	158	211
Connecticut—Hartford	23	64	79
Florida—Jacksonville	92	201	236
Georgia—Atlanta	49	112	134
Illinois—Springfield	36	94	117
Indiana—Indianapolis	29	75	90
Iowa—Des Moines	31	86	105
Kansas—Concordia	33	116	154
Kentucky—Lexington	28	114	151
Louisiana—New Orleans	104	121	141
Maine—Skowhegan	18	51	63
Maryland—Baltimore	41	109	133
Massachusetts—Boston	17	57	73
Michigan—East Lansing	19	43	51
Minnesota—Minneapolis	25	65	78
Mississippi—Vicksbury	57	136	161
Missouri—Columbia	43	93	107
Nebraska—North Platte	25	78	99
New Hampshire—Concord	18	62	79
New Jersey—Trenton	29	102	131
New York—Rochester	13	54	75
North Carolina—Raleigh	53	137	168
North Dakota—Devils Lake	19	49	59
Ohio—Columbus	27	77	94
Oklahoma—Guthrie	47	134	163
Pennsylvania—Harrisburg	19	43	51
Rhode Island—Providence	23	68	83
South Carolina—Columbia	41	106	132
South Dakota—Huron	19	50	61
Tennessee—Nashville	35	83	99
Texas—Austin	51	169	218
Vermont—Burlington	15	47	58
Virginia—Roanoke	23	61	73
West Virginia—Elkins	23	51	60
Wisconsin-Madison	29	77	95

Slope-Length Factor, L   Slope-length is the distance from the point of origin of overland flow to the point where the slope gradient decreases to the extent that deposition begins.   The soil loss per unit area increases as the slope length increases due to the greater accumulation of runoff.

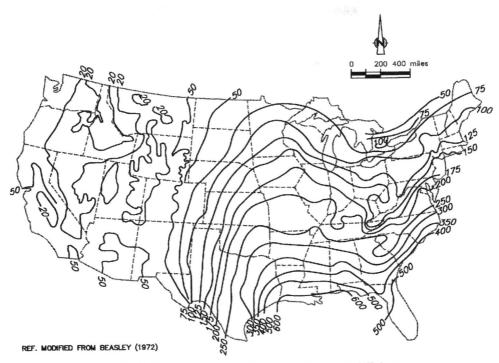

REF. MODIFIED FROM BEASLEY (1972)

Fig. 8.13 Average annual values of the rainfall factor.
(courtesy of Iowa State University Press)

Slope-Gradient Factor, S   As slopes increase, the velocity of the runoff water increases, which results in an increase in the ability of the runoff water to detach particles from the soil mass.

Values for L and S may be taken from Figure 8.14.

Cropping-Management Factor, C   The cropping-management factor, C, is the ratio of soil loss from land cropped under particular conditions, relative to that from continuously fallowed land. The factor estimates the effect of vegetative cover, soil conditions, and general management practices. On landfills, freshly covered and without vegetation or special erosion-reducing procedures of cover placement, C will usually be approximately unity. Typical values of C relevant to landfill design are noted in Table 8.17.

Erosion-Control Practice Factor, P   As applied to landfills, the P factor is similar to C except that it accounts for the additional erosion-reducing effects of land management practices. Typical values relevant to landfill applications are listed in Table 8.18.

E X A M P L E

Use the Universal Soil Loss Equation to calculate the soil loss for a northeastern Pennsylvania site of 5 acres. The design slope for the site is 14 percent and is 200 ft in length. Soils are 65 percent silt and 35 percent sand with slow to moderate permeability. The soil is bare of vegetation. The applicable factors are $R = 150$, $K = 0.30$, $LS = 3.3$, $C = 1.0$, $P = 1.0$.

S O L U T I O N

The soil loss is determined is determined to be

$A = 148.5$ tons/acre/yr, with seeding (grass), $C = 0.05$ after the first year, and

$A = 7.4$ tons/acre/yr.

Table 8.16 VALUES OF SOIL ERODIBILITY FACTOR, K, FOR MAJOR SOIL TYPES

Soil	K Value
Dunkirk silt loam	0.69
Keene silt loam	0.48
Shelby loam	0.41
Lodi loam	0.39
Fayette silt loam	0.38
Cecil sandy clay loam	0.36
Marshall silt loam	0.33
Ida silt loam	0.33
Mansic clay loam	0.32
Hagerstown silty clay loam	0.31
Austin clay	0.29
Mexico silt loam	0.28
Honeoye silt loam	0.28
Cecil sandy loam	0.28
Ontario loam	0.27
Cecil clay loam	0.26
Boswell fine sandy loam	0.25
Zaneis fine sandy loam	0.22
Tifton loamy sand	0.10
Freehold loamy sand	0.08
Bath flaggy silt loam with 2-inch surface stones removed	0.05
Albia gravelly loam	0.03

Table 8.17   VALUES OF *C* FOR VARIOUS CROP COVERS

Land Cover	"C" Value
Continous fallowed land	
(bare soil, no crop)	1.0
Mulch	
Heavy 1000 to 1500 lb / acre	0.2
Moderate 500 to 1000 lb / acre	0.4
Light 200 to 500 lb / acre	0.6
Grasses	
Newly Seeded, first month	0.6
Newly Seeded, during first year	0.05
Ground cover 95-100% as grass	0.003
80% as grass	0.01
60% as grass	0.04
Permanent pasture, turf-grass	0.03

As a general guideline, ranges of soil losses (all in tons/acre/year) are as follows:

0 < A < 5—frequently considered an acceptable loss/year

5 < A < 20—sedimentation retention will be required

A > 20—design changes will be required (e.g., terraces or changes in slope and depth)

Table 8.18   VALUES OF P FOR VARIOUS SURFACE PREPARATIONS

Erosion Control Practice	P Value
Surface condition with no cover; compact, smooth, scraped	1.30
Landfill surface	1.00
Rough irregular surface; equipment (tracks in all directions)	0.90
Small sediment basins (1 basin for 4 acres)	0.50

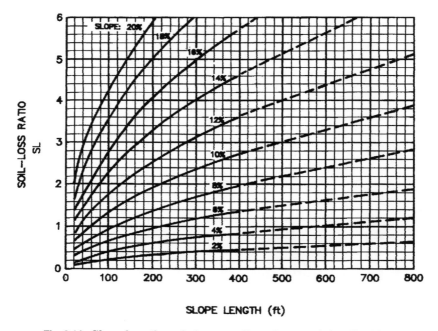

Fig. 8.14 Slope-length and slope-gradient factors. (after Beasley, 1972)
(courtesy of Iowa State University Press)

In the event of erosion, it is essential that any disturbed areas be revegetated as quickly as possible, otherwise, erosion will be quickly accelerated. Velocity controls/silt traps (e.g,. hay bales), erosion matting, sodded swales, lined channels, and drop pipes are all means of lowering erosion levels. Figure 8.15 illustrates a downslope flume that is used to convey the runoff rapidly down the sides of a landfill site.

8.9   WIND EROSION EFFECTS

8.9.1 Components

Dust is often a problem at landfill sites, especially in dry climates. If the soil is fine grained, it causes excessive wear of equipment, is a potential health hazard to personnel on the site, and is a nuisance to residences or businesses nearby.

Detailed computer models for calculating wind erosion losses are available, but the data and computational requirements for these models are substantial. Lutton et al. (1979) described a simple alternative model:

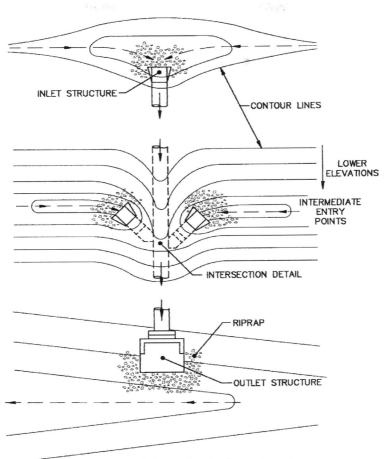

Fig. 8.15 Schematic of a downslope flume.

$$A' = f(K', C', L', T', V') \qquad (8.5)$$

where:

A'=the amount of erosion (tons/acre/yr),

K'=a soil erodibility index,

T'=a soil ridge roughness factor,

C'=a climatic factor,

L'=the open-area length along the prevailing wind erosion direction,

V'=an equivalent quantity of vegetative cover.

Thus, soil loss is a function of such features as cloddiness, surface roughness, and surface moisture of the soil. It is dependent on amount, type, and orientation of the vegetation; on wind velocity or force; and on the windward

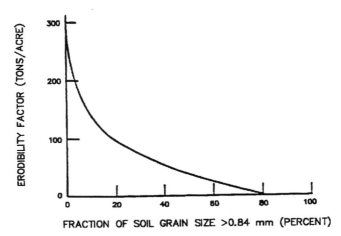

Fig. 8.16 Wind erosion versus percent coarse fraction.

distance across the open area. Thus, Equation 8.5 uses wind characteristics along with soil and vegetation factors. The features associated with the individual parameters are as follows:

Soil Erodibility, K' Soil erodibility, K', reflects the nature of the soil and an adjustment for knoll or hill configuration. The value of K' is determined from the product of the soil erodibility factor from Figure 8.16 (after Skidmore and Woodruff, 1968).

Climate Factor, C' The climate factor combines wind velocity and near-surface water content. Average monthly values for regions of the United States are indicated in Figure 8.17 (after Skidmore and Woodruff, 1968).

Field Length Factor, L' The field length factor is the unsheltered distance along the direction of prevailing wind erosion (see Figure 8.18).

Soil Ridge Roughness Factor, T' Soil ridge roughness reflects surface roughness beyond what is caused by clods or vegetation. Values are determined by making amplitude measurements from crest to trough on the ground, as indicated in Figure 8.19 (after Skidmore and Woodruff, 1968)).

Vegetative Cover Quantity, V' The vegetative cover quantity combines type and orientation effects with the tons per acre of vegetation cover. Values are indicated in Figure 8.20. For example, 800 lb/acre actual flat residue has an equivalent V' of 2500 lb/acre.

8.9.2 Design Aspects

Most wind studies have considered erosion from agricultural land where the fields are large. Landfills are generally smaller, and the cover material is

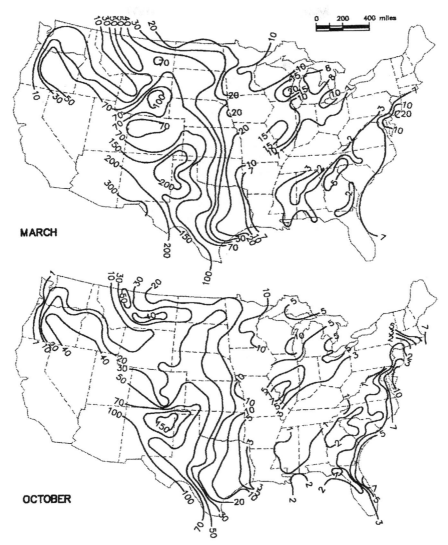

Fig. 8.17 Wind erosion climate factor C' in percent for March and October.

designed to reduce the erosion potential. Conversely, the soil is thoroughly remolded during placement, which reduces its erosion resistance. In addition, other features can be manipulated in the stages of landfill design.

The steps in solving Equation 8.5 are as follows:

1. Determine from Figure 8.16 an erodibility increment, $A_1' = K'$ that would occur for a wide, isolated, smooth, unsheltered, bare field having a calculated percentage of dry aggregates greater than 0.84 mm in diameter. Adjust K' for knoll configuration as necessary using Figure 8.21.

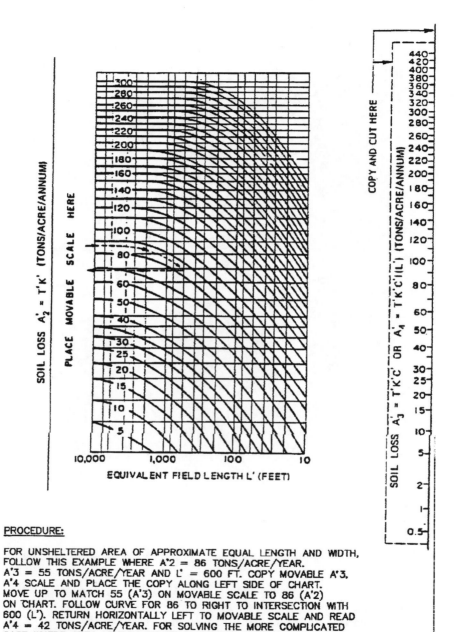

PLACE MOVABLE SCALE HERE

SOIL LOSS A'$_2$ = T'K' (TONS/ACRE/ANNUM)

EQUIVALENT FIELD LENGTH L' (FEET)

SOIL LOSS A'$_3$ = T'K'C' OR A'$_4$ = T'K'C'(IL') (TONS/ACRE/ANNUM)

PROCEDURE:

FOR UNSHELTERED AREA OF APPROXIMATE EQUAL LENGTH AND WIDTH,
FOLLOW THIS EXAMPLE WHERE A'$_2$ = 86 TONS/ACRE/YEAR.
A'$_3$ = 55 TONS/ACRE/YEAR AND L' = 600 FT. COPY MOVABLE A'$_3$,
A'$_4$ SCALE AND PLACE THE COPY ALONG LEFT SIDE OF CHART.
MOVE UP TO MATCH 55 (A'$_3$) ON MOVABLE SCALE TO 86 (A'$_2$)
ON CHART. FOLLOW CURVE FOR 86 TO RIGHT TO INTERSECTION WITH
600 (L'). RETURN HORIZONTALLY LEFT TO MOVABLE SCALE AND READ
A'$_4$ = 42 TONS/ACRE/YEAR. FOR SOLVING THE MORE COMPLICATED
CASE WHERE AREA LENGTH GREATLY EXCEEDS WIDTH, CONSULT THE
ORIGINAL REFERENCE (REF. SKIDMORE AND WOODRUFF, 1968)

Fig. 8.18 Chart for determining soil loss A'4, from A'2, A'3 and L.

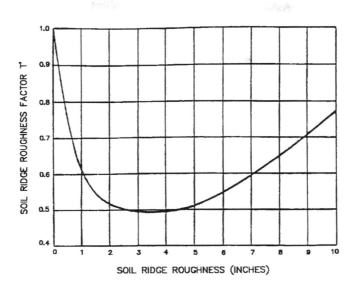

Fig. 8.19 Soil ridge roughness factor T' from actual soil ridge roughness.

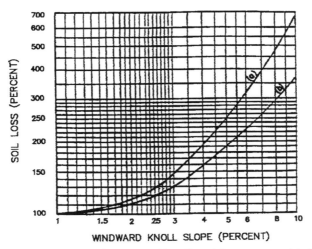

Fig. 8.21 Knoll adjustment from top of knoll and from upper third of slope. (Courtesy of Soil Science of America)

2. Account for the effect of roughness, T, from Figure 8.19 and determine the erodibility increment, $A_2' = A_1' \bullet T'$.

3. Account for the effect of local wind velocity and surface soil moisture, C', from Figure 8.17 and find the erodibility increment, $A_3' = A_2' \bullet C'$.

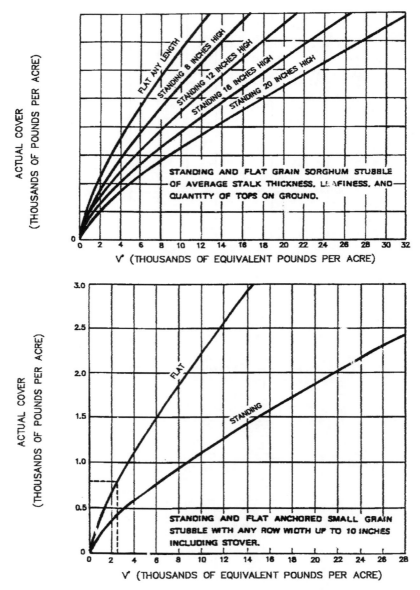

Fig. 8.20 Relationship of Factor V to quantity and type vegetative cover.

4. Account for the effect of length of field, L', and determine $A_4' = A_3' \cdot f(L')$. Calculation of A_4' is not a simple multiplication because L', A_3' and A_2' are interrelated (use Figure 8.18).

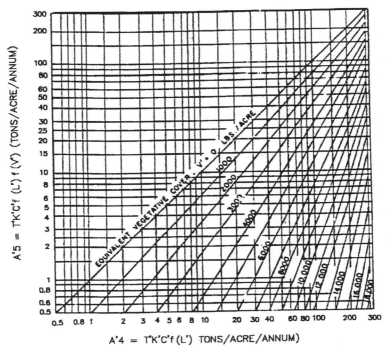

Fig. 8.22 Chart for determining soil loss A'5.

5. Account for vegetative cover in V (Figure 8.22) and determine the annual erosion as $A_5' = A_4'\ f(V)$, as per Figure 8.20 (after Skidmore and Woodruff, 1968).

Obviously a number of design features can be adjusted if necessary, including soil composition; use of mulch; wind barriers; vegetation cover, length, width, and orientation adjustment; and additives.

More localized dust problems, such as those from vehicular traffic (e.g., on an access road), can be temporarily controlled by wetting down the road with water or by using a deliquescent chemical. Calcium chloride, applied at 0.4 to 0.8 lb/yd^3 and admixed with the top 3 in. of the road surface is often effective. Frequent applications are usually required.

8.10   OFF-SITE RUNOFF CONTROL

Construction of a landfill generally involves a substantial alteration of site topography, increased slopes, and the redirection of drainage. The result is large increases in off-site flows and sediments.

Before the relatively impermeable final cover is placed on the completed landfill, a large percentage of the precipitation falling on the landfill will percolate through the intermediate cover and into the refuse. Once the final cover is placed, most precipitation gives rise to runoff, which must be controlled.

Most regulatory agencies require that virtually no changes be made to a system, which, in the case of a landfill, means that post-development runoff levels must not exceed pre-development levels. The types of controls that have been utilized include swales and storm-water recharge ponds constructed on the periphery of the site. Sedimentation controls are also often required.

A sedimentation facility can be as simple as a widening of a channel to slow the velocity of the water for a sufficient length of time for the sediment to settle out. Alternatively, a sedimentation pond may be used to prevent the peak flow level from leaving the site as well as to remove the sediment. Thus, it may be used to temporarily store excess peak runoff flow until it can be discharged at a lower controlled rate of release. Standard hydrologic procedures are followed in sizing the storm-water basins.

In general terms, the steps in designing a detention basin are as follows:

1. Determine the peak flow for pre-existing conditions for a specified recurrence interval.
2. Construct a hydrograph for the design storm for the completed landfill site arriving at the detention pond.
3. Calculate the storage needed, given the stage-discharge curves for the detention basin, for major storms such as the 25-year and 100-year storms. (The return period of the storm that needs to be controlled is typically specified by a government agency. However, it is prudent to consider the consequences of a more substantial rain.)
4. Check the performance of the basin during the more frequent storms, such as the 5- and 10-year storms.

The specifics of the pond design may vary considerably. An example of those designed for the Keele Valley Landfill in Toronto, Ontario are depicted in schematic form in Figure 8.23.

8.11 SUMMARY COMMENTS

In attempts to minimize leachate production, design initiatives such as increasing the surface slopes to enhance runoff and the placement of a low-permeability soil cover, among others, may be considered. The costs of these design initiatives must be weighed against the potential costs of leachate treatment. Significant concerns also exist relative to maintenance of the surface. Rainfall runoff is increased by increases in inclination of the surface slopes; accordingly, infiltration decreases. Because erosion also increases with increasing surface slopes, the balance between these opposing forces must be carefully evaluated.

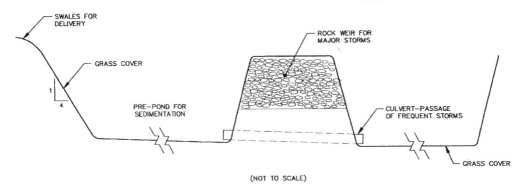

Fig. 8.23   Stormwater management ponds

There are limits to the effectiveness of surface control measures to limit leachate production.

On slopes of less than 3 percent, surface irregularities and vegetation commonly act as traps for runoff. At a 5 percent slope, the inclination is sufficient to facilitate runoff and normally will not incur excessive erosion. For very high slopes, a downslope flume such as depicted in Figure 8.15 may be needed.

References

Aldrich, H. P. and Paynter, H. M. 1953. *Analytical Studies of Freezing and Thawing of Soil*. T.R. #42, Arctic Construction and Frost Effects Laboratory, New England Division, U.S. Army Corps of Engineers, Boston, Mass.

Andersland, O. B., and Al-Moussawi, H. M. 1987. Crack Formation in Soil Landfill Covers Due to Thermal Contraction. *Waste Management and Research* 5: 445–452.

Beasley, R. P. 1972. *Erosion and Sediment Pollution Control*. The Iowa State University Press.

Bedient, P. B., and Huber, W. C. 1988. *Hydrology and Floodplain Analysis*. Addison-Wesley, Mass.

Chow, V.T., Maidment, D. R., and Mays, L.W. 1988. *Applied Hydrology*. McGraw-Hill, New York.

Edil, T. B., Ranguette, V. J, and Wuellner, W. W. 1990. *Settlement Of Municipal Refuse In Geotechnics of Waste Fills: Theory and Practice*. Ed: A. Landva and G. Knowles. ASTM STP 1070. American Society for Testing Materials.

Emberton, J. R., and Parker, A. 1987. The Problems Associated with Building on Landfill Sites. *Waste Management and Research* 5(4): 473-482.

Emcon Associates 1975. *Sonoma Solid Waste Stabilization Study*. U.S. Environmental Protection Agency EPA-530/SW-65d.1, Cincinnati, Ohio.

Flower, F. B., Leone, I. A., Gilman, E. F., and Arthur, J. J. 1978. *A Study of Vegetation Problems Associated with Refuse Landfills*. U.S. Environmental Protection Agency EPA-600/s-78-094, Cincinnati, Ohio.

Gee, J. R. 1981. "Prediction of Leachate Accumulation in Sanitary Landfills." In *Proceedings of the Fourth Annual Madison Conference on Municipal and Industrial Waste*, University of Wisconsin Extension at Madison, September 28–30.

Gilman, E. F., Leone, I. A., and Flower, F. B. 1981. *Critical Factors Controlling Vegetation Growth on Completed Sanitary Landfills*. U.S. Environmental Protection Agency EPA-600/S2-81-164, Cincinnati, Ohio, September.

Gilman, E., Flower, F., and Leone, I. 1983. *Standardized Procedures for Planting Vegetation on Completed Sanitary Landfills*. United States Environmental Protection Agency EPA 600/52 83-055, Municipal Environmental Research Laboratory, Cincinnati, Ohio, August.

Hatheway, A. W., and McAneny, C. C. 1987. An In-Depth Look at Landfill Cover. *Waste Age* 18 (August): 135–156.

Hartz, K. E., and HAM, R. K. 1983. Moisture Level and Movement Effects on Methane Production Rates in Landfill Samples. *Waste Management and Research* 1:139–145.

Holtan, H. N. 1961. *A Concept for Infiltration Estimates in Watershed Engineering*. U.S. Dept. of Agriculture. Agricultural Research Service.

Horton, R. 1933. The Role of Infiltration in the Hydrologic Cycle. *Transactions American Geophysical Union* 14:446–460.

Horton, R. 1939. Analysis of Runoff Plot Experiments with Varying Infiltration Capacity. *Transactions American Geophysical Union* 20:693–711.

Huggins, L. F., and Monke, E. J. 1966. *The Mathematical Simulation of the Hydrology of Small Watersheds*. Technical Report 1, Purdue University, Water Resources Center, Lafayette, Indiana, August.

Lambe, T., and Whitman, R., 1969. *Soil Mechanics*. John Wiley, New York.

Lee, C., Hoeppel, R., Hunt, P., and Carison, C. 1976. *Feasibility of the Functional Use of Vegetation to Filter, Dewater, and Remove Contaminant from Dredged Material*. Army Waterways Experiment Station, Report No. 18, WES-TR-D-76-4, Vicksburg, Miss. 88 pp.

Lutton, R. J. 1979. S*oil Cover for Controlling Leachate from Solid Waste, Municipal Solid Waste: Land Disposal*. U.S. Environmental Protection Agency EPA-600/9-79-023a, August.

Lutton, R. J., Regan, G. L., and Jones, L. W. 1979. *Design and Construction of Covers for Solid Waste Landfills*. U.S. Environmental Protection Agency EPA-600/2-79-165, Cincinnati, Ohio, August.

Mather, J. R., and Rodriguez, P. A. 1978. The Use of the Water Budget in Evaluating Leachate Through Solid Waste Landfills. PB80-180888, U.S. Dept. of the Interior, Washington D.C.

Richardson, G., and Koerner, R. 1987. *Geosynthetic Design Guidance For Hazardous Waste Landfill Cells and Surface Impoundments*. Contract No. 68-03-3338, Hazardous Waste Engineering Research Laboratory, Office of Research and Development, U.S. EPA, Cincinnati, Ohio.

Skidmore, E., and Woodruff, N. P. 1968. *Wind Erosion Forces in the United States and Their Use in Predicting Soil Loss*. U.S. Dept. of Agriculture, Handbook 346. Washington, D.C., 42 pp.

Soil Conservation Service. 1972. Hydrology. *SCS National Engineering Handbook, Section 4. U.S. Dept. of Agriculture,* Washington, D.C.

Soil Conservation Service. 1975. Urban Hydrology For Small *Watersheds. Tech, Report No. 55, U.S. Dept. of Agriculture*. Washington D.C.

Tchobanoglous, G., Theisen, H., and Eliassen, R. 1977. *Solid Waste Engineering Principles and Management Issues*. New York, McGraw-Hill.

Thornthwaite, C.W., and Mather, J.R. 1957. Instructions and Tables for Computing Potential Evapotranspiration and the Water Balance *Publications in climatology*. Vol. X, No. 3, Drexel Institute, New Jersey.

U.S. Department of Agriculture, Soil Conservation Service. 1972.   National Engineering

Handbook-Section 4: Hydrology. U.S. Government Printing Office.

U.S. Department of the Interior. 1982. *Surface Mining Water Diversion Design Manual*. September.

U.S. Department of Transportation, Federal Highway Administration. 1961. Design Charts for Open-Channel Flow, *Hydraulic Design Series No. 3*. Washington D.C., August.

Viessman, W., Lewis, G.L., and Knapp, J.W. 1989. *Introduction to Hydrology*. Harper Collins Publishers, New York.

Woodruff, N.P. and Siddoway, F.H. 1965. A Wind Erosion Equation. *Soil Science of America*. Vol. 29, pp. 602-608.

Yen, B.C. and Scanlon, B. 1975. Sanitary Landfill Settlement Rates. ASCE-Journal of the Geotechnical Engineering Division. Vol. 101. No. GT5.

8.12   PROBLEMS

8.1. A landfill can be designed with 3 different soil types as cover: fine sand, fine sandy loam, and clay loam. The cover is to be 1.6 m deep and vegetated with moderately deep-rooted plants. The costs to acquire and place the various soils are shown in the table below. It is presumed that February is the first month of the year in which leachate is produced. Monthly data on temperature and precipitation as well as runoff coefficients are also given below. Compare this total amount of leachate produced (assume no liner leakage) for the three cover soils. It costs $1.00 per 100 liter to dispose of the leachate collected. Compare disposal costs over a 5 year period against costs of cover placement.

Climatology Data:

Month	Feb	Mar	Apr	May	June
Total Monthly Precipitation (in)	3.15	2.92	3.62	3.01	2.84
Average Monthly Temperature (°F)	34.6	38.9	47.5	54.3	65.0

Soil Data:

Soil Type	*Runoff Coefficient*	*Cost of Placement ($ m^{-3})*
Fine sand	0.12	1.20
Fine sandy loam	0.19	2.10
Clay loam	0.28	3.60

8.2 Landfill sites are filled as a series of successive cells. Sketch the progressive cell completion and the consequent leachate production rates that you might expect, where the site will be completed in twenty years.

8.3 Develop a water balance for the surface cover of a landfill, indicating the fate of precipitation for a landfill in your community.

8.4 Assume you must design a soil cover related to a 25 year storm of 30 minutes duration (use Figure 8.7). Using the data provided below, quantify which one soil type would have the highest infiltration rate; storage capacity; lowest percolation rate and which would allow the least runoff assuming very poor grass cover. State any assumptions you need to make. If you had to use one soil type for a cover, which would you use?

Soil Type	Total Porosity	Field Capacity	Hydraulic Conductivity
Sandy Loam	.50	.40	10^{-5} cm/s
Clay	.70	.60	10^{-8} cm/s
Coarse Sand	.30	.15	10^{-2} cm/s
Loam	.70	.55	10^{-4} cm/s
Fine Sand	.35	.25	10^{-3} cm/s
Loamy Clay	.65	.50	10^{-6} cm/s

BARRIER LAYERS

9.1 INTRODUCTION

A low-permeability material or combination of materials is placed
at the base of the landfill to reduce the discharge to the underlying hydrogeologic
environment. The liner is designed as a barrier to intercept leachate and to
direct it to a leachate collection system, as illustrated in Figure 9.1. A low-permeability layer may also be placed as part of a cover system to reduce percolation into the refuse and/or as part of a gas control system. The focus of this
chapter is the function, performance, and design of these barrier layers.

Liners or barrier layers may be constructed of a single material or as a composite of more than one material. They may also be created as a system of more
than one liner separated by a drainage plane to evaluate the performance of the
liner above and to reduce the hydraulic head on the liner below.

Liners are frequently constructed with natural materials serving as the primary barrier to liquid movement. Soils comprising a high percentage of
clay-sized particles are most frequently used. The in situ soil at a landfill site
may be suitable for use as a bottom liner; however, the likely existence of fractures and root holes near the soil surface will require that the soil be worked and
recompacted. It is often the case that the soils in place at a disposal location are
not suitable for liner construction, thus requiring the importation of soils for use
in liner construction or the use of materials to amend the existing soils to reduce
their permeability. Bentonite is frequently used for this purpose, and fly ash and
cementing materials have been used experimentally.

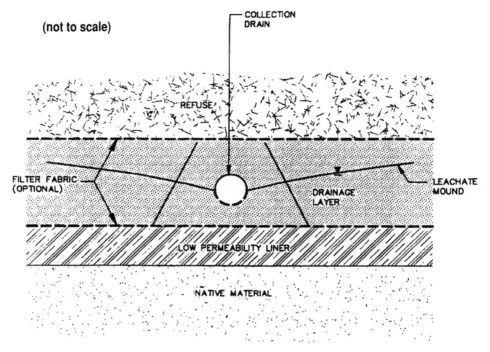

Fig. 9.1 Components of the leachate collection system.

(FMLs) constructed mainly of plastics such as polyethylene and polyvinyl chloride. FMLs may be utilized on their own or as part of a composite system, with clay acting as a secondary or backup barrier system. Many different types of FMLs exist, with varying costs and capabilities. The determination of the best FML for a particular application depends on a number of considerations, not the least of which is the nature of the leachate to which it will be exposed.

The primary concern about liners is their permeability and increases in permeability that may occur in the very long term. Faults in liners that might cause the in situ permeability to exceed design expectations are discussed later but can include cracks and fissures in soil liners caused by improper placement procedures or punctures and incomplete sealing of the seams in the FML.

Compatibility between liner materials and leachate components resulting in long-term deterioration has been a major concern in the use in landfills and is addressed later in this chapter.

Liner function, design, and construction are discussed in the following sections. Section 9.2 considers liners constructed with soil materials. Section 9.3 addresses FMLs. Section 9.4 examines double liners and composites. Thus, the major part of the chapter deals with barrier layers as a bottom liner. Section 9.5 discusses barrier layers as elements of landfill covers.

9.2 Design of Liners Using Natural Materials

9.2.1 Introduction

Twenty years ago, the few liners in existence were constructed of natural materials; mainly clay tills and other soils with a substantial clay content. Availability and cost were key factors involved in material selection. High clay content was considered useful not only for low permeability but also for potential contaminant retardation; clay has an ability to attenuate many of the chemical constituents found in leachates. Nevertheless, there is concern about the appropriateness of using clay liners for MSW landfills (at least by themselves) due to the existence of some liner failures, the difficulties in estimating hydraulic conductivity in the field, and questions about long-term liner integrity in contact with leachate.

9.2.2 Materials Utilized in Natural Soil Liners

Different types of clay (e.g., montmorillonite, kaolinite, illite) exist, and the differences among them determine how successfully they can be used as landfill liners. The crystalline structure of a clay is composed of two basic building blocks, a silicate tetrahedron (SiO_4) and an octahedron consisting of aluminum $[Al_2(OH)_6)]$ or magnesium $[Mg_2(OH)_6]$. Different clay minerals are formed as the sheets of the basic units stacked on top of each other with different ions bonding them together.

Although there are many types of clay minerals with various substituted and exchangeable ions, they are generally classified into three groups, as listed in Table 9.1. Of particular note is the sizable magnitude of the expansion index associated with montmorillonite; water molecules enter easily between the layers, which expand or swell considerably. As a consequence, montmorillonite clays have been widely used for drilling mud and slurry wall construction because swelling helps to reduce hydraulic conductivity at the soil interface.

A mix of at least 50 percent clay- and silt-sized particles provides a wide moisture range within which the clay can be compacted to obtain the desired low-permeability (and thus low hydraulic conductivity). This mix also favors soils with low frost and erosion susceptibility. In some cases a 25 percent clay-sized fraction (<0.001 mm) has been specified to ensure high chemical-attenuation characteristics.

Sometimes, mixtures of soils are employed. Hoeks et al. (1987) used a mixture of sand plus 5 to 15 percent (w/w) European bentonite compacted in laboratory permeameters and demonstrated hydraulic conductivities less than 5×10^{-8} cm/s with water and 0.01 N $CaSO_4$. Bentonites are clay minerals of the smectite group; naturally occurring bentonite is a type of montmorillonite, the most common mineral of the smectite group. Water is easily absorbed between the layers of smectite, causing swelling of the clay and thus an apparent low permeability. However, a major difficulty with sand-bentonite liners is the potential incompatibility with MSW leachate (Farquhar and Parker, 1988; Mad-

Table 9.1: CHARACTERISTICS OF CLAY MATERIALS

Clay Groups	Cation Exchange Capacity (meg/100g)	Specific Surface (m^2/g)	Expansion Index
Kaolinite	3 - 15	10 - 20	
Sodium			0.20
Calcium			0.06
Illite	10 - 40	65 - 100	
Sodium			0.15
Calcium			0.21
Montmorillonite (smectite group)	80 - 150	700-840	
Sodium			2.5
Calcium			0.80

From Grim, 1968, and Lambe and Whitman, 1969.

sen and Mitchell, 1989). Depending on the leachate composition and the bentonite, the tendency exists for Ca to replace Na in the montmorillonite (Wyoming bentonite) structure causing shrinkage and the development of cracks (Farquhar, 1988)., which greatly increase permeability. Thus, high sodium swelling clays should be avoided in landfill applications (Gray, 1989; Madsen and Mitchell, 1989). On the other hand, Ca-bentonite or calcium-treated Na-bentonite experience much less shrinkage and are suitable for landfill applications. Alternatively, the effect of pore fluid on the permeability of bentonite can be reduced by treating the bentonite with polymers.

Soils with high non-swelling clay mineral (kaolinite and illite) content are most suitable for compacted clay liners in depths of 0.5 m and greater. Calcium-based swelling clays are used for amending coarse-grained soils to reduce the permeability.

9.2.3 Placement Techniques

As indicated in the previous section, the most important property of a liner is its hydraulic conductivity. Even soils with the best constituents, if poorly constructed in place, can be expected to leak significantly. For clay liners, this uncertainty can be minimized by:

- Properly selecting soil in terms of clay content
- Use of thicker liners to compensate for construction variability
- Carefully supervising construction methods and quality control testing
- Establishing an effective monitoring program to reveal construction flaws before serious problems occur
- Using lysimeters in the field to monitor performance.

After placement, the permeability of a clay material is variable and is not a unique function of the water content or dry density; other factors also affect the permeability. Important variables related to compaction include the type of compaction, compactive effort, size of clods, and the bonding between lifts. There is no universal code to specify the unique set of conditions that will guarantee a specifically targeted hydraulic conductivity. These concerns are compounded in a field situation where large expanses of the liner must be constructed and remain exposed to weather prior to placement of MSW.

Strict control of moisture content is necessary since it is a major consideration in determining the permeability. Major increases in moisture content reduce both the density of the clay and the ability of machines to "work" the clay surface. Hydraulic conductivity can be decreased by increased compaction or by changes in the dry unit weight of the soil. It is easier to compact clay with less than the optimum water content (water content at maximum density for a constant compaction effort). However, since a primary interest is in minimizing the hydraulic conductivity, soil must be compacted with more than the optimum water content using a kneading type of compaction.

The response to different moisture contents and different compaction efforts is depicted in Figure 9.2, which shows the dependence of clay liner hydraulic conductivity on degree of compaction and moisture content (after Mundell and Bailey, 1985; Daniel, 1987; and Madsen and Mitchell, 1989). Measurements of hydraulic conductivity made on specific compacted samples are shown in Figure 9.2(a). For an individual compaction effort, hydraulic conductivity decreases substantially at moisture contents greater than the optimum value.

Optimum water content corresponds to a degree of saturation between 0.8 and 0.9. On the dry side, or dry of optimum, the soil is dry and hard. It is difficult to compact the soil into a dense, uniform mass. At the optimum water content (peak), the soil is wetter, has a lower strength, and is easily compacted into a dense structure. On the wet side, or wet of optimum, the soil is very wet and soft and has low strength. It is easily deformed into a homogeneous macrovoid-free mass. Dry unit weight drops because water displaces the soils. A variation in hydraulic conductivity of three orders of magnitude can occur in a soil at a given density depending on compaction procedures (Quigley et al., 1988).

Compaction, wet of optimum, produces low K because the soil particles are arranged in a dispersed pattern, whereas a dry-side compaction produces a floc-

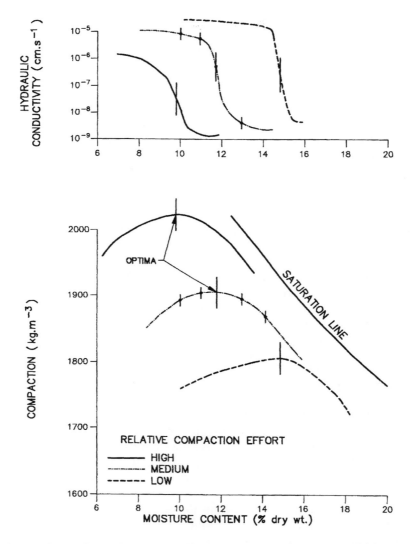

Fig. 9.2 Interrelation between hydraulic conductivity, degree of compaction and
moisture content. (After Farquhar, 1989)

culated pattern, which offers better paths for the flow of water and, consequently, higher K. Wet of optimum, at high degrees of saturation, the clay is soft
and has low strength. The result is that the soil is more easily remolded, and
clods and macropores are broken up as the clay is squeezed together. Since the
hydraulic conductivity through clay micropores is very low, the overall permeability is also low.

Fig. 9.3 Disking of clay line to break up clods, Keele Valley Landfill, City of Vaughan, Canada.

A reasonable goal for compaction of liners, is 0.90 percent of maximum dry density. Figure 9.2(b) shows the increased density with compaction effort and also the nonlinear dependence of density on moisture content resulting in an optimum moisture content to produce maximum density. The curves such as those in Figure 9.2 are specific to the clay mixture being tested.

As a result of the tradeoffs demonstrated in Figures 9.2 (a) and (b), most designs for compacted clay liners recommend moisture contents from 0 to 3 percent wet of optimum and at compactions equivalent to 95 percent Standard Proctor or 90 percent modified American Society of State Highway Transportation Organization (ASSHTO) (Mundell and Bailey, 1985).

Figures 9.3 and 9.4 show equipment preparing the liner at Keele Valley Landfill in Vaughan. Figure 9.3 shows the disking of the clay to break up the clods. The tendency to form clods increases as the plasticity of the clay increases (Foreman, 1984). Figure 9.4 shows a water truck distributing the water, closely followed by the sheep-footed dozer to create the desired compaction.

Plastic clays are more difficult to break and compact in the field at the optimum moisture content, are more sensitive, and take relatively less time to dry when above optimum, or to moisten when below optimum. Although clays with higher plasticity index are usually less permeable, a reasonable compromise

Fig. 9.4 Additions of water and compaction efforts on clay liner,
Keele Valley Landfill, Vaughan, Canada.

between permeability and ease of installation is to use clay with a plasticity
index less than 20 (Oweis and Khera, 1990).

Additional features that can affect hydraulic conductivity of the natural soil
liner include desiccation and freezing/thawing. Desiccation can cause significant
increases in K. Self-healing with elevated overburden pressure can occur, but it
is best to avoid the problem from the outset. Avoiding desiccation problems can
be accomplished in part by adding cover soil (>0.3 m or 1 ft) or keeping the soil
moist by applying water during the periods in which it is exposed. A clay with a
high liquid limit tends to develop more desiccation cracks ($\geq$ 30–50 percent).
Clay with a low plasticity index is less workable.

If properly maintained, clay liners remain wet for the life of the landfill. It
is therefore the water-wet or leachate-wet behavior that is related to the performance of the liner.

Freezing/thawing, which causes irreversible increases in K, can often be
avoided by adding cover soil or refuse in excess of the frost depth.

Whether or not hydraulic conductivity measurements made in the laboratory are representative of field conditions has been a longstanding question.
This is due in part to physical differences between the laboratory test conditions

and those present in the field. This concern is exacerbated by the spatial variability of soil properties and placement methods in the field. .

9.2.4 Evaluation of Hydraulic Conductivity and Permeability

Laboratory Testing It is not uncommon for the hydraulic conductivity measurements of a specific soil to differ by one to two orders of magnitude depending on the experimental procedure used. For example, rigid-wall permeameters often produce values higher than confined flexible-wall unit, especially with swelling clays.

In addition, Daniel (1981) demonstrated the significance of the sample diameter, as indicated in Table 9.2 The average permeability of the in-place liner was back-calculated from measured leakage rates and found to be 1×10^{-5} cm/s, a value similar in magnitude to the sample of 243.8 cm diameter. Conclusions drawn from tests involving smaller diameters could clearly be in error.

If the structure of the clay changes during permeation, shrinkage, cracking, and/or piping can occur; this can result in increased flow in a rigid-wall permeameter, producing high values. The confining pressure provided in a flexible-wall permeameter can cause the clay to "heal" despite the structural changes, so that little change in hydraulic conductivity is observed (Madsen and Mitchell, 1989).

In addition to the test apparatus, the permeant employed in the test can influence the test results obtained. Ideally, the experiments should be run with landfill leachate. If the landfill is new, no leachate exists, so the use of the leachate is infeasible. However, it is expected that differences between permeability measurements using leachate and $0.01\,N$ $CaSO_4$–based tests will be small for non-swelling clays (Farquhar, 1989). Whether to use rigid- or flexible-wall permeameters is still a subject for debate. However, in cases where leachate-soil compatibility is not expected to be a problem (leachate from MSW and non-swelling clay materials, the use of flexible wall permeameters with confining pressures similar to those expected in the field is preferred.

Differences between Lab and Field Measurements of Hydraulic Conductivity As noted

Table 9.2: RESULTS SHOWING EFFECT OF SAMPLE DIAMETER
ON PERMEABILITY MEASUREMENTS

Sample Diameter (cm)	Permeability (cm/s)
3.8	1×10^{-7}
6.4	8×10^{-9}
243.8	3×10^{-5}

Source: After Daniel, 1981.

in the previous subsection, the use of different instruments and leachates can influence hydraulic conductivity measurements. However, an even larger concern is whether the laboratory-determined values are representative of field conditions. Laboratory experiments are more easily controlled and much less costly, but they are of short duration, utilize high hydraulic gradients, and do not reflect spatial soil heterogeneities.

As an example, laboratory hydraulic conductivity tests are often run with a hydraulic gradient several hundredfold higher than that expected in the field (normally very close to unity). The high gradients are utilized to minimize test duration; however, they may generate unnatural flow conditions that can adversely affect the estimate of field hydraulic conductivity.

Other reasons for variations between the laboratory and field measurements include the following:

- Laboratory tests are too short for leachate/clay interactions and biomass/solids buildup that occur over time in the field. Thus, additional mechanisms for change in the field situation are not replicated in the laboratory tests.
- Desiccation cracks may occur if compacted soil is not protected.
- Higher confining pressures exist in the laboratory than in the field.
- The field soil matrix (clods, cracks, fissures) is not reproduced in the laboratory tests. Few samples are sufficiently large to account for field heterogeneities such as clods and cracks, which produce increased field hydraulic conductivities.
- The soil compaction methods used in the field and laboratory are different (Lambe and Whitman, 1969). Laboratory compaction methods generally result in permeabilities lower than those attained using field equipment (Dunn and Mitchell, 1984; Day and Daniel, 1985).

Determination of field-scale hydraulic conductivity is possible through seepage measurements in the field using lysimeters. These field lysimeters can be constructed in and below the liner or in a trial liner (e.g., a section of 30 m by 30 m). Figure 9.5 shows the construction of a field lysimeter in the middle of a clay liner at the Keele Valley Landfill; such a lysimeter is costly (approximately $50,000) and can be used to determine the performance of the liner only after the fact. Alternatively, double-ring infiltrometers may be employed during construction, but even these instruments require months for completion of the field test.

Construction Techniques To achieve desired field-scale hydraulic conductivity, careful attention must be paid to construction techniques. For example, many designs call for construction of the liner as a series of layers. This involves placing individual layers in 15 cm (6 in.) lifts at 2 to 3 percent wet of optimum moisture content, to create a 0.7 to 1.3 m (2 to 4 ft) overall thickness in a total of four to eight lifts. Each of the 15 cm lifts is compacted by a heavy-footed roller making several passes on each lift. The roller feet need to

Fig. 9.5 Lysimeter being used within the clay liner, Keele Valley Landfill, Vaughan, Canada. (photograph courtesy of Metropolitan Toronto Public Works Department)

be sufficiently long to work the layer being placed into the layer below. The intent is to ensure good bonding between the lifts of soil to minimize the formation of lateral flow channels between the lifts. If lateral flow occurs, significant increases in hydraulic conductivity will result.

Experimentation has shown that in situ liner hydraulic conductivities may exceed laboratory measurements by one to two orders of magnitude, partly due to the variability in the compacted soil fabric as a result of clay texture and compaction method. Daniel (1987), Herrmann and Elsbury (1987), Gray (1989), and Quigley et al. (1988) all refer to the presence of clay clods as being detrimental to the compaction process. If these clods are not broken up during placement, localized zones of increased hydraulic conductivity around the clods are created that can lead to excess seepage through the liner. Improperly worked clay and failure to bond a layer to the one below during placement will also produce higher secondary hydraulic conductivities due to cracks and other volume defects. Thus, proper placement techniques are essential.

For example, the liner at the Keele Valley Landfill was placed by a dozer in 15 cm layers, with each layer being disked to break up clods. Each layer was then worked in four to six passes with a 30 tonne heavy-footed roller, as shown Figure 9.4.

Liner Thickness   The total vertical thickness of in-place liners must be sufficient to avoid construction irregularities (e.g., a minimum of 0.75 m). Early experience with clay liner construction was that thinner liners were not adequately constructed. A safety factor is needed to compensate for construction irregularities and the difficulty of obtaining the needed gradient (slope) over the large areal extent of the liner. Many landfills have been constructed as containment facilities with compacted clay liners with a minimum thickness of 1 m and a leachate collection system.

In many situations it is becoming the norm to use composite liners with an FML over a barrier of natural material, often a compacted clay layer. However, the primary protection role is often assigned to the compacted clay barrier because the long-term performance of the FML is difficult to estimate.

Increases in the liner thickness result in a decreased hydraulic gradient through the liner, which thereby reduces the seepage rates. In addition, the effects of construction defects and liner heterogeneities on hydraulic conductivity are reduced as the thickness increases. However, increases in both costs and landfill volume utilization result as the thickness increases.

9.2.5 Quality Control and Quality Assurance

Because clay liner permeability is so dependent on the moisture content, compaction density, and kneading to homogenize soil fabric and thickness, these parameters must be frequently monitored during placement. As a result, in situ measurements of density and moisture content are required (Mundell and Bailey, 1985; Daniel, 1987; Gordon, 1987).

Daniel (1987) documented the actual hydraulic conductivities of four clay liners in Texas to be generally 10 to 1000 times higher than values measured in the laboratory.

Korfiatis et al. (1987) stressed the need for statistical analyses when interpreting and reporting liner hydraulic conductivity data. An outlier (high- or low-permeability) may be attributable to measurement errors such as improper sample collection or lab analyses. The variations in hydraulic conductivity are demonstrated by the example results graphed in Figure 9.6 (Donald and McBean, 1994).

9.2.6 Contaminant Transport Through Clay Liners

As examined in Chapter 6, contaminant migration through a clay liner may occur through the processes of advection and diffusion. However, because of the low permeability of a well-constructed liner, Quigley et al. (1988) showed that, of these two processes, diffusion is the more important mechanism; diffusion may transport contaminants at rates greater than permeability-dependent advective flow. Concern is frequently expressed that the permeability of the barrier soils may increase during long-term exposure to landfill leachate. In fact, evidence to the contrary has been published for several MSW Sites (Finno and Schubert, 1986; Griffin et al., 1976; Farquhar and Constable, 1978; Biene and Geil, 1985; Wuellner, et al., 1985). It appears that precipitate formation and the develop-

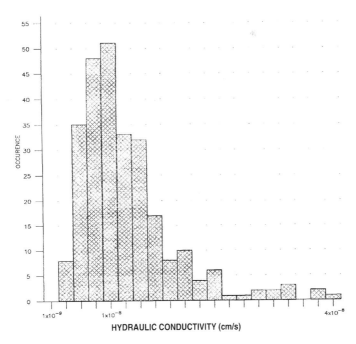

Fig. 9.6 Histogram of hydraulic conductivity of clay liner, Keele Valley Landfill Site.

ment of active biomass near the upper surface of the barrier produce a clogging action that reduces the effective hydraulic conductivity of the barrier.

The barrier may also retard contaminants through the processes of sorption, precipitation, biodegradation, and filtration. Over time, these processes may affect the hydraulic conductivity of the liner.

Research on quantifying the capacity of clay materials to attenuate leachate contaminants began in the early 1970s with the work of Griffin et al. (1976) and Fuller and Korte (1976). Some investigators found that exposure of clay materials to landfill leachates tended to increase permeability (Anderson and Jones, 1985). However, most investigators (preceding references) found that permeabilities decreased because of the biomass and precipitate accumulation at the clay surface.

Thus, changes in both the hydraulic conductivity of the barrier soil and leachate composition can occur over time due to numerous processes. Design procedures should recognize these various mechanisms, notwithstanding the difficulties in quantifying them. The importance of most of these mechanisms is difficult to determine since they take a long time and are leachate-specific. Most laboratory experiments are not of sufficient duration to account for longer-term interactions between the liner materials and the leachate and for the buildup of biomass and other solids along the liner surface. The more important mechanisms are discussed next.

Carbonate Dissolution   The rate of carbonate dissolution depends on the chemical character of the leachate and its temporal variation, on the amount and size of particles of carbonates, and on the leachate percolation rate. Dissolution is likely to be modest because the time during which the leachate is strongly acidic (pH <6) is brief.

Cation Removal   Cations are removed by exchange of resident ions. Alkaline earth metals are released in most cases and are possibly discharged from the liner. Elevated hardness of downflow from natural attenuation landfills is commonly a result. Many heavy metals tend to be removed quickly and extensively in clay through precipitation as hydroxides and carbonates and by coprecipitation with Fe and Mn hydrous oxides (Griffin et al., 1976; Fuller and Korte, 1976). The removal efficiency tends to improve with increased pH.

Anion Removal   Attenuation of anions in clay soils is not so well documented as it is for cations (Griffin, 1983). Given "typical" MSW leachate, extensive retardation of phosphate can be expected. Some contaminants are poorly attenuated in clay soils. Examples include chloride, monovalent cations, some organic acids (Weinberg et al., 1985), and neutral nonpolar organics (Brown and Anderson, 1983).

Clay has a limited capacity to absorb many ionic species. Thus, if the adsorbent capacity of the liner is exhausted with respect to these contaminants, contaminant breakthrough will occur.

Other Inorganic Contaminants   It seems likely that inorganic liquids, including salts, acids, and bases in the pH range of 1 to 13, do not adversely affect the permeability of clay barriers (Farquhar and Parker, 1989).

Peterson and Gee (1985) used laboratory columns to evaluate the impact of leachate from acidic mine tailings on the permeability of a clay liner. After 3 years of experimentation and the collection of 30 pore volumes of column effluent, they found that the permeability of the clay was reduced by two orders of magnitude. This was attributed to precipitate formation in response to acid neutralization.

Attenuation of Biodegradable Organics   Biodegradable organics are normally removed, although often only following an initial breakthrough that allows the biomass to develop. Farquhar and Parker (1989) reported that MSW landfill leachates do not appear to adversely affect the permeability of natural barriers.

Although many of the processes discussed will certainly slow the breakthrough of contaminants through the liner, there is one process that will create contaminant flux. Specifically, diffusion will cause contaminants to travel at velocities exceeding the theoretical fluid velocity and eventually to break through the low-permeability barrier. Contaminants that are poorly attenuated in the barrier will diffuse at velocities that can exceed the fluid velocity by an order of magnitude in clay soils. Contaminant diffusion that is driven by large concentration gradients through the barrier can be substantial.

Impact of Organic Solvents   Prediction of the impact of organic solvents on liners is difficult because there are many types and widely varying concentrations, which make generalizations difficult. The results of lab testing published during the 1980s indicated that concentrated organic liquids could alter the structure of clay soils and increase their hydraulic conductivities 100- to 1,000-fold. Some of the findings follow:

- Pure organic liquids of several types, but solvents in particular, can substantially increase the permeability of clay barriers. This situation is obviously one to avoid. Dilute aqueous solutions of many organic liquids do not appear to affect the permeability of clay barriers, especially the nonswelling clays. It would appear that concentrations up to 0.1 percent by weight (total solvents) can be tolerated, but knowledge of the interactions is limited.
- There is a need for caution when dealing with wastes such as benzene, xylene, or carbon tetrachloride, since these chemicals tend to cause shrinkage and cracking of some clays (Green et al., 1981).
- Anderson and Jones (1985) compared double-ringed permeameters with flexible- and rigid-wall units using bentonite slurry mixtures. Methanol and xylenes increased permeability from less than 5×10^{-8} cm/s to nearly 10^{-5} cm/s for most slurry preparations.
- Acetone first reduced permeabilities but then increased them (in excess of 10^{-6} cm/s) by displacing water from clay and causing it to crack.
- Xylene caused structural changes in the form of blocks and cracks through an expanded lattice structure. Permeabilities increased from 10^{-8} to greater than 10^{-6} cm/s (Farquhar and Parker, 1989).
- Brown and Anderson (1983) summarized their work into four major organic liquid groups, as listed in Table 9.3. Weinberg et al. (1985) reported that some combinations of linear materials produced high-capacity, nearly irreversible retardation of certain organics (e.g. hexachlorobenzene in fly ash and ground salt). Some organics, however, such as phenoxy acetic acid, were poorly retarded.

Laboratory-scale experiments to study the impact of leachates on barrier materials have been preferred to field experimentation because of the comparative economies in time and money. Also, many experiments are accelerated through the application of hydraulic and concentration gradients that are often

Table 9.3: SUMMARY EFFECT OF CONTAMINANT ALTERATION OF ORGANIC LIQUID GROUPS

Organic Liquid Group	Representative Liquids	Comment on Potential Interaction
Organic acids	Acetic acid	Multiple potential attack mechanisms Poor adsorption
Organic bases	Aniline	Rapid adsorption onto clays
Neutral polar	Acetone, methanol	Compete with H_2O ethylene glycol to wet clay Reduce fluid viscosity Adsorption inversely proportional to solubility
Neutral nonpolar	Xylene, heptane	Do not compete with H_2O to wet clay Adsorb poorly onto clay

Source: After Brown and Anderson, 1983.

much higher than field conditions. These factors make the extrapolation of these data somewhat tenuous.

In summary, barriers constructed from natural materials with significant clay content can be expected to retard the transport of some contaminants through interaction with the barrier materials and/or biodegradation. Retardation generally improves when the barrier exhibits alkaline pH, significant cation-exchange capacity, and organic and hydrous oxide content. Nevertheless, reliance on barrier retardation must acknowledge several limitations.

1. Not all contaminants will be retarded to the extent necessary to meet groundwater quality standards.
2. Retardation may have a capacity limit, so contaminant loading must be considered in designing the barrier thickness to avoid breakthrough.
3. The movement of some constituents through a clay liner can increase the effective permeability.
4. Ideal conditions of homogeneous barrier materials, bulk intergranular flow, and equilibrium reactions will not occur in the field.
5. Interferences and interactions among contaminants affect retardation, so leachate- and barrier-specific data are required to design a specific system.

9.2.7 Field Performance of Compacted Clay Liners

Compacted clay liners consisting mainly of nonswelling clays have been used extensively for MSW landfills. Cases have been reported where compacted clay liners have failed to exhibit field values of hydraulic conductivity (K) of less than 10^{-7} cm/s (Herrmann and Elsbury, 1987; Williams, 1987). However, these situations were usually the result of inadequate design and installation procedures. In the State of Wisconsin, by contrast, the use of strict design, installation, and quality assurance guidelines has produced nearly 30 MSW landfill liners with hydraulic conductivities of less than 10^{-7} cm/s. Similar controls at the Keele Valley Landfill in metropolitan Toronto have resulted in a liner with field hydraulic conductivity measurements consistently less than 10^{-8} cm/s (Lahti et al., 1987). A landfill in Boone County, Kentucky, was constructed with a 0.6 m thick liner; values of K_{lab} of 2×10^{-8} cm/s and K_{field} of 4×10^{-7} cm/s were reported initially, with lower values after 9 years.

Quigley and Rowe (1986) reported that heavy metals were retained within the upper 15 cm of a clay liner, with Fe, Zn, Cu, and Pb at background levels beyond the 15 cm depth. The upper 20 cm of the clay was black and oily, suggesting anaerobic microbial activity as the mechanism responsible for the reduction of the leachate organic matter. Diffusional transport of Cl and Na reached 1.5 m into the clay in 15 years.

To date, published performance data of actual leakage rates from full-scale landfills have been limited. Using lysimeter data, Gordon et al. (1989) reported very low leakage rates, with an average value of 10 L/1000 m^2/d, from three landfills constructed in Wisconsin with 1.2 to 1.5 m thick clay liners. They reported that leachate breakthrough at one of these sites had yet to occur, even after 8 years.

In Eau Claire County, Wisconsin, mobile chloride ions moved downward, as did divalent metallic ions, either from the leachate or as a result of exchange mechanisms within the clay (Farquhar and Parker, 1989).

Incompatibility between MSW landfill leachate and the compacted clay liners has not been a problem when the swelling clay mineral content has been kept to a minimum (Gordon, 1987). In fact, liner hydraulic conductivity often decreases in the field because of sealing due to precipitate formation, solids accumulation, and biomass growth along the upper face of the liner and into cracks and fissures (Quigley and Rowe, 1986; Daniel, 1987).

9.2.8 Diffusion Through Clay Liners

Fick's second law, which is derived from the continuity of mass equation, can be used to relate the concentration of a diffusing substance to space and time. Specifically, one-dimensional transient diffusive flow is represented by the following:

$$\frac{\partial C}{\partial t} = D_s \frac{\partial^2 C}{\partial x^2} \qquad\qquad (9.1)$$

where:

C = solute concentration,

$\dfrac{\partial C}{\partial t} \dfrac{\partial^2 C}{\partial x^2}$ = concentration gradients,

D_s = apparent diffusion coefficient
= WD,
W = empirical coefficient (usually 0.5– 0.01),
D = diffusion coefficient.

The solution of Fick's second law using the error function (after Ogata, 1970) becomes:

$$\frac{C}{C_0} = \text{erfc}\left[\frac{x}{2\,(D_s t)^{1/2}}\right] \qquad\qquad (9.2)$$

where:

C_0 = initial concentration (M/L^3)

Ulchrin and Lewis (1988) have modeled both advection and diffusion by:

$$\frac{C}{C_0} = \frac{1}{2}\left[(\exp)\left(\frac{xu'}{2D'} - \frac{x}{\sqrt{D'}}\sqrt{\frac{u'^2}{4D'} + K'}\right)\text{erfc}\left(\frac{x}{2\sqrt{D't}} - \sqrt{\langle\frac{u'^2}{4D'} + K'\rangle t}\right) + \right.$$

$$\left. \exp\left(\frac{xu'}{2D'} + \frac{x}{\sqrt{D'}}\sqrt{\frac{u'^2}{4D'} + K'}\right)\text{erfc}\left(\frac{x}{2\sqrt{D't}} + \sqrt{\frac{u'^2}{4D' + K'}t}\right)\right]$$

$$\qquad\qquad (9.3)$$

where:$erfc\ (x)$=complementary error function,
x = length,
u' = $\dfrac{\text{pore velocity}}{R}$,
R = retardation coefficient (dimensionless),
D' = D/R
K' = first-order decay reaction rate coefficient.

9.2.9 Summary

Hydraulic conductivities less than 10^{-7} cm/s are reasonably attainable when only natural materials with significant clay content are used as the low-permeability liner. The tendency toward reduced hydraulic conductivity due to biomass and other solids buildup at the surface of the liner has been observed. Permeability is generally not adversely impacted by MSW leachate but is affected by many pure organic liquids.

Quigley et al. (1987, 1988) showed clearly that in low-permeability clay soils, transport through a liner by diffusion releases contaminants at rates much greater than permeability dependent or advective flow. Thus, diffusion must be quantified by compatibility experiments in the laboratory and reflected in the design.

Low hydraulic conductivities may be achieved by using (1) water content wet of optimum, (2) a large compactive effort, and (3) a reduced clod size. The liner must also be protected from desiccation and freezing thaw.

Among the three major types of clay liners that have been reported in the literature, compacted clay liners may be the most versatile, with hydraulic conductivities of less than 10^{-7} cm/s attainable if the depth of clay is greater than 1 m, the water content is 2 to 3 percent wet of optimum, and compaction to 95 percent Standard Proctor is achieved. Considerations during placement should include avoiding or breaking down clods, placing and kneading the soil in thin layers (e.g., 15 cm), and maintaining quality assurance and quality control. In situ clay deposits can function well as natural barriers; however, surface preparation may be necessary to reduce the effect of fractures and root holes. Availability of suitable sites with significant in situ clay deposits are limited in some regions. Soils amended with Ca-bentonite may also serve as suitable liners.

9.3 GEOSYNTHETICS

9.3.1 Overview

The utilization of geosynthetics is a new and rapidly emerging industry with potential applications that go far beyond the landfill application of interest here. Here we discuss geosynthetics as concerned with landfills, primarily MSW landfills.

In 1982, the U.S. EPA banned reliance on clay liners alone for hazardous waste sites and stated that landfills should have a liner that "prevents migration of leachate during its active life." Implicit in this was the (correct) determination that after landfill closure, leachate generation is governed by the permeability of the cover. The EPA came to the conclusion that a synthetic membrane leads to "virtually 100 percent removal efficiency" and therefore specified use of single or double liners using "impermeable" synthetic membranes.

The degree to which synthetic membranes are impermeable is questionable. For example, Brown (1982) pointed out that there was no long-term field experience as to the performance of synthetic membranes, and Haxo (1981) stated that no synthetic is suitable for all wastes. Regardless, since that time, there has been widespread utilization of geosynthetics.

The term *geosynthetic* is actually a general term that includes geotextiles, geomembranes, geonets, and geogrids. The selection of the geosynthetic appropriate for a specific circumstance depends on its required function. Examples of specific functions include:

- filtration—to retain soil while allowing the passage of water
- transmission—to enhance lateral drainage (note that if the geosynthetic is to be used as a drain, the geosynthetic must possess sufficient stability to retain its thickness under pressure as well as retain high transmissivity once refuse is placed above it)
- isolation—to isolate two constituents from each other
- barrier—to decrease the transmission of water.

Virtually all geosynthetics are manufactured of noncellulosic extruded synthetic fibers or slit yarns. Nylons (polyamides), olefins (polypropylene, polyethylene), and polyesters comprise nearly all the commercially available products. Therefore, in response to the various types of functions that geosynthetics are expected to perform, a series of different types of geosynthetics have been developed (after Fluet, 1984). They include the following:

- Geonets—for drainage;
- Geogrids—for slope stability (to reinforce the supporting soil or to prevent excessive deformation)
- Geomembranes—for isolation
- Geotextiles—for reinforcement, separation, filtration, and drainage
- Geomats—for prevention of erosion of exposed slopes such as landfill caps.

Geotextiles are the most flexible; however, geomembranes are an essential component for serving as a barrier to intercept leachate. In fact, a system or combination of these individual components is necessary.

9.3.2 Geonets

Geonets are used for lateral drainage by providing a medium through which the planar flow of fluids can occur. The geonet has minimal depth, but the gridlike character provides extensive flow opportunity. It is illustrative to utilize hydraulic conductivity to develop some comparisons. Recall from Chapter 6 that transmissivity is defined as the hydraulic conductivity multiplied by the thickness. As depicted in Figure 9.7, the hydraulic transmissivity is the same for the drainage profiles of two different media, but the velocity is much greater for the geonet, and the volume requirements are vastly different.

EXAMPLE
───

For the drainage net, K (flow direction) = 0.2 m/s. For a typical value of thickness of 4.5 mm = 4.5×10^{-3} m, the transmissivity is 9×10^{-4} m^2/s.

For a sand layer, $K = 3 \times 10^{-3}$ m/s. For a thickness of 0.3 m (1 ft) of sand, the transmissivity is 9×10^{-4} m^2/s.
───

The result is that the use of a geonet will reduce the volume requirements while maintaining the necessary transmissivity and be much easier to place.

9.3.3 Geotextiles

Geotextiles are employed in the landfill as filters to prevent the movement of soil fines into drainage systems, to provide planar flow for drainage, or to act as a cushion to protect geomembranes. Because geotextiles have relatively high permeability, they allow the movement of liquid through the geotextile while preventing the movement of adjacent soil particles. Thus, geotextiles are permeable textiles used with geotechnical materials.

Typical values of hydraulic conductivity are 10^{-3} to 10^{-2} cm/s for geotextiles (in comparison with approximately 10^{-11} cm/s for geomembranes). Thus, geotextiles are comparable with sands and gravels in terms of hydraulic conductivity.

It is generally agreed that leachate drainage pipes should not be wrapped directly in a geotextile because of the potential for clogging. However, differences of opinion exist as to whether geotextiles should be used on top of the drainage media or as underlay or drainage underblanket. Geotextiles will minimize the ingress of fines into the drainage media and pipe, which may otherwise become blocked due to the buildup of fines. However, the fabric itself may become clogged due to the buildup of fines or clogged with biomass given the presence of leachate. Waste itself has some prefiltering capability for removing material in suspension, so the geotextiles may not be necessary. It is important to ensure that the initial lifts of waste being placed on top of the leachate collection system have a low fines content.

9.3.4 Geomembranes or Flexible Membrane Liners (FMLs)

Geomembranes are very low-permeability membrane liners used with any geotechnical engineering materials so as to minimize fluid flow across them.

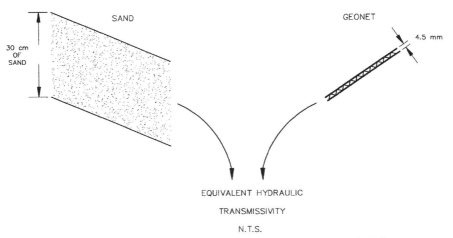

Fig. 9.7 Two different media with equivalent hydraulic transmissivity.

Thus, the intent of geomembranes in landfill design is to limit the movement of leachate. Geomembranes range in thicknesses from 0.75 mm (30 mil) to 3.00 mm (120 mil) and are delivered to a site in lengths up to 500 m.

Some smooth-surface membranes have failed along the shear plane. Therefore, the trend is to include several methods of increasing friction by texturing the geomembrane to minimize slippage of the geomembrane/soil interface.

Leakage Rates   Although geomembranes have very low permeability, they still allow some leakage. Moisture moves through the membranes by diffusion which is driven by concentration gradients (Fick's first law) rather than advective flow (in accord with Darcy's law). The flow rates through conventional, fixed- or falling-head permeameters are small. Thus, either evaporation would negate the leakage measurements, or extremely high gradients would be required to produce measurable flows.

Leakage through the geomembrane can also occur due to pinholes and large holes. Pinholes can be defined as holes having a dimension significantly smaller than the geomembrane thickness. Pinholes are often caused by grit that is inadvertently included in the manufacturing process. Leakage calculations can be made for pinholes. Large holes are holes having a dimension significantly larger than the geomembrane thickness. Bernoulli's equation for flow through an aperture can then be used to quantify the leakage.

The following three equations (after Giroud, 1984) are used to calculate leakage:

Darcy's equation for geomembrane permeability:

$$\frac{Q}{A} = K\left(\frac{Z}{T}\right) \qquad\qquad 9.4$$

Poiseuille's equation for pinholes:

$$Q = \frac{\pi \rho g \bar{z} d^4}{128 \eta T} \qquad\qquad 9.5$$

Bernoulli's equation for large holes:

$$Q = ca\sqrt{2gZ} \qquad\qquad 9.6$$

where:

Q = discharge (m^3),
Q/A = discharge per unit area (m/s),
Z = depth below water level (m),
K, T = hydraulic conductivity (m/s) and thickness of the geomembrane (m)
d = pinhole diameter (m),
a = hole surface area (m^2),
ρ and η = density (kg/m^3) and dynamic viscosity (kg/m·s) of water,

g = gravity (9.81 m/s^2),
c = dimensionless coefficient (e.g., c = 0.6).

The leakage flux through geomembranes is a function of the head (see Figure 9.8(a)), and the leakage rate is proportional to the interface condition (see Figure 9.8(b)) (e.g., does clay lie below the FML.)

Giroud et al. (1989) calculated theoretical leakage rates based on a 1 cm^2 hole per 4000 m^2 (one hole per acre) and a 0.3 m leachate head. They estimated leakage rates of 1000 L/1000 m^2/d for a geomembrane in poor contact with relatively permeable underlying soils and less than 1 L/1000 m^2/d for a correctly installed composite liner. Giroud (1991) indicated typical leakage rates for different layer sequences as listed in Table 9.4.

Note that Equation 9.6 is valid only if the medium located underneath is free draining. Thus, any estimated values are conservatively high values in that the discharge through the holes will be much smaller if the medium located under the geomembrane becomes saturated with leaking liquid. Leakage through a geomembrane with a hole is drastically reduced if the geomembrane is

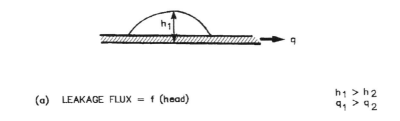

(a)   LEAKAGE FLUX = f (head)

$h_1 > h_2$
$q_1 > q_2$

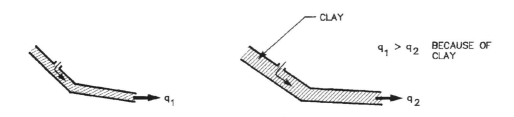

(b)   LEAKAGE RATE IS PROPORTIONAL TO INTERFACE CONDITION

Fig. 9.8 Leakage rates through geomembranes.

Table 9.4: TYPICAL RATES OF LEAKAGE THROUGH THE PRIMARY LINER
(liters/hectare/day)

	Hydraulic Head	
	3 cm	30 cm
Geomembrane/clay	1	10
Geomembrane/silt	10	100
Geomembrane/sand	300	1,500
Geomembrane/gravel or geonet	600	2,000

Notes: 5 holes/ha, each hole 2 mm diameter

Source: Giroud (1990)

placed on a low-permeability soil such as clay.

Composite liners made of a geomembrane placed on a layer of compacted clay are sometimes used to decrease the risk of leakage. Such composite liners are effective only if the geomembrane and clay are in close contact over their entire surface. This requirement is difficult to fulfill due to geomembrane wrinkles and clay surface irregularities and cracks. As a result, a leak in the geomembrane may travel a lengthy distance between the two liners instead of being stopped by the clay (Giroud, 1984).

Using a combination of models, Geoservices (1987) demonstrated that key factors in geomembrane leakage rates are the number of holes in the geomembrane, hole size, leachate head, permeability of the underlying soil, and the degree of contact between the geomembrane and soil.

To minimize the leakage, it is necessary to:

1. Limit the number of holes and the head of liquid over the holes
2. Limit the number and size of holes by using an effective quality assurance/quality control (QA/QC) procedure (to be discussed later)
3. Minimize tears, punctures, cracking, faulty seams, and chemical attack on the liner.

The likelihood that leaks will occur and the ability of the geomembrane to resist chemical attack is, in part, dependent on the flexible membrane liner (FML) employed.

Geomembrane Types   A series of different compositions for geomembranes exist. Selecting a liner for a given application involves defining the site requirements, the length of storage desired, and the wastes to be contained. Among the alternatives are the following:

- ***Thermoplastics*** — for example, polyvinyl chloride (PVC), and thermoplastic nitrile (TN–PVC). Some become soft when heated and can be molded, extended, and shaped.

- *Crystalline thermoplastics* — for example, low-density polyethylene (LDPE), high-density polyethylene (HDPE), polypropylene (PP), and elasticized polyolefin. Some of the polymeric chains are ordered in a crystal lattice.

- *Thermoplastic elastomers* — for example, chlorinated polyethylene (CPE), chlorosulfonated polyethylene (CSPE or Hypalon), and thermoplastic ethylene-propylene diene monomer (T-EPDM).

- *Elastomers* — for example, butyl rubber or isoprene isobutene rubber, ethylene propylene diene monomer (EPDM), neoprene, and polychloroprene.

There are also a number of additives that are sometimes introduced during the process of manufacturing geomembranes to improve processing, physical properties, and resistance to weather and soil exposure. These additives include:

- *Fillers* — For example: mineral particles, metallic oxides, fibers, and reclaimed polymers. The small mineral particles (typically 1 to 200 µm) are used to reduce the cost and increase the stiffness of the geomembrane without altering the permeability;

- *Processing aids* — Used to reinforce or soften the compound during the manufacturing process;

- *Plasticizers* — Used to impart flexibility to the compound, although some plasticizers attract microorganism attack;

- *Carbon black* — Used to impart a black color to the compound which retards aging by ultraviolet light from the sun and increases the stiffness of elastomeric compounds;

- *Fungicides and biocides* — Used to prevent fungi and bacteria from attacking the polymer;

- *Antioxidants* — Used to reduce the aging effect of ultraviolet light and ozone;

- *Scrim reinforcement*— Used to increase the strength and improve tear and puncture resistance; typically nylon or polyester.

Except for highly compounded FMLs such as polyvinyl chloride, the base polymer mainly determines the liner's ultimate chemical resistance. Matrecon (1980) details many of the manufacturing aspects of the various geomembrane types.

Most geomembranes are manufactured by one of the following processes:
- *Extrusion* — whereby a molten polymer is extruded into a non-reinforced sheet. It can be laminated with a fabric to reinforce it.

- *Spreading* — in which a fabric is coated. The geomembrane is thus reinforced by the underlying fabric.

- *Calendering* — in which a heated polymeric compound is passed through a series of heated rollers. Sometimes two sheets of compound are simultaneously run, to minimize the risk of pinhole formation through the entire thickness. In calendered reinforced geomembranes, the reinforcing (support) is achieved by using open fabrics such as scrims (loosely woven strong fabric) of nylon or other lightweight fibers (Oweis and Khera, 1990).

Indications of Seam-Related Concerns   Field seaming is a critical factor for minimizing leakage, since most geomembrane leaks occur at the seams. Because many of the FMLs arrive on site in 6 to 15 m wide rolls, the length of field seaming on a large project is enormous. Seams that bind panels of geomembranes together must not leak and must be physically strong in both shear and peel. They must also retain their integrity over the long term. Exam les of field seams are illustrated in Figure 9.9 (after Fluet, 1991).

Many tests of the integrity of FML seams are required, and the variety of tests is considerable. We simply provide an overview of these tests, since this is a substantial, evolving field of research.

The required durability/polymeric properties that are tested for include ultraviolet stability, chemical stability, freeze-thaw stability, biological stability, specific gravity, color, and water sorption. In addition, the required mechanical properties for which tests are carried out include tensile modulus, ultimate tensile strength, ultimate elongation, burst strength, puncture strength, and creep strength.

The FMLs are installed in the form of sheets, the sizes of which are limited by handling weight and dimension. The seams must then be joined together at the site. Note also that some materials, such as HDPE, must be allowed time to relax and temperature-adjust prior to seaming. Geomembrane placement is usually restricted to temperatures between 5° and 40°C.

Seaming Procedures   The seaming procedure to be used depends on the composition of the geomembrane. The integrity of the field seam is determined by many factors, the most important of which is compatibility with the liner material.

Thermal Seaming   Thermal seaming procedures involve heat only (>260°C). These procedures are preferred in many facilities because no solvents are required. These procedures are applicable only to geomembranes made with base products sensitive to heat (thermoplastics, crystalline thermoplastics, and thermoplastic isomers). The procedures include the following:

- Hot air bonding
- Hot wedge (or knife) bonding
- Dielectric bonding. Dielectric bonding uses a high-frequency electrical current to agitate the molecules within the FML to generate the heat required for the melt. This procedure is not utilized in field applications because it is highly sensitive to dust and humidity.

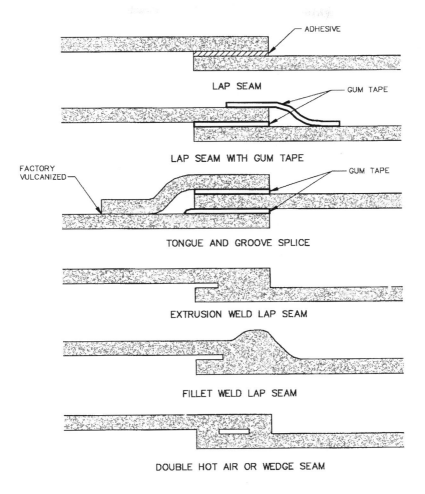

Fig. 9.9 Examples of alternative field seams for geomembranes.

Chemical Seaming   Chemical seaming procedures use the following substances:

- Cement
- Solvent — the selection depends on the ability of a solvent to dissolve the FML
- Bodied solvent
- Vulcanizing adhesive—used on elastomers that will not go into solution with solvents and have poor thermal bonding properties.

The chemical used to seam a particular geomembrane depends on the composition of the membrane and the environment in which the membrane is to be

placed. For hazardous waste disposal facilities, the general practice is to avoid
any bonding method that will leave a residue of volatile organic solvents that
might end up being confused with leachate. There is also concern about the
long-term integrity of the seam materials in contact with leachate. Haxo et. al.
(1983) have conducted experiments in this regard; the reader is referred to their
work for guidance.

Extrusion or Fusion Welding   Extrusion seaming methods, used only on
HDPE, involve supply of a hot base product. Specialized welders extrude a rib-
bon of molten HDPE that melts and then bonds to the two HDPE sheet surfaces,
forming a fillet weld or an insertion (flat) weld.

Mechanical Methods   Mechanical seaming methods utilize taping or
sewing. The presence of moisture can interfere with the curing and bonding
characteristics of the seaming. The presence of dirt or foreign material can jeop-
ardize seam strength and provide a path for fluid to migrate through the seam or
to initiate stress cracks. Thus, quality assurance/quality control procedures are
important considerations.

Once the seam has been completed, it is essential to test the integrity of the
seam. Non-destructive seam testing methods include visual, vacuum, air pressure, ultrasonic, air launch and probe. Visual examination is possible if adjoining sheets are welded together. The vacuum method employs a transparent
suction box approximately 1 m in length placed over a seam portion wetted with
a soap solution. The ultrasonic technique verifies the thickness of the weld to
ensure the pressure of the weld. Non-destructive tests are described in
Table 9.5.

An important aspect of the quality assurance of geomembrane installation
is the complete documentation of seaming operations, which must include a
record for each section of seam with the name of the seamer, the equipment
employed, the date, the weather, and the like.

To the extent feasible, seams should be 100 percent non-destructively
tested for continuity, and the degree of destructive testing should be limited,
since the geomembrane liner is damaged in the process of taking the test samples.

The difficulty of seaming and repair has led to a decline in the popularity of
elastomers. In contrast, some of the HDPE seaming methods (e.g., double-seam
fusion welding) give greater confidence. As noted previously, there are a number
of nondestructive methods for evaluating synthetic membrane seams. However,
the vacuum chamber method should be restricted to FMLs 30 mil or thicker, due
to deformation, and an air lance should be restricted to thicknesses less than
45 mil. For more details on seam testing and quality assurance, see Pacific
Northwest Laboratory (1984, 1985) and USEPA (1985).

Spray-Applied Geomembranes   It is important to note that the preceding procedures apply to FMLs that are brought to a site in large rolls. Geomem-

branes that are made on site are usually continuous (i.e., no seaming is necessary) and are made by spraying the sealant onto a substrate. Some geomembranes are manufactured in situ by spraying asphalt, bitumen, or a molten polymer directly onto a carrier surface (typically a nonwoven geotextile). Thus, these FMLs are referred to as *spray-applied* geomembranes. This technique has the advantage of no seams; however, the quality control of thickness and temperature is difficult, particularly on slopes. The sprayed material must penetrate the fabric and adhere to it after curing to provide a consistent reinforced spray-applied geomembrane. With this type of FML it is difficult to control the thickness of the spraying; thus, the thickness may vary significantly from one location to another (e.g., from 3 to 7.5 mm, or 120 to 300 mil).

The speed of installation of a geomembrane has been characterized in Fluet (1990) as follows:

- Small project (100,000 ft^2) at the rate of 5000 to 10,000 ft^2/d
- Medium project (1,000,000 ft^2) at the rate of 20,000 to 30,000 ft^2/d
- Very large project (10,000,000 ft^2) at the rate of 100,000 ft^2/d.

For geomembranes, calculation of the collection efficiency is difficult because leakage occurs primarily through liner imperfections (including the seam) as opposed to flux rate across the entire liner.

Additional concerns that must be addressed as part of the QA/QC implementation include the following:

- A liner can be punctured by falling objects, equipment moving on the liner, abrasion, and movement against sharp objects. As soon as the leachate drainage blanket is complete, it is recommended that a 2 ft thick layer of fine soil be spread over the membrane to cushion the system against damage from placement of wastes.
- Tearing typically results when a puncture is subjected to a tensile stress.
- Wind blowing over a geomembrane exerts varying amounts of uplift force depending on the velocity of the wind and the roughness of the surrounding land. When not adequately sandbagged, the membrane will lift off the ground and exert tear stresses on the sheet and seams.
- Landfill-generated gases can cause uplift of a membrane. Venting locations must be developed to alleviate tear stresses.
- The physical and mechanical properties of the geomembrane must be compatible with the expected behavior of the underlying material. Circumstances can require geomembranes with high tensile strength or with large elongation (Giroud and Goldstein, 1982).
- Folds should be eliminated and seams should be prohibited in cold weather.
- Membranes must be installed with appropriate slack to accommodate

Table 9.5 NONDESTRUCTIVE TESTS USED TO EVALUATE SEAM CONTINUITY

Nondestructive Test	Description	Applicability	Comments
Vacuum box	A soapy solution is applied to the geomembrane. A box with a transparent window is sealed against the geomembrane and vacuum is established in the box. Soap bubbles will form if there is a leak.	Mostly for stiff geomembranes	Most commonly used test with stiff membranes such as HDPE whose thickness exceeds 0.75 mm (30 mil). Cannot be used in corners or around small radii without special apparatus. Rrelatively slow process since testing area is limited by size of vacuum box.
Air pressure	A double seam with intermediate open channel is made. Pressurized air is blown into the channel.	Any type of geomembrane if seamed with double seam with intermediate channel	Used only with double seams with intermediate open channel (i.e. seams made with double hot wedge or double hot air). More severe loading than vacuum test, but tests only a small fraction of seam strength. Causes some damage to geomembrane because leading hole must be cut. Quite efficient method since long sections of seam (up to 100 m) may be tested at one time. When defects are found, a vacuum box is often used to locate the defect.
Ultrasonic	Several types of ultrasonic techniques are used to assess the continuity of a seam by (i) comparing the measured thickness of the seam to the thickness it should have; and/or (ii) directly detecting voids in the seam.	Geomembranes which may be fused (not applicable to adhesive seams)	Reliable test when conducted by very experienced operator over small areas. Difficult to interpret readout over long periods of time due to operator fatigue
Spark testing	A conducting wire is placed in the seam during seaming. A spark can be established between the wire and an electric device if the wire is exposed (i.e. if a portion of seam is missing).	All geomembranes, but requires conducting wire inserted in seams	Difficult to set up accurately over large areas. Applicable in areas where vacuum cannot be used (corners, etc.). Results not always reliable
Air lance	A pipe with a nozzle is used to blow pressurized air at the edge of a seam. If a portion of seam is missing, air flows under the geomembrane and inflates it or causes the geomembrane to vibrate, often audibly.	Mostly for flexible geomembranes	Qualitative test only, results not very reproducible
Probe	A stiff probe such as a blunt screwdriver is used to verify mechanically if the seam is continuous.	All geomembranes and all seams with well-defined edge	Qualitative test only, results not very reproducible

Source: Giroud & Fluet, 1986.

Table 9.6: FAILURE CATEGORIES

Physical	Biological	Chemical
Puncture	Microbial attack	Ultraviolet attack
Tear		Ozone attack
Creep		Hydrolysis
Freeze/thaw cracking		Ionic species attack
Wet/dry cracking		Extraction
Differential settling		Ionic species incompatibility
Hydrostatic pressure Abrasion		Solvents

Source: Matrecon, 1980.

shrinkage.

- Long, steep slopes may be broken by berms, but anchor berms must be installed at the top of the slope.
- Quality control of installation must be maintained, particularly in relation to proper connections to appurtenances.

Matrecon (1980) defined the failure categories listed in Table 9.6.

Selection of Geomembrane Materials   Selection of the best geomembrane material for a particular application must consider several different factors. A list of properties relevant to this decision are included in Table 9.7. Table 9.8 compares the properties of three geomembrane types, and Table 9.9 lists additional comparisons between HDPE and PVC.

The cost per square meter of an installed flexible membrane liner, which is generally proportional to the thickness, can range from $3/m^2 for a single 20 mil PVC liner to $14/m^2 for a single 100 mil HDPE. Alternatively, a "rule of thumb" is 1 cent per mil per square foot. Although these costs include fabrication, installation, and field seaming, they do not include the cost of subliner preparation, leachate collection layers, engineering design, or QA/QC.

Interactions of Leachate with Synthetic Barrier Materials   The resistance of a liner material to chemical attack and permeation are vital considerations in liner selection, as leaks may occur from chemical attack. No consensus yet exists on what characterizes a nonresistant liner. Changes in weight, dimensions, tear strength, tensile strength, percent elongation at break, modulus of elongation, and hardness are all used (Schwope et al., 1986).

Table 9.7: CONSIDERATIONS UTILIZED IN MATERIALS SELECTION FOR GEOMEMBRANE

Durability/Polymeric Considerations	
Ultraviolet stability	Photochemical attack of polymers is caused by ultraviolet light that fosters oxidation of the polymer
Chemical stability	Of foremost importance
Thermal expansion and contraction	FMLs may be stretched tight in locations. The strains tend to be very localized and can lead to significant fabrication problems and failure of seams
Biological stability	Once the bacteria or fungi have attached themselves to the synthetic or natural material adjacent to the liner, they will eventually use it for a food source
Color	
Water adsorption	Polymeric materials exposed to water will react over long time periods

Mechanical Property Considerations	
❖ Tensile modulus	❖ Burst strength
❖ Ultimate tensile strength	❖ Puncture strength
❖ Ultimate elongation	❖ Creep strength

Tests are frequently carried out under accelerated conditions to examine the compatibility of polymeric liner materials with specific wastes. Samples of the liner materials are immersed in a representative sample of the waste for these tests. Schwope et al. (1986) summarized the results of waste chemical interactions with the polymeric FMLs.

Haxo and Nelson (1984) carried out tests with liner exposed to municipal solid waste for nine years. They found that beneath the leachate, the following occured:

- Membranes tended to sorb leachate
- Changes in physical properties were modest
- Seams made with adhesives tended to lose strength but not leak
- Seams heat-sealed or made with solvent suffered no loss of strength.

Chemical compatibility may be determined from historical and/or test data, although for major projects the latter is certainly preferred. Lab chemical interaction tests on FMLs are carried out by specialized consultants. Such tests include gas permeability, water vapor transmission, water sorption, and water vapor sorption. These tests and alternatives for conditioning the membranes (complete, immersion, pouch, or tub exposure) are discussed in Haxo and Nelson (1984), Lord and Koerner (1984), and Haxo et al. (1985).

Synthetic materials have gained acceptance as barriers at landfills because they exhibit permeabilities of less than 10^{-11} cm/s for many fluids, they are highly resistant to numerous chemicals, and in many cases they can be installed at lower cost than clay liners. (It should be noted that geosynthetics with permeabilities of 10^{-11} cm/s are not necessarily better than clay liners at 10^{-8} cm/s

Table 9.8: A COMPARISON OF THREE FLEXIBILE MEMBRANE LINERS
FOR LANDFILL APPLICATION

Property	CSPE	PVC	HDPE	References
Heat resistance	Excellent	Poor above 140°C	Excellent	Neal (1989)
Microbial resistance	Good	—	Good	
Chemical resistance	Very good (poor to aromatic hydrocarbons)	Very good	Excellent	Neal (1989)
Ultraviolet resistance	Excellent	Poor	Excellent	Neal (1989)
Puncture resistance	Fair to good	Good	If thick, good If thin, fair	Neal (1989)
Field seams	Best made on warm day		Fusion or extrusion welding used in field - no glues needed. Easy for non-destructive testing.	
Ease of placement	Good	Good, but tendency to lose plasticizer and shrink - must be installed loosely.	Fair	Neal (1989)
Cost	High	Low	Moderate	Neal (1989)
Tensile Strength	—[a]	High	Excellent	
Ozone resistance	Good	Poor	—	
Cold weather difficulties	Good resistance to cracking	Stiff and brittle in cold weather	—	

[a]Interpret as "information not available."

Table 9.9: COMPARISONS BETWEEN HDPE AND PVC(FLUET, 1991

HDPE	PVC
Stiffness and thermal expansion resulted in wrinkles (difficult to install and wrinkles act as water barrier in direction of preferred drainage)	Flexibility (easy to install, negligible water barrier)
Yield (tensile failure possible if large elongations)	Elastic behavior even under very large elongation
Low friction coefficient likely to cause soil cover instability	Medium friction coefficient causing some concern regarding soil cover stability
Excellent durability	Questionable durability

because hydraulic gradients for the FMLs are much higher). In contrast, however, synthetic materials are vulnerable to damage and are difficult to seam in the field. Some liners are susceptible to attack from certain chemicals and field experiences with them are limited (Reades, et al., 1987).

Several studies of examination of landfill liners have been reported. These include Haxo et al. (1985), who showed no leachate leakage after 56 months as a result of MSW leachate. Haxo et al. (1983), Haxo and White (1976), and Smith (1977) demonstrated that some synthetic liner materials are quite resistant to MSW leachate, whereas others are considerably less resistant. It is noteworthy that changes in material properties did occur even though leachate leakage did not. Seam strength loss occurred for CPE, CSPF, and EPDM. Hot-sealed and body solvent-welded seams performed best. Results from the tests showed that leachate contaminants had diffused through the liner materials and into the ion-free water. Emcon (1983) found results similar to those of Haxo et al. (1985), with the net effect being that the results are encouraging for the use of synthetic liners at MSW landfills.

Certain organic liquids have a substantial impact on the integrity of the synthetic liners. Specifically, chlorinated solvents, including chloroform, trichloroethylene, tetrachloroethylene, and tetrachloroethane, have significant effects on most synthetic materials, producing increased permeabilities and swelling (Haxo et al., 1985). Polyethylene exhibits high permeation rates for chlorinated solvents but performs well with acids, bases, and hydrocarbons in general. CPE and CSPE perform nearly as well as HDPE but show less resistance to solvents and hydrocarbons in general. Neoprene demonstrates good resistance to hydrocarbons but is difficult to seam properly and is expensive in comparison with other synthetic materials.

Quality Assurance/Quality Control (QA/QC) Because geomembranes are thin, the quality of the liner installation is an extremely important factor in liner performance. Since the 1983–84 quality assurance investigations sponsored by the USEPA, the state of the art of geosynthetic construction quality assurance has received considerable attention and has undergone significant improvements (Giroud and Fluet, 1986). Nevertheless, it is essential to establish a QA/QC program at the outset.

Installation of a large geomembrane is a complex process that requires experienced construction personnel and seamers, as well as detailed inspection and testing. In essence, the two components of a QA/QC program are as follows:

1. *Quality assurance* — a planned and systematic pattern of all means and actions designed to provide adequate confidence that items or services meet contractual requirements and will perform satisfactorily in service
2. *Quality control* — those actions that provide a means to measure and regulate the characteristics of an item or service to contractual and regulatory requirements.

The specifics of QA/QC requirements can assume many forms; we simply provide some indications of the contents of such a program. The first step should be a third-party review of the design. The QA/QC should specify the following important considerations regarding placement of a membrane liner (modified from Schultz, 1985).

1. Prepare a subgrade that is firm, flat, and free of sharp stones, gravel, or debris.
2. Use a qualified installation contractor with experience, preferably experience with the geomembrane being utilized.
3. Follow manufacturers' recommended procedures for the adhesive system and seam overlap.
4. Conduct installation during dry, moderately warm weather (between 5°C and 40°C).
5. Plan and implement a quality control program that will help ensure that the liner meets specifications.
6. Document inspections for review and record keeping.

The importance of a thorough QA/QC program for design and installation is demonstrated by data from actual projects, which indicate one defect per 10 m (30 ft) of seam without quality assurance, and one defect per 300 m (1000 ft) of seam with quality assurance. Lane and Miklas (1990) reported finding a typical range of 1 to 22 leaks/4000 m^2 with an overall average of 14 leaks/4000 m^2 during tests of 61 new or in-service geomembrane-lined lagoons. They reported a frequency of one hole/130 m^2 without quality assurance and one hole per 4000 m^2 with quality assurance.

Obviously, QA/QC programs have an impact on the field efficiency and integrity of the liner, but the program does involve a cost. The cost of a QA/QC program, expressed as a percentage of the cost of lining materials plus installation, has been indicated as 7 percent for a single liner with minimum QA/QC, and 12 percent for a double liner with typical QA/QC (Giroud, 1991).

9.3.5 Summary

Selection of the proper synthetic material as a barrier is a complex decision that involves numerous considerations, including liquid/material compatibility, resistance to weathering, ease and quality of seam preparation, physical properties, and cost.

9.4 DOUBLE LINERS AND COMPOSITE LINERS

Over the last several years, the trend has been toward the use of double liners and composite liners. A double liner consists of two liners with a highly permeable drainage layer or leachate detection system between the liners. The intent of this design is for a majority of leachate to be contained by the upper or primary liner. The lower or secondary liner is intended to contain that fraction of leachate that leaks through the upper liner. It can also act as a primary liner should catastrophic failure of the upper liner occur.

Double liners are required by the EPA for hazardous wastes and are increasingly being required for municipal wastes. Single FMLs are required beneath municipal landfills in Germany, Holland, Italy, Taiwan, Austria, and Saudi Arabia (Cadwallader, 1986).

Composite geosynthetics are a new trend. Composites may consist of, for example, a needle-punched nonwoven material laminated to a geogrid. Thus, the intent is to combine the advantages of the hydraulic properties with the high-density properties. A composite liner is not a double liner.

A state-of-the-art liner system may consist of combinations of various components, the functions of which are as follows:

1. *Optional filter material* — (soil or geotextile) separates the bottom portion of waste from the leachate collection and drainage medium, to reduce clogging of the drainage system. The filter fabric may still clog due to suspended solids, biological growth, and precipitates. Therefore, a high-permeability filter fabric is recommended for use, if one is used at all. Note that geotextiles are sensitive to ultraviolet light degradation if left exposed and therefore must be protected from accidental damage during installation.

2. *Drainage layer* — must have high transmissivity and resist plugging. Gravel is normally used on the bottom and geocomposite for side slopes due to the ease of installation. The gravel should be specified to have a grain diameter larger than 38 mm to minimize biogrowth effects and be meant to resist degradation when exposed to low pH. These concerns are discussed

further in Chapter 10.

3. *Protector layer* — prevents materials in the drainage layer from puncturing the primary geomembrane liner. This layer is usually a thick, needle-punched geotextile (filter fabric).

4. *Barrier layer* — frequently a geomembrane or a natural soil liner or a combination of both. Geosynthetic clay liners (GCL), or prefabricated clay blankets are factory-manufactured dry bentonite clay layers sandwiched between geotextiles can also be used (Daniel and Koerner, 1991).

5. *Leak detection system* — identifies leakage from the primary liner system and enables it to be collected and removed. A geonet is preferable to granular materials because it is easier to place on side slopes, and granular materials can puncture a geomembrane. A geonet also offers faster detection of leaks. The secondary containment system is designed so that leachate passing through defects in the primary FML is detected in the secondary leachate collection system (i.e., the detection system) and removed. The secondary leachate collection system is commonly referred to as a "witness drain," since it bears witness to the integrity of the primary FML.

6. *Secondary barrier (geomembrane)* — The last defense against leachate escape. Technical requirements are generally the same as for the primary layer.

7. *Subgrade soil* — native material.

9.5 NATURAL MATERIAL VERSUS SYNTHETIC AS PART OF THE COVER SYSTEM

Natural soils and geomembranes are also used in landfill caps to prevent fluid flow into the landfill. A minimum 2 ft thick clay cap is frequently required above the venting and drainage layer to provide a low hydraulic conductivity barrier to percolation. Clay soil should be used for this layer and should have a minimum of 50 to 70 percent by weight that passes the 200 sieve and a saturated hydraulic conductivity of 10^{-7} cm/s or less. The layer should be constructed in maximum 6-in. lift heights after the soil is compacted to at least 95 percent Standard Proctor Density with successive layers worked together by a sheeps foot roller with at least 6+ inch feet, and then proof-rolled flat.

At municipal sanitary landfills with clay covers, it is commonplace to see damage to the cover caused by animals burrowing holes for shelter or to access potential food sources. (There are not many reported cases in which bottom liners have been damaged because this type of damage generally occurs only during the construction phase.)

There are more reported instances of animals damaging clay liners or covers than synthetic liners or covers. Rigatuso (1991) found only seven isolated incidents of animals damaging synthetic liners during a survey of all 50 states in

Table 9.10: DESIRABLE PROPERTIES OF A CAP MATERIAL

- ❖ Resist biological attack

- ❖ Be "critter" resistant

- ❖ Retard gas permeation

- ❖ Be impermeable to percolated surface waters

- ❖ Have sufficient friction characteristics to prevent cover soils from slipping and cracking open

- ❖ Resist diminishing physical properties due to long-term soil burial

- ❖ Should not impede vegetative growth; be nonherbicidal and resist root penetration

- ❖ Not deteriorate under ultraviolet light if exposed for any significant length of time

- ❖ Retain physical properties over a broad range of temperatures

- ❖ Retard random permeation; an

- ❖ Accommodate localized settlement without rupture

After Kriofske et al., 1990.

the United States. Rigatuso reported that PVC liners with plasticizers were attacked more frequently than HDPE. Only one incident was identified of an animal's chewing through an HDPE liner, and that occurred only when no escape route was left.

Landfill caps must withstand substantial deformation in areas of localized settlement without cracking. Cap construction materials must be minimally affected by radiation and ultraviolet rays unless rapidly covered by soil following placement. These materials must exhibit the necessary friction characteristics to avoid slippage since slopes on landfill caps may result in long inclined planes of soil. The best liner is that which transmits friction loads from the cover soil directly through the liner to the subsoils. Additional desirable properties for cover material are listed in Table 9.10.

An example of material modification to reduce slippage is the process of texturing HDPE. However, this significantly reduces its physical properties, making it less capable of conforming to localized settlements, less puncture resistant, and inferior in tensile properties (Kriofske et al., 1990).

The landfill cap is also designed to trap and properly vent gases generated during decomposition of organic waste. Gas control is a concern for a municipal site but not necessarily for a hazardous waste site. Gas will collect under the geomembrane cover and move to collection points. HDPE has the best resistance

Table 9.11: GAS TRANSMISSION RATES THROUGH GEOMEMBRANES

30 mil Liner	Methane cc/100 in^2/day	Hydrogen Sulfide cc/100 in^2/day
HDPE	3.5	5 - 15
VLDPE	14.0	40 - 60
PVC	18.0	70 - 110

Source: Kriofske et al.,, 1990.

to gas permeation followed by VLDPE and PVC (Kriofske et al., 1990, as summarized in Table 9.11).

In a modern waste landfill the compacted clay layer is constructed on a compacted layer of protective soil that provides a uniform foundation. An FML incorporated into the cover design will be exposed only to surface water infiltration, so chemical compatibility is not normally of concern. The FML may lie within the frost zone in northern regions and thus may be exposed to considerable temperature ranges. Also, surface settlement may lead to large strains. In addition, FML systems must be designed to provide for venting of gases generated within the cell and are therefore subject to more designed penetrations. An example of a very comprehensive cap and liner system is depicted in Figure 9.10.

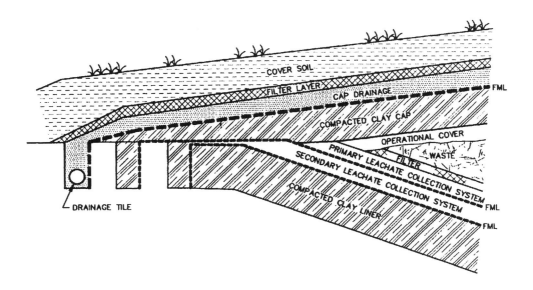

Fig. 9.10 Cap and liner profile.

If infiltration of rainfall is reduced to 100 mm/yr by a good landfill cap, it will take approximately 100 years for 10 m^3 of rain to infiltrate each square meter of cap, that is, before one "bed volume" of rainfall enters the waste and is available for dilution. Lower infiltration rates will expand this period still further and thereby influence the rate at which the processes described in Chapters 4 and 5 will occur. Since movement of liquid through the wastes will involve a good deal of channeling, it may be the case that many centuries of leachate management will be required at a modern landfill before chemical concentrations fall to low concentrations. The nature of these processes and the long time period over which they occur will need to be taken into account in setting criteria for good landfill practice and particularly for issuing certificates of completion.

Summary

Although both the cap and bottom liner systems perform identical functions, the design criteria for selection of the two membrane systems and details of the system are significantly different. Liners and caps can be effective as low-permeability barriers. The permeability of the liners may be adversely affected by pure organic liquids, but the degree of interaction is problem-specific. Permeability of the liners is generally not affected by MSW leachate or dilute solutions of salts, acids, or bases. Stringent QA/QC programs, although they are relatively expensive, are likely worth their investment.

References

Anderson, D., and Jones, S. 1985. "Clay Barrier – Leachate Interaction." In *Proceedings of National Conference on Management of Uncontrolled Hazardous Waste Sites*. Washington, D.C., October 13–November 2.

Biene, R., and Geil, M. 1985. "Physical Properties of Lining Systems Under Percolation of Waste Liquids and Their Investigations, in Contaminated Soil." Ed. J. W. Assink and W. J. Van Den Brink. *Proceedings of the First International TNO Conference on Contaminated Soil*. Utrecht, The Netherlands, 11–15 November. Martinus Nijhoff Publishers, Dorchecht.

Brown, K. W. 1982. *Testimony before the House Site Committee on Natural Resources*. Agriculture Research and Environment of the Committee on Science and Technology. November 30.

Brown, K. and Anderson D. 1983. *Effects of Organic Solvents on the Permeability of Clay Soils*. U.S. Environmental Protection Agency Report EPA-600/52-83-016.

Cadwallader, M.W. 1986. *Synthetic Flextile Membrane Liners in Waste Disposal Facilities: Guidelines for Proper Selection, Specification, and Compliance with Government Regulations. HazPro '86.*

Daniel, D. 1981. "Problems in Predicting the Permeability of Compacted Clay Liners." In Symposium on Uranium Mill Tailings Management, Fort Collins, Colo.pp. 665–675.

Daniel, D. 1984. Predicting Hydraulic Conductivity of Clay Liners *Journal of Geotechnical Engineering* 110 (2): 285–300.

Daniel, D. 1987. "Earthern Liners for Land Disposal Facilities." In *Geotechnical Practice for Waste Disposal '87*. Ed. R. D. Woods.ASCE, N.Y., pp. 21–39.

Daniel, D., and Koerner, R. 1991. Landfill Liners from Top to Bottom. *Civil Engineering*

December: 46–50.

Day, S. R., and Daniel, D. E. 1985. Hydraulic Conductivity of Two Prototype Clay Liners. *ASCE Journal of Geotechnical Engineering* 111 (8): 957–970.

Donald, S., and McBean, E. 1994. Statistical Analyses of Compact Clay Landfill Liners. *Canadian Journal of Civil Engineering* 21:184–196.

Dunn, R. J., and Mitchell, J.K. 1984. Fluid Conductivity Testing of Fine-Grained Soils, *Journal of Geotechnical Engineering* 110 (11):1648–1665.

Emcon Associates. 1983. *Field Verification of Liners from Sanitary Landfills.* U.S. Environmental Protection Agency EPA-600/2-83-046. Cincinnati, Ohio.

Farquhar, G. 1989. "Overview of Landfill Liners Using Natural Materials." *Second International Landfill Symposium*, Sardinia, Italy, Oct 9–13.

Farquhar, G., and Constable, T. 1978. *Leachate Contaminant Attenuation Soil.* Waterloo Research Institute, Project no. 2123, University of Waterloo, Waterloo, Canada.

Farquhar, G., and Parker, W. 1989. "Interactions of Leachates with Natural and Synthetic Envelopes: Lecture Notes in Earth Sciences." Vol. 20, *The Landfill Reactor and Final Storage.* ed. P. Baccini. Springer-Verlag, Berlin, pp. 174–200.

Finno, R. J., and Schubert, W. R. 1986. Clay Liner Compatibility in Waste Disposal Practice. *ASCE Journal of Environmental Engineering* 112, no. 6(December) 1070–1084.

Fluet, J. C. 1984. "Geosynthetic Products and Applications." Presented at *Sanitary Landfill and Leachate Management.* University of Wisconsin at Madison.

Fluet, J. C. 1990. Quality Assurance: Planning Your Work and Working Your Plan. *Geosynthetics World* 1 no. 1(May):20–25.

Fluet, J C. 1991. Geomembrane Applications in Sanitary Landfills. Presented at *Sanitary Landfill and Leachate Management.* University of Wisconsin at Madison.

Foreman, D. E. 1984. "The Effects of Hydraulic Gradient and Concentrated Organic Chemicals on the Hydraulic Conductivity of Compacted Clay." Master's thesis, Geot. Eng. Thesis GT 84-2, University of Texas at Austin.

Fuller, W. H., and Korte, N. 1976. "Attenuation Mechanisms of Pollutants Through Soil." In *Gas Leachate from Landfills*, ed E. J. Genetelli and J. Cirello, US Environmental Protection Agency EPA-600/9-76-004, Cincinnati.

Geoservices Inc. 1987. *Background Document on the Proposed Liner and Leak Detection Rule.* U.S. Environmental Protection Agency EPA-530-87-015.

Giroud, J. 1991. *Implementing Geomembrane Systems in Sanitary Landfills.* Report of GeoSyntec Consultants. Boynton Beach, Fl.

Giroud, J., and Goldstein, J. 1982. Geomembrane Liner Design. *Waste Age* September:27–30.

Giroud, J. P., and Fluet, J. E., Jr. 1986. Quality Assurance of Geosynthetic Lining Systems. *Geotextiles and Geomembranes* 3: 249–267.

Giroud, J. P. 1984. "Impermeability: The Myth and a Rational Approach." Paper presented at International Conference on Geomembranes, Denver, Colo. pp. 157–162.

Giroud, J. P., Khatami, A., and Bater-Tweneboah, K. 1989. "Evaluation of the Rate of Leakage Through Composite Liners." Technical note in *Geotextiles and Geomembranes* 8:337–340.

Giroud, J. P. 1990. *Geosynthetic Lining Systems.* Report of GeoSyntec Consultants, Boynton Beach, Florida.

Giroud, J. P., Badu-Tweneboah, K., and Bonaparte, R. 1992. Rate of Leakage through a Composite Liner due to Geomembrane Defects. *Geotextiles and Geomembranes* 11:1–28.

Gordon, M.E. 1987. "Design and Performance Monitoring of Clay-Lined Landfills." Geotechnical Special Pub 13, ASCE New York, pp. 500–514.

Gordon, M., Huebner, P., and Miazga, T. 1989. Hydraulic Conductivity of Three Landfill Clay Liners. *Journal of Geotechnical Engineering* 115 (8):1148–1160.

Gray, D. H. 1989. "Geotechnical Engineering of Land Disposal Systems." *Lecture Notes in Earth Sciences.* Vol. 20, *The Landfill Reactor and Final Storage*, ed. P. Baccini. Springer-Verlag, Berlin, pp. 201–251.

Green, W. J., Lee, G. F., and Jones, R. A. . 1981. Clay-Soils Permeability and Hazardous Waste Storage. *Journal of Water Pollution Control Federation* 53, no. 8, (August) 1347–1354.

Griffin, R. A. 1983. "Mechanisms of Natural Leachate Attenuation." *Notebook for Sanitary Landfill Design Seminar.* Department of Engineering and Applied Science, University of Wisconsin at Madison.

Griffin, R., Cartwright, K., and Shimp, N. 1976. Attenuation of Pollutants in Municipal Landfill Leachate by Clay Minerals. *Environmental Geology Notes* no. 78. Illinois State Geological Survey, November.

Grim, R. E. 1968. *Clayey Mineralogy.* 2d Ed. McGraw-Hill, New York.

Haxo, H. E. 1981 *Durability of Liner Materials for Hazardous Waste Disposal Facilities.* U.S. Environmental Protection Agency EPA 600/9-81-0026.

Haxo, H. E., Jr., and White, R. M. 1976. *Evaluation of Liner Materials Exposed to Leachate.* Second Interim Report. U.S. Environmental Protection Agency EPA 600/2-76-255.

Haxo, H. E., Jr., White, R. M., Haxo, P. D., and Fong, M. A. 1983. Liner Materials Exposed to Municipal Solid Waste Leachate. U.S. Environmental Protection Agency. EPA 60/2-82-097A.

Haxo, H. E., Jr., Haxo, R. S., Nelson, N. A., Haxo, P. D., White, R. M., Dakessian, S., and Fong, M. A. 1985. *Liner Materials for Hazardous and Toxic Wastes and Municipal Solid Waste Leachate.* Noyes, Park Ridge, N. J.

Haxo, H. E., and Nelson, N. A. 1984. "Factors in the Durability of Polymeric Membrane Liners." International Conference on Geomembranes. Denver, Colo. 1984.

Herrmann, J. G., and Elsbury, B R. 1987. "Influential Factors in Soil Liner Construction for Waste Disposal Sites." *Geotechnical Practice for Waste Disposal 1987.* ed., R. D. Woods. Geotechnical Special Publication, no. 13, ASCE, New York, pp. 522–536.

Hoeks, J., Glas, H., Hotkamp, J., and Ryhiner, A. H. 1987. Bentonite Liners for Isolation of Waste, Disposal Sites. *Waste Management Research* 5:93—105.

Korfiatis, G. P., Rabah, N., and Lekmine, D. 1987. "Permeability of Compacted Clay Liner in Laboratory Scale Models." Ed. R. D. Woods. Geotechnical Practice for Waste Disposal. Geotechnical Special Publication, no. 13, ASCE. pp. 640–654.

Kriofske, K. P., Yazdani, G., and Nobert, J. 1990. "VLDPE: State-of-the-Art Synthetic Membrane for Capping Landfills." Presented at 13th Annual Madison Waste Conference. University of Wisconsin at Madison, Sept. 19–20.

Lahti, L. R., King, K. S., Reades, D. W., and Bacopoulos, A. 1987. "Quality Assurance Monitoring of a Large Clay Liner." In *Proceedings, Geotechnical Practice for Waste Disposal. Geotechnical Division ASCE.* Ann Arbor, Mich., pp. 640–654.

Lambe, T. W., and Whitman, R. V. 1969. *Soil Mechanics.* John Wiley, New York.

Lambe, t. 1958. The Structure of Compacted Clay. ASCE *Journal of the Soil Mechanics and Foundations Division* T. (SM2): 34.

Lane, D. L., and Miklas, M. P. 1990. Finding Leaks in Geomembrane Liners using an Electrical Method: Case Histories. *Hazardous Materials Control* 3 (3): 29–37.

Lord, A. E., Jr., and Koerner, R. M. 1984. "Fundamental Aspects of Chemical Degradation of Geomembranes." In *Proceedings of the First International Conference on Geomembranes*, Denver, Colo. June 20–24.

Madsen, F. T., and Mitchel, J. K. 1989. "Chemical Effects on Clay Fabric and Hydraulic Conductivity." *Lecture Notes in Earth Sciences.* Vol. 20, *The Landfill Reactor and Final Storage*, ed. P. Bacinni. Springer-Verlag, Berlin, pp. 201-251.

Matrecon, Inc. 1980. *Lining of Waste Impoundment and Disposal Facilities.* EPA-SW-870, Office of Water and Waste Management, Washington, D.C.

Mundell, J. A., and Bailey, B. 1985. The Design and Testing of a Compacted Clay Barrier Layer to Limit Percolation Through Landfill Covers. *Hydraulic Barriers in Soil and Rock.* ASTM STP 874, pp. 246–262.

Neal, W. C. 1989. How to Pick a Geomembrane. *APWA Reporter* 56, no. 6 (June):8–9.

Ogata, A. 1970. Theory of Dispersion in Granular Medium. *USGS Professional Paper*, no. 41-I.

Oweis, I. S., and Khera, R. P. 1990. *Geotechnology of Waste Management.* Butterworths, Cambridge,

England.

Pacific Northwest Laboratory. 1984. *Technology of Uranium Mill Ponds Using Geomembranes.* NUREG/CR-3890, PNL-5164.

Pacific Northwest Laboratory. 1985. *Field Performance Assessment of Synthetic Liners from Uranium Tailings Ponds.* NUREG/CR-4023, PNL-5005.

Peterson, S. R. and Gee, G. W. 1985. "Interactions Between Acidic Solutions and Clay Liners: Permeability and Neutralization." In *Hydraulic Barriers in Soil and Rock.* ed. A. I. Johnson, R. Frobel, N. Cavalli, C. Pettersson. ASTM STP 874 American Society for Testing and Materials, Philadelphia, Pa.

Quigley, R. M., and Rowe, R. 1986. "Leachate Migration Through Clay Below a Domestic Waste Landfill, Sarnia, Ontario, Canada: Chemical Interpretation and Modelling Philosophies." In *Hazardous and Industrial Solid Waste Test and Disposal: Sixth Volume.* ed. D. Lorenzem, R. Conway, L. Jackson, A. Hamza, C. Perket, and W. Lacy. ASTM STP933, ASTM, Philadelphia, pp. 93–103.

Quigley, R., Fernandez, F., and Rowe, R. 1988. Clayey Barrier Assessment for Impoundment of Domestic Waste Leachate (Southern Ontario) Including Clay-Leachate Compatibility By Hydraulic Conductivity Testing. *Canadian Geotech Journal* 25:574–581.

Quigley, R. M., Yanful, E. K., and Fernandez, F. 1987. "Ion Transfer by Diffusion Through Clayey Barriers." In *Geotechnical Practice for Waste Disposal '8,* ed. R. D. Woods. Geotechnical Special Publication, no. 13, ASCE, New York, pp. 137–158.

Reades, D. W., Pohland, R J., Kelly, G., and King, S., 1987. Discussion on Hydraulic Conductivity of Two Prototype Clay Liners by Day and Daniel. *Journal of Geotechnical Engineering* 113(7):809–813.

Rigatuso, G. 1991. Rat Tales. *Environmental Information Ltd.* September:6–7.

Schultz, D. W. 1985. "Field Studies of Geomembranes Installation Techniques." International Conference on Geomembranes. Denver, Colo.

Schwope, A. D., Costas, P. P., and Lyman, W. J. 1986. *Resistance of Flexible Membrane Liners to Chemicals and Wastes.* U.S. Environmental Protection Agency EPA-/600/52-85/127. January.

Smith, A. J. 1977. *Comparison of Potential Landfill Liner Materials by Exposure to Leachate.* Harwell Laboratory, Oxfordshire, England, IG 985.

Ulchrin, C. G. and Lewis, T. E. 1988. A Basic Encoded Model for One-Dimensional Ground Water Systems Incorporating First-Order Degradation Kinetics. *Journal of Environmental Science and Health* A23(5):469–482.

U.S. Environmental Protection Agency. 1985. *Draft Minimum Technology Guidance in Double-Liner Systems for Landfills and Surface Impoundment Design, Construction and Operation.* EPA-/530-SW-85-012.

Weinberg, R., Heinze, E., and Forstner, O. 1985. "Experiments on Specific Retardation of Some Organic Contaminants by Slurry Trench Materials." In *Contaminated Soil.* ed. J. Assink and W. Van den Brink. *Proceedings of the First International TNO Conference on Contaminated Soil.,* Utrecht, The Netherlands, 11–15 November. Martinus Nijhoff Publisher, Dorchecht.

Williams, C. E. 1987. "Containment Applications for Earthern Liners." *Proceedings of the 1987 Specialty Conference on Environmental Engineering, ed.* J. D. Dietz.

Wuellner, W., Wierman, D. A. and Koch, H. A. 1985. "Effect of Landfill Leachate on the Permeability of Clay Soils." *Proceedings of the Eighth Annual Madison Waste Conference.* September 18–19, Wisconsin.

9.6 PROBLEMS

9.1. Your laboratory has been asked to test the compatibility of the following four landfill liner materials and four test liquids. Describe briefly how you would perform these tests, and express the anticipated results as either an increase or decrease in permeability. Identify the major shortcomings of your methods. Using any combination of the liner materials you wish, sketch the construction of a liner system that you would recommend for a mixture of municipal solid wastes (MSW) and limited (no non-aqueous-phase liquids) hazardous wastes.

Liner materials	Liquids
compacted clay (clay minerals mostly kaolinite and illite)	young MSW leachate
sand-bentonite mix	trichloroethylene
HDPE (60 mil)	acid sludge (pH = 2)
asphalt	brine (3.7% solids)

9.2 Develop and design a leak detection system that is cost-effective for a hazardous waste site of 5 ha.

DESIGN OF LEACHATE COLLECTION SYSTEMS

10.1 INTRODUCTION

Concern with the environmental impact of landfills has resulted in criteria for new landfills to be designed as containment facilities. In a containment facility, the waste material, products of waste decomposition, and free moisture must be prevented from traveling beyond the limits of the disposal site. The containment requirement thus necessitates the prevention of landfill gas and leachate from migrating from the site in significant quantities.

An essential component of the containment facility is the leachate collection system. The purposes of the leachate collection system are to collect leachate for treatment or alternative disposal and to reduce the depths of leachate buildup or level of saturation over the low-permeability liner.

The leachate collection system may contain many individual components, each of which plays a specific role. Thus, the leachate collection system is not a single entity but is instead a multicomponent system. Two main leachate collection systems may be necessary, namely, an underdrain system and a peripheral system. The underdrain system is constructed prior to landfilling and consists of a drainage system that removes the leachate from the base of the fill. The peripheral system is constructed around the edge of the disposal area and is used to control leachate seeps through the face of the landfill. The peripheral system can be installed after landfilling has occurred and as such is commonly used as a remedial method. The majority of the discussion to follow focuses on design of the underdrain system.

(a) A SIMPLE COLLECTION SYSTEM

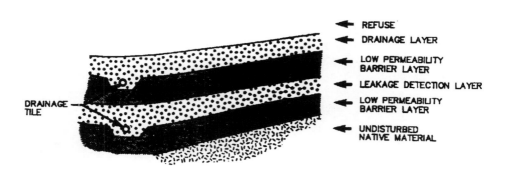

(b) A MORE SOPHISTICATED COLLECTION SYSTEM

Fig. 10.1 Typical leachate collection systems.

The underdrain system, or what we refer to as the leachate collection system for simplicity, may include a drainage layer of high-permeability granular material and drainage tiles to collect the flow diverted essentially laterally toward them, as depicted in Figure 10.1(a).

For the leachate collection system to function, there must be a low-permeability liner (as per the discussion in Chapter 9) underneath. A more elaborate, double-liner system may consist of (from the bottom up) a sloping, low-permeability barrier layer; a leachate detection layer consisting of a drainage layer of high-permeability granular material; another sloping low-permeability barrier

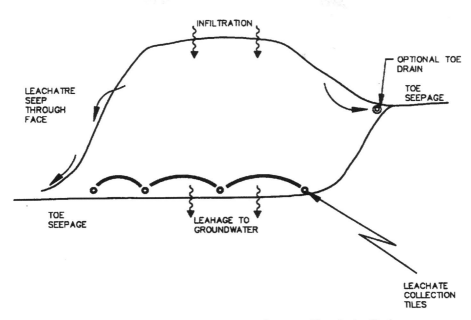

Fig. 10.2 Schematic of various pathways of leachate discharge.

layer; and a drain blanket, as depicted in Figure 10.1(b).

Thus, in terms of preventing leachate migration, the intent is to retard the leachate that percolates vertically through the unsaturated zone of refuse with the low-permeability boundary. The landfill liner is sloped, and lateral flow is facilitated via the drainage layer, thereby minimizing the percolation of the liquids from the waste down into the undisturbed native material.

At the point where the leachate meets the low-permeability boundary, saturated depths of leachate will develop, and leachate flow will be governed by hydraulic gradients within the drainage layer (and if the mound is sufficiently high, also within the refuse). The net result of the leachate collection system is that resulting discharge may include (1) primarily discharge into the leachate tile collection system, (2) seepage through the low-permeability material beneath the site, and (3) seepage through the refuse face above grade on the outside of the landfill, or (4) a combination of these. Figure 10.2 schematically indicates these various pathways.

In this chapter we discuss the interdependencies of the design variables associated with the leachate collection system (e.g., liner slope, drainage tile spacing, and hydraulic conductivity of the drainage material) and give equations describing the depth of mounding for different design configurations.

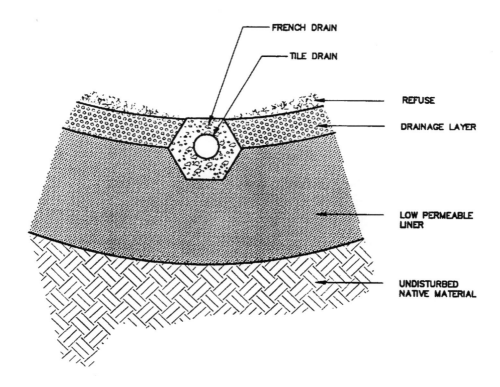

Fig. 10.3 Collection tile and surrounding components.

10.2 COMPONENTS OF THE LEACHATE COLLECTION SYSTEM

When the leachate first reaches the containment and collection system, it flows downward through the drainage layer and impinges on the upper surface of the liner. Leachate begins to pond, or build up, on the upper liner surface, filling the pore spaces of the drainage layer. Simultaneously, horizontal leachate drainage flow begins through the drainage layer, toward the collection pipes, and leachate begins to infiltrate into the unsaturated liner.

The detailed configuration in the immediate vicinity of the drain is depicted in Figure 10.3. Typical design specifications include the *tile drain*, which may or may not be keyed into the low-permeability liner, and a *french drain*, which is utilized so that in the event of pipe failure or clogging, the gravel pack has the hydraulic capacity to transport the leachate, even without the flow-carrying capacity of the tile drain.

Obviously, a number of other components may be introduced, each providing an additional aspect of containment and/or leak detection but involving additional cost. In fact, the cost of the leachate collection system for even a moderately sized landfill may be millions of dollars. The liner/drain system is one

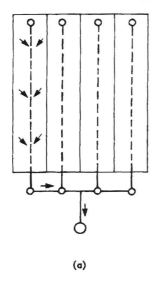
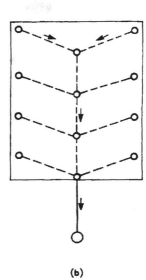

Fig. 10.4 Examples of leachate collection pipe layouts.

of the principal elements over which there can be a high degree of control during construction.

The layout of the leachate collection tiles may assume many different forms, two of which are indicated in Figure 10.4 (from Stecker, 1992). The circles on the schematics indicate cleanout access points. The maintenance features are described in Section 10.5. The leachate collection system being utilized at the Keele Valley Landfill is shown in Figure 10.5. The individual pipes are placed in a dendritic pattern, draining by gravity to one or more locations where the leachate is then removed from the landfill and dealt with by means that are discussed in Chapter 11. A gravel drainage layer is also utilized.

The drainage layer consists of a high-permeability gravel drain to provide a uniform and continuous connection between the waste and the leachate collection system and a modestly sloped pathway along which liquid collected on the clay liner is transmitted to the collector drain pipes. Typically, the material design specifications require the hydraulic conductivity of the drainage layer to be a minimum of 10^{-3} cm/s; 10^{-2} cm/s is desirable. The granular material of the drainage layer should be pre-washed to remove fines. Stipulations on the materials also specify that limestone-based aggregate not be utilized in the drainage media because of the danger that the acidic pH of the leachate (to be discussed in Chapter 11) will cause dissolution followed by subsequent precipitation and encrustation within the rock/gravel media. Chemical precipitation could have

Fig. 10.5 Leachate collection tile network, Keele Valley Landfill,
City of Vaughan, Canada.

serious consequences on the long-term permeability. Natural-washed gravel is
generally appropriate.

As soon as the leachate drainage blanket is complete, a 0.7 m (2 ft) thick
layer of fine soil or refuse is placed over it to cushion the engineered system
against damage and to act as a filter to prevent clogging of the drainage layer.
Table 10.1 summarizes the roles of the individual components of the leachate collection system.

10.2.1 Design Considerations for Tile Spacing

Infiltration through the cap and the resulting percolation through refuse
creates a mound of leachate on top of the barrier layer. To control the height of
this mound, a leachate collection system is used. If collection drains are used, a
mound of leachate will develop between the drains. Thus, design considerations
must involve examination of the interrelationships among the series of components. This includes:

- The flow rate or flux of the leachate impinging on the barrier layer
- The spacing between the tiles
- The slope of the liner
- The thickness and hydraulic conductivity of the drainage layer.

Table 10.1 ROLES OF INDIVIDUAL COMPONENTS OF THE LEACHATE COLLECTION SYSTEM AND LOW PERMEABILITY LINER

System Component	Comments
Barrier Layer	A very low-permeability synthetic or natural soil liner to restrict and control the rate of vertical downward flow of liquids.
Drainage Layer	A high-permeability gravel drainage layer to laterally drain the liquid to the collector drain pipes. The layer lessens the contact time of the leachate with the waste by conducting the percolate away from the waste.
	Should be at least 30 cm (12 inches) thick with a minimum hydraulic conductivity of 1×10^{-3} cm/s.
Slope	To encourage lateral migration, the bottom final slope should be 2% minimum after long-term settlement.
French Drains and Tiles	Maximize the amount of leachate diverted to, and collected by, the tile drains. The material utilized in the french drain should consist of rounded to subangular gravel and have a uniformity coefficient of less than 4 and a maximum particle diameter of 2 in. A minimum depth of 15 cm (6 in.) of gravel is placed prior to placement of the leachate tile drains, and a minimum of 15 cm (6 in.) of gravel above the top of the pipe. Typical pipe perforations consist of two or more rows of holes at the 2 and 10 o'clock positions. A minimum slope of 0.5% is typically employed, using a minimum diameter of 6 in.
Filter Layer	Granular or synthetic, used above the drainage layer to prevent clogging. A filter layer (granular or synthetic) is frequently used above the drainage layer to reduce the potential for migration of fines into the drainage layer..
Refuse	Hydraulic conductivity of 1×10^{-4} cm/s. This value is typical of unmilled, poorly compacted refuse (Ham et al., 1978).

Note: There is no provision for a geotextile to be wrapped around the pipe. The presence of a geotextile would interfere with the movement of any fines from the trench, into the pipe, where the fines would be removed as part of routine cleanout operations.

   If the tiles are separated by too large a distance, the leachate mound will penetrate back up into the refuse. As a result of larger mound heights, increases in the hydraulic gradient occur. Consequently, seepage of leachate through the low-permeability liner is increased (and the efficiency of capture of the leachate is decreased). Significant quantities of leachate will eventually migrate through the liner if the mounding process is allowed to go unchecked, resulting in increases in the contaminant flux to the groundwater. In addition, if the mounding depths become excessive, seepage through the exterior face or side slope of the landfill may occur, affecting the quality of surface water and resulting in increased odors and local effects to ecology.

Flow toward a collection point is essentially due to the gradient of the phreatic surface. The surface profile will change in accord with the liner and drainage tile configuration.

Figure 10.6 shows the mounding between adjacent tiles for two types of tile and liner slope configurations: Figure 10.6(a) is referred to as the continuous-slope configuration, and Figure 10.6(b) is referred to as the sawtooth configuration.

10.3   ANALYTIC FORMULATIONS FOR DESIGN OF THE DRAINAGE TILE SPACING

The mathematical models developed in the following sections will be utilized to develop methodologies to examine a series of design considerations, including:

- The depth, hydraulic conductivity and slope of the drainage layer
- The thickness of the low-permeability barrier layer
- Two measures of hydraulic performance: the maximum saturated depth over the barrier, and the amount of leakage through the barrier.

10.3.1 Configuration of the Low-Permeability Liner

Leachate mounding is dependent on the slope of the liner, the infiltration (percolation) rate of the leachate, the permeability of the drainage layer and the barrier layer, and the drainage tile spacing. Placing the tiles in close proximity minimizes the leachate mounding; however, it increases the cost. A series of mathematical formulations have been presented in the technical literature for relating the design variables to the depth of mounding. These formulations range from simple analytic models to relatively complicated numerical models. However, for applications, calculations for leachate quantity flux are normally adequately addressed using relatively simple models.

The assumptions employed in the mathematical formulations include the following:

1. Flow in the drainage layer occurs primarily in one direction (lateral). Many of the formulations ignore the leachate quantity that passes through the liner.
2. Saturated steady-state flow conditions exist. The models assume saturated Darcy flow occurs within the drainage layer. This assumption means that the wetted volume of the sand grain itself is assumed large relative to the

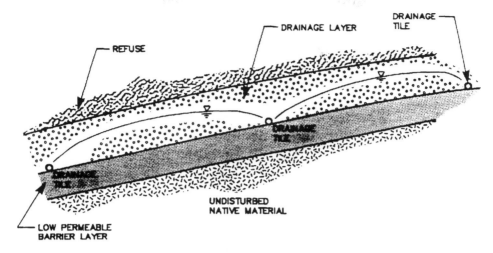

(a) CONTINUOUS SLOPE CONFIGURATION FOR LOW PERMEABLE LINER SYSTEM

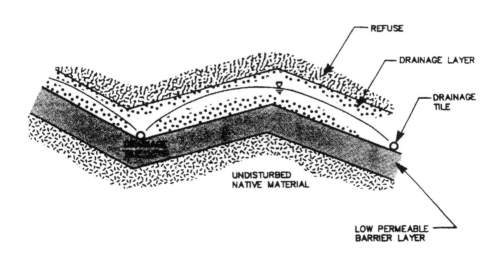

(b) SAW-TOOTH CONFIGURATION FOR LOW PERMEABLE LINER SYSTEM

Fig. 10.6   Leachate mounding in liner systems.

diameter of the grain itself. Violation of this condition will be of concern only when coarse gravels are used for the drainage layer.

3. The drainage media are homogeneous and isotropic.

Generally, these assumptions are reasonable. Given the difficulty in data assignments for use in the models, there is considerable uncertainty in estimates of the mound profiles. Thus, ignoring the leachate, flux through the liner has minimal effect on the mound profile (since the leachate flux is small). Note that this is not to suggest that the effects on water quality as a result of the leachate flux to the groundwater beneath the liner are unimportant. In addition, ignoring the flux through the liner makes the final design of the leachate collection system conservative.

Different analytic models are applicable in response to different liner configurations, namely, the *continuous-slope configuration* and the *sawtooth configuration*.

A commonly employed configuration consists of contiguous V-shaped (or sawtooth) elements constructed from a material of very low-permeability and equipped with a leachate collection system at the bottom of the V. The two branches of the V form a small angle with the horizontal.

The types of drainage tile configurations divide the landfill into discrete segments, so that the models are applied between adjacent collection pipes. More sophisticated models exist (e.g., McEnroe, 1989a and Donald and McBean 1994), but because of their increased mathematical complexity, these models are hard to use. For example, leachate profile drawdown is actually influenced in three dimensions; however, in normal practice, design of leachate collection systems for MSW landfills ignores the three-dimensional feature and considers a simple cross section as shown in Figure 10.6(a) or (b).

Of interest, then, are analytic solutions for design of the leachate collection systems, as such solutions are simple and easy to use. To design a leachate collection system, the designer must be able to understand how different variables such as hydraulic conductivity and configuration of the system can affect its performance in terms of mound depth, lateral drainage, and percolation rate.

An important design feature is the drain separation distance, D. The apex of the mound profile is defined as the point at which the flow divides, with portions of the flow draining to each of the adjacent tile drains. The distance L measures the drainage length contributing to the downstream drain as shown Figure 10.7. The sawtooth configuration has a fixed drainage length L, equal to one-half the drain separation ($D/2$), whereas the continuous slope configuration has a variable drainage length that is greater than that of the sawtooth. The two systems have identical values of L when the liner slope equals zero.

10.3.2 Continuous-Slope Formulation

Consider the problem of drainage on an impervious sloping barrier with drains at both the upper and lower ends. Following the derivation of McBean et al. (1982), for the drainage configuration indicated in Figure 10.6(a), leachate mounding will occur between parallel drains, which serve as flow boundaries. The static head at any point between the two parallel drains in the saturated flow region above the low-permeability boundary can be defined as

$$z_x = sx + y_x \tag{10.1}$$

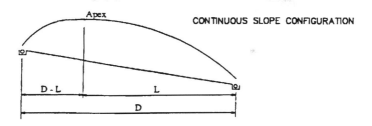

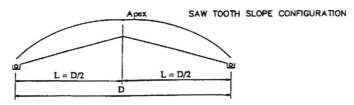

Fig. 10.7 Comparison of drain separation versus drainage length for continuous and sawtooth systems.

where:

z_x	=	static head at location x,
s	=	slope of the liner (radians),
x	=	horizontal distance from the point of interest to the drain centerline,
y_x	=	the depth of flow at location x.

Refer to Figure 10.8 for a schematic representation of the variables.

From Darcy's equation the lateral flow at location X, Q_x, can be written as

$$Q_x = -KA\left(\frac{dz}{dx}\right) \tag{10.2}$$

where:

K	=	hydraulic conductivity of the media,
A	=	cross-sectional area of flow
dz/dx	=	gradient of static head in the direction of flow.

Assuming the system is at a steady state (no changes with time), gives the vertical arrival of water via infiltration as equal to the lateral flow:

$$Q_x = (L - x)\,p \tag{10.3}$$

where:

p =rate of infiltration of moisture per unit area.

Assuming a unit width of aquifer and combining Equations 10.2 and 10.3 provides

$$(L-x)\,p \; = \; Kys + Ky\!\left(\frac{dy}{dx}\right) \tag{10.4}$$

Defining $\alpha = p/K$ and $w = L - x$, further transforming w, and introducing the variable v so that $y = vw$ results in the differential equation

$$\frac{-dw}{w} \; = \; \frac{v\,dv}{v^2 - sv + \alpha} \tag{10.5}$$

Solving the preceding equation and invoking the boundary condition $y(0) = y_0$, yields three conditional cases:

Case I: $4\alpha > s^2$

$$x = L\left[1 - \frac{\left(\left(\frac{y_0}{L}\right)^2 - s\left(\frac{y_0}{L}\right) + \alpha\right)^{\frac{1}{2}}}{\left(\left(\frac{y}{L-x}\right)^2 - s\frac{y}{(L-x)} + \alpha\right)^{\frac{1}{2}}}\exp\left(\frac{s}{\left(4\alpha - s^2\right)^{\frac{1}{2}}}\left\{\tan^{-1}\left(2\frac{\frac{y_0}{L} - s}{\left(4\alpha - s^2\right)^{\frac{1}{2}}}\right)\right.\right.\right.$$

$$\left.\left.\left. - \tan^{-1}\left(\frac{2\frac{y}{L-x} - s}{\left(4\alpha - s^2\right)^{\frac{1}{2}}}\right)\right\}\right)\right] \tag{10.6}$$

In physical terms, in Case I the infiltration/permeability ratio dominates the square of the slope. This means that the leachate tends to accumulate or resist flow, and thus the profile is high. This is the only case where the apex of the mound lies between the flow boundaries, as in Case I in Figure 10.9.

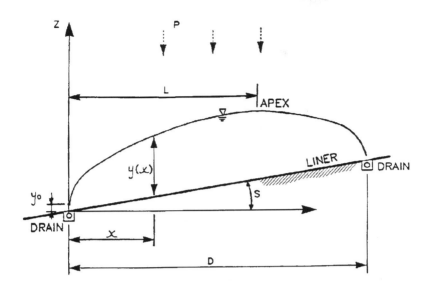

Fig. 10.8 Schematic of variable definitions for the continuous-slope configuration.

Case II: $4\alpha = s^2$

$$x = L\left(1 - \left(\frac{\left(s - \frac{2y_0}{L}\right)}{\left(s - \frac{2y}{(L-x)}\right)}\right)\exp\left\{\frac{2s\left(\frac{y_0}{L} - \frac{y}{(L-x)}\right)}{\left(s - \frac{2y_0}{L}\right)\left(s - \frac{2y}{(L-x)}\right)}\right\}\right) \qquad (10.7)$$

The physical situation of Case II would be highly unusual in that the percolation rate/hydraulic conductivity is precisely equal to the square of the slope. This situation is unlikely to occur in practice, but in the event that it does occur, the equation is provided for completeness.

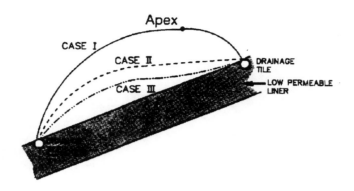

Fig. 10.9 Physical flow profiles the continuous-slope configuration.

<u>Case III</u>: $4\alpha < s^2$

$$
x = L\left[1 - \frac{\left(\left(\frac{y_0}{L}\right)^2 - s\frac{y_0}{L} + \alpha\right)^{\frac{1}{2}}}{\left(\left(\frac{y}{L-x}\right)^2 - s\frac{y}{(L-x)} + \alpha\right)^{\frac{1}{2}}} \exp\left\{ \frac{-s}{\left(s^2 - 4\alpha\right)^{\frac{1}{2}}} \tanh^{-1} \right. \right.
$$

$$
\left. \left. \left[\frac{2\frac{y_0}{L} - s}{\left(s^2 - 4\alpha\right)^{\frac{1}{2}}} \right] - \tanh^{-1}\left[\frac{2\frac{y}{L-x} - s}{\left(s^2 - 4\alpha\right)^{\frac{1}{2}}} \right] \right\} \right]
$$

(10.8)

In case III, the slope term dominates the stability factor 4α, which indicates that leachate will flow readily, resulting in a low profile as depicted by case III in Figure 10.9. Under Case II and III conditions, there is no flow divide per se; all the percolation rate, p, that impinges on the low-permeability liner between the two drains flows to the lower drain.

The solution of Equation 10.6 is dependent on the location of the apex, L. An initial estimate of L allows forward and backward flow profiles to be calculated from the adjacent drains. Successive substitutions for L are required until a common apex is determined, and a smooth, continuous profile is generated by calculation from the adjacent tile drains. Although solution of Equation 10.6 requires successive substitution, convergence typically occurs very quickly in three or four iterations.

Nevertheless, the use of Equations 10.6 through 10.8 is relatively onerous. Nomographs for typical design situations are provided. Figures 10.10 through 10.12 give examples of the magnitudes.

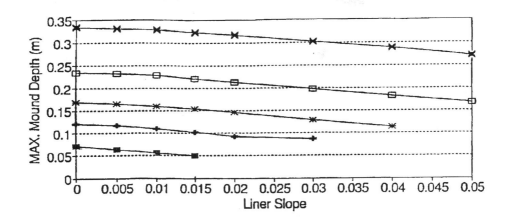

(DRAINAGE LAYER K = 10^{-3} cm/s AND TILE SEPARATION DISTANCE OF 15 m)

Fig. 10.10 Maximum mound depth versus liner slope.

Flow toward a drainage tile is essentially due to the gradient of the phreatic surface. The slope of the liner has little effect, if any, in driving the leachate toward the drainage tile, but a nonzero slope is essential, to prevent ponding.

EXAMPLE
_____

Given an infiltration (percolation) rate of 15.2 cm/yr (6 in./yr), a hydraulic conductivity of the drainage layer of 10^{-3} cm/s, a maximum allowable mound depth of 0.3 m, and a drainage tile spacing of 30 m, what is the necessary minimum slope of the liner?

SOLUTION

From Figure 10.11, the minimum acceptable slope is 0.015.

EXAMPLE
_____

For a 7.6 cm/yr infiltration rate, a drainage tile spacing of 30 m, and a drainage layer with hydraulic conductivity of 10^{-3} cm/s, determine how the maximum mound changes for different liner slopes.

SOLUTION

From Figure 10.11, the 7.6 cm/yr line shows that relatively small changes in maxi-

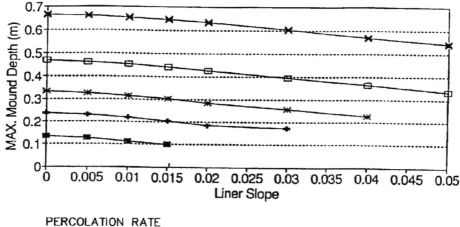

(DRAINAGE LAYER $K=10^{-3}$ cm/s AND TILE SEPARATION DISTANCE OF 30 m)

Fig. 10.11 Maximum mound depth versus liner slope.

mum mound depth occur for sizeable changes in liner slope.

Towner (1975) modified the development by Childs (1971) in developing a complex analytic solution for the three preceding cases. Towner found an approximate solution based on a simplification of his more complex original equations. The result was a two-part drainage profile approximation procedure.

Assignment of the Boundary Value, y_0

The conditions that apply in the immediate vicinity of the drainage tile are free drainage conditions. A solution by Numerov for unconfined flow in the vicinity of a bottom drain (Harr, 1962) indicates that the hydraulic gradient at the brink of the drain is approximately equal to -1. This solution is based on a system with a horizontal barrier and zero recharge and leads to the assumption (after McEnroe (1990)

$$y_0 = \frac{p}{K}\frac{D}{2} = \alpha\frac{D}{2}$$

EXAMPLE
___

Estimate the value of y_0 for a drainage tile spacing of 30 m, a percolation rate of 15.2 cm/yr, and hydraulic conductivity of the drainage layer of 10^{-3} cm/s.

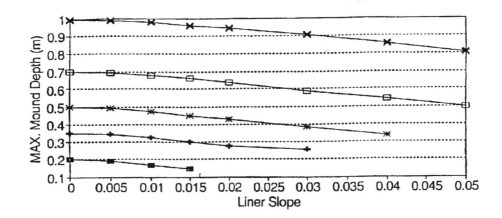

PERCOLATION RATE

—■— 2.5 cm/yr	—+— 7.6 cm/yr	—*— 15.2 cm/yr
—□— 30.0 cm/yr	—✱— 61.0 cm/yr	

(DRAINAGE LAYER K=10^{-3}cm/s AND TILE SEPARATION DISTANCE OF 45 m)

Fig. 10.12   Maximum mound depth versus liner slope.

SOLUTION

$$y_0 = \frac{(15.2\text{cm/yr})\ (15\text{'m})}{\left(10^{-3}\text{cm/s}\right)\left(3.15 \times 10^{7}\text{s/yr}\right)} = .7\ddot{\text{c}}\text{m}$$

10.3.3 Flat-Slope Configuration

Equations 10.6 through 10.8 are implicit in form, requiring an iterative numerical evaluation (i.e., successive substitutions). However, when the slope of the liner system equals zero, Equation 10.6 becomes

$$y = \left[y_0^2 + 2\alpha\left(\frac{Dx}{2} - \frac{x^2}{2}\right)\right]^{\frac{1}{2}} \tag{10.9}$$

which is easily solved. The maximum mound depth occurs at $x = D/2$.

Equation 10.9 is a slightly more general formulation than those presented by Harr (1962), Cedergren (1967), and Matrecon (1980). The maximum leachate mound, y_{max}, is

$$y\max = \frac{D}{2}\sqrt{\alpha} = \frac{D}{2}\sqrt{\frac{p}{K}} \tag{10.10}$$

where the boundary of depth, y_0, is taken as zero.

The flat-slope configuration is the simplest design case and gives a conservatively high estimate of the mound, relative to the nonzero slope situation (as easily seen in Figures 10.10 through 10.12).

EXAMPLE

Using Equation 10.10, determine the maximum mound depth for a 30 m tile drain spacing, a drainage layer hydraulic conductivity of 10^{-3} cm/s, a percolation rate of 7.6 cm/yr, and zero liner slope. What flow can be expected in the leachate tile, assuming a pipe run length of 40 m?

SOLUTION

The maximum mound depth is

$$y = \frac{30m}{2}\sqrt{\frac{7.6\text{cm/yr}}{\left(10^{-3}\text{cm/s}\right)(3600\text{s/hr})\,(24\text{hr/day})\,(365\text{day/yr})}}$$

$$= 0.23\ m$$

The flow, Q, in the drainage tile arrives from both sides of the tile and thus

is

$$
\begin{aligned}
Q \quad &= (7.6\ cm/yr)\,(30\ m)\,(40\ m) \\
&= 91.2\ m^3/yr \\
&= 2.9 \times 10^{-6}\ m^3/s
\end{aligned}
$$

10.3.4 Sawtooth Formulation

An alternative landfill liner configuration, the sawtooth configuration, is depicted in Figure 10.6(b). The basis of equations for mounding associated with the sawtooth configuration is essentially equivalent to that of the continuous-slope configuration with the exception that the upper flow boundary is a different type (i.e., no drain is present) and was used as the origin of calculation by McEnroe (1989a). In addition, McEnroe transforms the contributing variables differently than McBean et al. (1982) and starts profile calculations at the upper boundary.

Based on the Dupuit assumption for unconfined flow, the differential equation governing the steady drainage on a sloping barrier with the variable as defined in Figure 10.13 is

$$Ky\left(\frac{dy}{dx} - s\right) + Ix = 0 \tag{10.11}$$

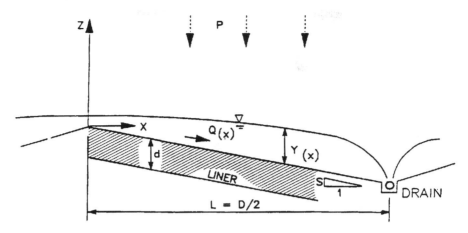

Fig. 10.13 Schematic of variable definitions for the sawtooth configuration.

This is equivalent to Equation 10.4 with transformation of the origin (i.e., $x_{sawtooth} = L - x_{continuous}$). Transforming Equation 10.11 by substituting the expressions $x_0 = x/L$, $y_0 = y/L$, and $y_{0*} = y_0/L$, defining $u^* = y_0/x_0$, substituting u^*x^* for y^*, and then separating variables leads to the equivalent form

$$\frac{-dx^*}{x^*} = \frac{u^* du^*}{u^2* - su^* + \alpha}$$ (10.12)

Invoking the boundary condition $u^* = y_0^*$ at $x^* = 1$, the solution of Equation(10.12) yields three conditional equations:

Case I: $4\alpha > s^2$

$$x^* = \left[\frac{\alpha - sy_0^* + y_0^{*2}}{\alpha - su^* + u^{*2}}\right]^{0.5} \exp\left\{\frac{s}{\left(4\alpha - s^2\right)^{\frac{1}{2}}}\left[\tan^{-1}\left(\frac{2u^* - s}{\left(4\alpha - s^2\right)^{\frac{1}{2}}}\right)\right.\right.$$

$$\left.\left.-\tan^{-1}\left(\frac{2y_0^* - s}{\left(4\alpha - s^2\right)^{\frac{1}{2}}}\right)\right]\right\}$$ (10.13)

Case II: $4\alpha = s^2$

$$x^* = \frac{s - 2y_0{}^*}{s - 2u^*}\exp\left[\frac{2s\,(y_0{}^* - u^*)}{(s - 2y_0{}^*)\,(s - 2u^*)}\right]$$

(10.14)

Case III: $4a < s^2$

$$x^* = \left[\frac{\alpha - sy_0{}^* + y_0{}^{*2}}{\alpha - su^* + u^{*2}}\right]^{\frac{1}{2}}\left[\frac{(2y_0{}^* - s - A)\,(2u^* - s + A)}{(2y_0{}^* - s + A)\,(2u^* - s - A)}\right]^{\frac{s}{2A}}$$

(10.15)

where: $A = (s^2 - 4\alpha)^{1/2}$.
and

$$
\begin{aligned}
x^* &= & x/L \\
y^* &= & y/L \\
y_0{}^* &= & y_0/L \\
u^* &= & y^*/x^*
\end{aligned}
$$

Equations (10.13) and (10.14) are identical with Equations (10.6) and (10.7), although the variables are transformed differently.

Cases I, II and III are illustrated in Figure 10.14.

It is noteworthy that Equations (10.13) through (10.15) are implicit (since x is on both sides of the equation) but are simpler than the equations for the continuous-slope configuration (10.6) through (10.8), since the distance L is known to be one-half the tile separation distance (i.e., $L = D/2$). Note that Equation 10.13 simplifies to Equation 10.9 when the slope of the liner equals zero. Thus, the various equations are consistent.

Several additional mathematical formulations for calculating the maximum depth of mounding for the sawtooth configuration have been presented, including those of Moore (1983) and Richardson and Koerner (1987). These alternative equations are listed in Table 10.2. A comparison of the mounding depths using the three models for the sawtooth configuration is provided in Table 10.3. The mound depths are very similar at small slopes but diverge at the higher slopes.

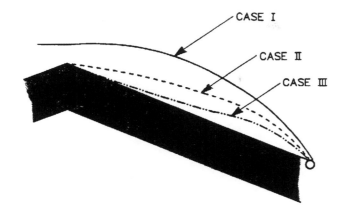

LEGEND

—————   CASE I

- - - - - -   CASE II

.—..—..—   CASE III

Fig. 10.14 Physical flow profiles for the sawtooth configuration.

Table 10.2 ALTERNATIVE MATHEMATICAL EQUATIONS FOR DETERMINING MAXIMUM MOUNDING DEPTH FOR SAWTOOTH CONFIGURATIONS

$$y_{max} = \frac{D}{2}\left[\sqrt{\alpha + \tan^2 s} - \tan s\right]$$

After Moore, 1983.

$$y_{max} = \frac{D}{2}\left[\frac{\tan^2 s}{\alpha} + 1 - \frac{\tan s}{\alpha}\sqrt{\tan^2 s + \alpha}\right]\sqrt{\alpha}$$

After Richardson and Koerner, 1987.

Table 10.3 COMPARISON OF CALCULATED MAXIMUM MOUNDING DEPTHS USING DIFFERENT FORMULATIONS

		Maximum Mound Depth (M)		
Tile Spacing (m)	Slope (%)	McEnroe (1989)	Moore (1983)	Richardson and Koerner (1987)
100	0	1.545	1.542	1.542
	1	1.225	1.121	1.179
	2	1.010	0.838	0.999
	3	0.855	0.651	0.909
	4	0.740	0.526	0.861
	5	0.650	0.437	0.833
50	0	0.772	0.771	0.771
	1	0.612	0.561	0.589
	2	0.505	0.419	0.499
	3	0.427	0.326	0.454
	4	0.370	0.263	0.430
	5	0.325	0.219	0.417

Notes: Assuming infiltration $p = 30$ cm/yr, $K = .001$ cm/s and $y_0 = 0$

10.3.5 Reliability Questions

The extent of mounding is important in terms of the propensity for leakage through the liner. It is interesting, therefore, to compare magnitudes for the alternative liner configurations, as illustrated in Figure 10.15. It is readily apparent that larger maximum mound depths are encountered when the continuous-slope configuration is utilized rather than the sawtooth configuration.

Nevertheless, the situation changes in the event of a drain tile failure. For reasons that will be explored in Section 10.5, drain tile failure may occur, and the resulting mounding profiles assume the magnitudes indicated in Figure 10.16(a) and (b). As Figure 10.17 shows, the higher mounding profiles are now associated with the sawtooth configuration.

10.3.6 Additional Configurations

As discussed in Section 10.3.2 for a zero bottom slope the equations assume a different and much simpler form. Generalizing, Harr (1962) presented an equation for the configuration indicated in Figure 10.18 for the shape of a water table between two parallel open drainage ditches for a constant recharge, *p:*

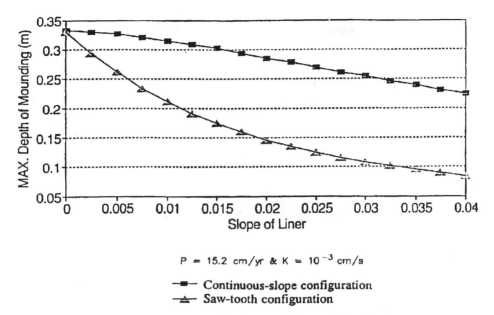

Fig. 10.15 Maximum mounding depth versus slope of liner for two liner configurations.

$$y = \sqrt{h_1^2 + \left(\frac{h_2^2 - h_1^2}{D}\right)x + \frac{p}{K}(D - x)\,x} \qquad (10.16)$$

10.4   PERFORMANCE MEASURES OF THE LINER

Several measures are frequently used in evaluating the effectiveness of a liner, including residence time in the mound, efficiency of capture, and time of first leachate.

10.4.1 Residence Time in the Mound

The maximum residence time in the mound can be estimated using the estimated distance of travel (from knowledge of the location of the apex of the mound) and an estimate of the velocity, from Darcy's equation. Given L, the distance of necessary movement of the leachate from the apex, and Darcy's equa-

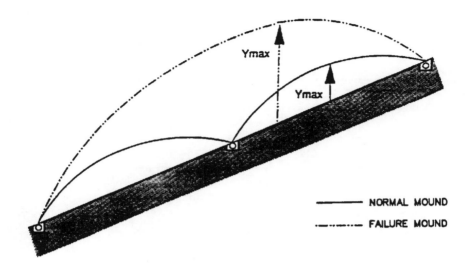

(a) IMPACT OF DRAIN TILE FAILURE FOR CONTINUOUS-SLOPE
    CONFIGURATION

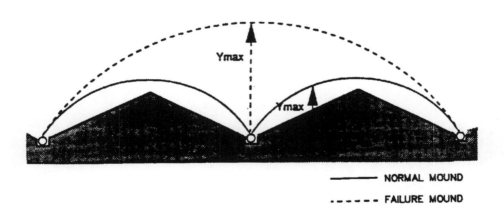

(b) IMPACT OF DRAIN TILE FAILURE FOR SAW-TOOTH CONFIGURATION

Fig. 10.16 Impact of drain tile failure on two liner configurations.

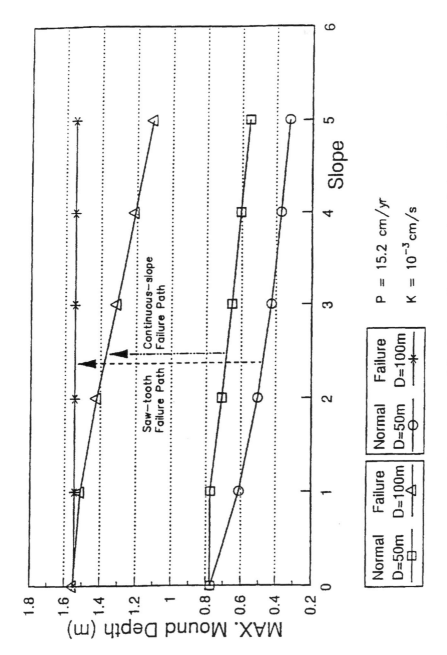

Fig. 10.17 Maximum mound depth versus liner slope, given drainage tile failure.

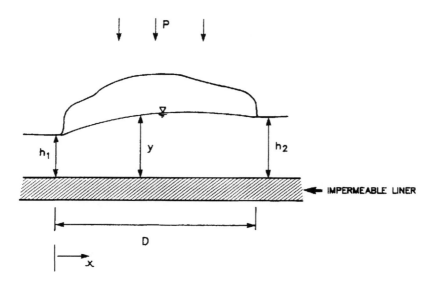

Fig. 10.18 General characterization of leachate mounding for different boundary depths.

tion, the average linear velocity, $\bar{v}$ from Equation (6.2), is estimated from,

$$\bar{v} \approx \frac{Ks'}{n}$$

(10.17)

where:

K = hydraulic conductivity of the drainage layer,
s' = approximated by the bottom slope, and
n   = the porosity,

then the residence time T of the mound is estimated from,

$$T = \frac{L}{\bar{v}} = -\frac{Ln}{Ks'}$$

(10.18)

10.4.2 Efficiency of Capture

The efficiency of capture relates the proportion of leachate that is diverted to horizontal flow, to the tile drains, relative to the proportion that continues to seep downward.

The analytic models described in Section 10.3 all assumed that no percolate was lost due to percolation through the liner. From a practical point of view, this assumption of leachate mounding calculations is valid for any barrier material exhibiting a permeability that is at least two orders of magnitude tighter than the drainage layer above it. A typical design may provide for a clay liner with a hydraulic conductivity of 1×10^{-7} cm/s and a sand blanket with an hydraulic conductivity of 1×10^{-3} cm/s, a difference of four orders of magnitude. However, restrictions on leakage flux through the liner are still very relevant for ground-water quality protection.

The physical laws governing liquid moving downward through a low-permeability clay liner are more complex than those governing liquid moving in sand and gravel drain layers. Because of the micropores that exist in clay soils, water moves not only by the hydraulic head caused by gravitational forces but also by capillary forces that tend to draw the liquid into the soil. Thus, it may be erroneous to predict the time required for leachate to appear at the bottom of a clay liner using calculations based only on the saturated coefficient of permeability, K, of the barrier layer. The reason for this is that Darcy flow assumes that the gravitational potential is the only force moving liquid through the soil. Flow through partially saturated soil occurs both as a result of the gravitational potential and capillary potential. During initial wetting, the capillary potential can greatly exceed the gravitational potential in the soil. Thus, the flow process during initial wetting usually occurs much faster than the gravity-induced Darcy flow that governs the process after the clay liner has become saturated. The smaller the pore radius, the larger the capillary attraction force.

The magnitude of the capillary action is most conveniently measured in terms of a capillary potential, Ψ. The units of Ψ are in centimeters because it expresses the contribution of capillary attraction forces to the hydraulic head tending to move the liquid through the soil.

The capillary potential may pass through seven orders of magnitude as the soil goes from dry to saturated. When the soil is very dry, the capillary forces are so large that they dominate the gravitational forces; however, as moisture content nears saturation, the capillary forces become so small that gravitational forces become dominant.

The importance of the capillary potential should be minimal, since during placement the clay liner is saturated. This being the situation, travel times and flux rates are usually calculated using Darcy's equation.

The total volume of leachate, V, passing through the liner in time Δt, is given by

$$V = -K\frac{dh}{dz}A\Delta t \qquad (10.19)$$

where:

K	=	saturated hydraulic conductivity of the low-permeability liner.
dh	=	change in the total hydraulic head,
dz	=	distance over which the head change occurs,
A	=	cross-sectional area through which the flow occurs.

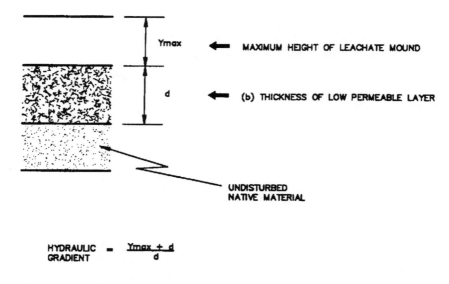

HYDRAULIC = $\dfrac{Y_{max} + d}{d}$
GRADIENT

(GRADIENT ASSUMES NO CAPILLARY SUCTION BELOW SOIL LINER)

Fig. 10.19 Calculation of hydraulic gradient.

For a saturated liner of thickness d with no leachate mound, $dh/dz = d/d = 1.0$.

An upper bound can be calculated for the seepage volume in a liner/drain module where liquid mounds within the drainage layer to a depth y_{max}. In this case,

$$V = \left(\frac{d + y_{max}}{d}\right)KA\Delta t \tag{10.20}$$

Figure 10.19 depicts the gradient information.

Gordon et al. (1989) published data indicating that leakage through a clay liner decreases over time (likely due to a decrease in permeability of the liner). However, it is difficult to reflect this in design of a liner because of the limited field experience obtained to date.

Although theoretically a thin low-permeability layer can be constructed, a number of factors must be considered:

- Potential for loss of strength and integrity due to exposure to leachate
- Construction related issues
- Deterioration in the liner due to freezing/thawing before the liner is covered with refuse.

Theoretically, total containment is not possible, although leakage through the base of a double-lined site approaches zero.

The first transient model (which calculates leakage flux and lateral collection of flows) was proposed by Wong (1977), who presented a simple but approximate solution for time-varying drainage through a liner. Wong's model was amplified by Kmet et al. (1984). Demetracopoulos and Korfiatis (1984) used Wong's solution to explore the sensitivity of collection system performance to various factors. They extended Wong's model to general input conditions, assuming quasi-steady-state behavior for successive time steps. More recently, Korfiatis and Demetracopoulos (1986) solved the complete transient flow equations numerically and compared their results with Wong's approximate solution.

The mathematical models that include losses through the liner become considerably more numerically cumbersome. For example, McEnroe and Schroeder (1988) and Petyon and Schroeder (1990) investigated steady-state drainage and leakage on a low-permeability sloping barrier. The HELP model, described in Section 7.3, also reflects leakage losses.

10.5   LEACHATE REMOVAL ALTERNATIVES

Schematics of alternatives for leachate removal are depicted in Figure 10.20. Local circumstances determine which type of system will be utilized. The above-grade approach creates some site-operation difficulties, whereas the below-grade approach penetrates through the liner, always a cause for concern.

10.6   SYSTEM RELIABILITY AND MAINTENANCE REQUIREMENTS

Clogging problems of drainage systems occur in agricultural irrigation, weeping tile systems, sanitary landfills, septic system leachate fields, and the like. However, system reliability in landfills is particularly important in relation to leachate collection systems, since excavation and replacement are generally not feasible.

The clogging of a leachate collection system can be caused by one (or more) of the following factors: sedimentation, biological growth, chemical precipitation, biochemical precipitation, pipe breakage, pipe separation, and pipe deterioration.

The severity of clogging will depend to a great extent on the waste materials deposited in the landfill due to the large number of physical, biological, and chemical factors involved in the deposition of sludges in the collector pipes. Annual cleaning programs are frequently recommended.

Preventive and remedial measures can be used to address problems. Preventive measures are intended to interrupt the sequence of causal steps neces-

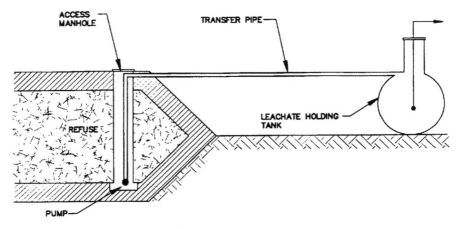

(A) LEACHATE REMOVAL/STORAGE ABOVE
GRADE ALTERNATIVE

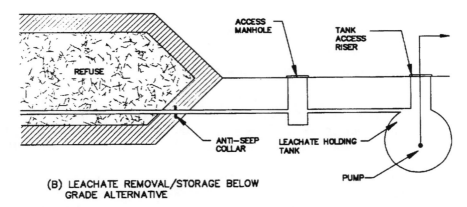

(B) LEACHATE REMOVAL/STORAGE BELOW
GRADE ALTERNATIVE

Fig. 10.20 Alternatives for leachate removal and storage

sary for a clogging mechanism to occur. Remedial measures are intended to
eliminate the clogging problem once it has occurred.

In general, the equipment required for cleaning out collector tile drains is
similar to that used for municipal sewer cleaning activities. Both rodding and
cable machines can be utilized to clear collector lines. Cables are normally used
for smaller-diameter lines (15 to 30 cm), and rodding equipment for lines greater
than 30 cm in diameter. In all cases, the collector run should be designed so that
access ports or manholes are provided at intervals spanning no more than

300 m. Hydraulic jet cleaning can be done for 200 to 300 m distances relatively easily, yet becomes difficult for distances in excess of 300 m.

Removal mechanisms include:

- Mechanical procedures —roto-routers, pigs, sewer balls, snakes, and buckets
- Low-pressure jets — 70 to 140 psi at nozzle
- High-pressure jets — 410 to 1300 psi at nozzle but may cause damage to drain envelope
- Chemical methods — such as SO_2 gas; this procedure poses some danger to personnel and the environment.

Inspection utilizing a closed-circuit television camera requires a "pull cord" for pipes over 100 m in length. The use of television camera inspection allows the opportunity to complete routine pipe deflection testing and hydraulic jet cleaning. Rodding machines are commonly used to insert the pull cord and have a length limitation of 500 m. However, if entry is possible from both ends of the pipe, a length of 1000 m is functional.

As a means of minimizing the problems, preventive measures are generally recommended. Examples include:

- Ensuring that the slope of the collectors is of a suitable grade to promote self-cleaning of the system to the greatest possible extent (However, the flows in most collection systems are relatively small, thus limiting this opportunity)
- Ensuring that the collector pipe diameters are of sufficient size to accept conventional sewer cleaning equipment
- Ensuring that the system layout facilitates the entry and retrieval of sewer cleaning equipment and displaced sludges.

An example of a clean-out system is depicted in Figure 10.21.

However, all the cleaning procedures presume that the clogging has occurred within the leachate collection tiles themselves. There is increasing evidence that clogging also occurs within the surrounding gravel matrix, such as occurred in Peterborough, Ontario (McBean et al., 1993).

Biological clogging occurs when organisms grow in the drain envelope and interfere with the normal flow of leachate. For such growth to occur, appropriate environmental conditions must exist. For example, heavy metals often present in hazardous waste landfills may be toxic or inhibiting to the clog-forming species (Bass et al., 1984). Factors that influence growth include carbon/nitrogen in the leachate, nutrient supply, concentration of polyuronides, and temperature.

Biological clogging of drainage media is greatly reduced under anaerobic conditions in comparison with aerobic conditions because anaerobic bacteria do not generally produce the aggregations and slimes that are characteristic of aerobic bacteria. These findings suggest that biological clogging is more likely to

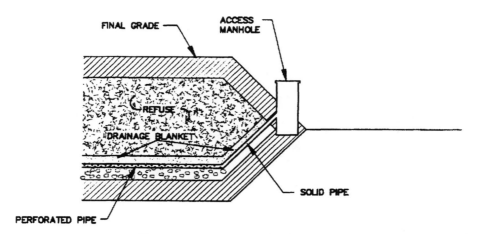

Fig. 10.21 Schematic of clean-out system

occur in toe drains as opposed to the underdrain systems.

Also, it is almost certain that microorganism growth of bacteria and fungi will affect the filtration capability of geotextiles and the drainage capability of geonets and geocomponents. The initial phenomenon is clearly one of blocking and/or clogging rather than degradation, but remediation of microorganism clogging is very difficult.

In addition to biological clogging, physical/chemical reactions may also be a source of clogging. For example, air accessing into the drainage layer will mainly affect the solubility of iron. There are reducing conditions within the landfill. If the iron-containing leachate gets into contact with more oxidizing conditions in the leachate collection system, the soluble Fe^{2+} will be transformed into the insoluble Fe^{3+} and will cause iron precipitate.

In addition, a drop in the partial pressure of carbon dioxide caused by access of air could result in the precipitation of $CaCO_3$. At a CO_2 partial pressure of 0.4 atm the calcium solubility is approximately 10-fold higher than under normal atmospheric conditions.

Temperature influences the solubility of different salts, as well. Usually, the solubility of a salt decreases at lower temperatures. Temperature gradients between the landfill and drainage layer may cause precipitation of leachate components.

Table 10.4 gives the composition of clogging material found in leachate collection pipes as determined by a series of investigators.

Table 10.4 COMPOSITION OF CLOGGING MATERIAL FOUND IN DRAIN PIPES AND DRAINAGE LAYERS OF SANITARY (g/kg) CLOGGING MATERIAL

	Bass et al (1984)	Essig et al. (1981)		Ramke (1987)			
Mg	29.2	4.5	21	2.8	4.9	8.2	18.0
Ca	241	117	240	137	153	307	223
Fe	156	208	8.5	157	125	9.0	48
Mn	5.7	0.6	3.7	3.3	4.7	5.5	3.0
CO_3^{2-}	NA	360	560	207	294	282	396
GR	NA	700	900	844	870	968	853
Drain/Filter	F	D	D	D	D	D	D

If blockage occurs, leachate heads build up. With increased hydrostatic pressure above the bottom liner, higher leachate losses will occur. The stability of the landfill will also decrease, and slope failures may occur, as indicated by Ramke (1987).

Precipitates and sludges may also be a problem in the leachate pumping station, in sumps and in temporary storage reservoirs. As a result, attention should be given to providing extra storage capacity, to sloping and configuring of the reservoir floor to assist removal by pumping, and to making clean-out accessible and safe.

Once collector spacing has been determined, design considerations include:

- Collector sizing and type—of at least 150 mm (6 in.) diameter to facilitate the access of mechanical cleaning equipment. This diameter is likely to have sufficient flow-carrying capacity. However, to reduce the effects of silting and to facilitate inspection and cleaning, a minimum pipe diameter of 225 mm and preferably 300 mm is usually recommended. Frequently, specifications call for schedule 80 PVC pipe or HDPE.
- *Collector slope*—2% if practical but not less than 0.5%.
- *Collector perforations*—at 2 and 10 o'clock positions. The drain slits must be as wide as possible to prevent clogging. A flushing jet has a better flow through a wide slot than through a small channel. Ramke (1987) indicated that the relation of width of holes or slots to wall diameter should be at a minimum 1:1, but 1.5:1 is recommended. The size of the slots or holes is dependent on the grading curve of the drainage layer material. The smallest grain of the drainage layer must not pass the drain slots. Generally, the

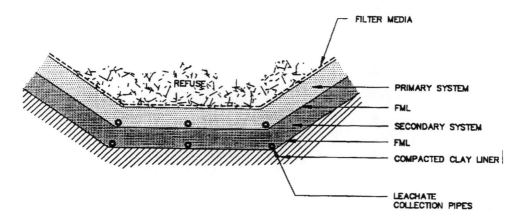

Fig. 10.22 Double-liner configuration using a lower composite liner.

diameter of drain slots should not be less than 10 mm.

- *French drain around the collector pipe*—38 to 50 mm washed stone.
- *Attention to field construction practices*—within pipes, accumulation of deposits may occur in areas of hydraulic perturbation such as where pipe joints have been poorly installed.

10.7 LEAK DETECTION SYSTEMS

10.7.1 Multiple Liner/Drain Modules

In November 1984 the Resource Conservation and Recovery Act was amended by the Hazardous and Solid Waste Amendments (HSWA). HSWA requires that new units and lateral expansions of existing units at hazardous waste landfills have two or more liners and a leachate collection system above and between the liners.

Incorporating multiple liner/drain modules into a landfill (such as depicted in Figure 10.22) somewhat increases the efficiency of leachate collection. However, the primary purposes of the secondary liner system are:

1. To act as a leak detection system—sizable quantities of leachate collection from the secondary system demonstrates that the primary liner has failed
2. To act as a backup system, in the event of failure of the primary system.

Because the intention is to allow minimal leakage through the primary system, that which does occur will not generally be captured with high efficiency by

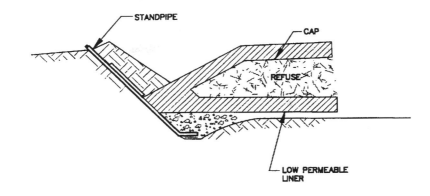

(a) FOR A BELOW-GRADE CELL

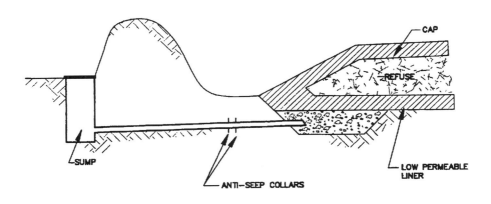

(b) FOR A ABOVE-GRADE CELL

Fig. 10.23 Examples of leak detection monitoring systems.

the second liner system (i.e., the second liner will display a very low individual efficiency).

Leak detection monitoring facilities include those depicted in Figure 10.23. Typically, an 8 to 10 inch HDPE pipe is employed. Note that submersible pumps do not fit down less than a 4 in. diameter casing. A vertical standpipe is generally not employed, since it would have to penetrate the liner. Although it is possible to design a secure penetration of the liner, it is best to avoid when possible.

If standpipes are used in the refuse, the potential for large down-drag forces must be considered. Down-drag forces acting on a standpipe are caused by the differential settlement that occurs between the compressible waste and the rigid standpipe.

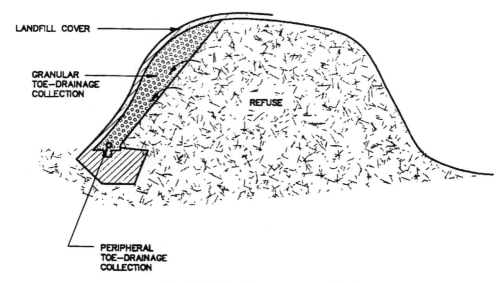

Fig. 10.24 Leachate seep remediation.

10.8 REMEDIATION OF TOE LEAKAGE PROBLEMS

Toe leakage and leachate seeps may be the result of perched water table levels or of biological slimes in the vicinity of drainage tiles.

    If an underdrain system — the focus of much of the discussion in this chapter — was not placed prior to placement of the wastes, or in the event of a failure of the collection system for reasons such as outlined in Section 10.5, surface seeps may develop. The seeps must be dealt with, or they will simply increase in quantity and cause surface water contamination. Remediation may be accomplished by placement of a peripheral toe-drain system such as depicted in Figure 10.24.

    It may also be possible to decrease toe leakage and mounding depths by implementing a series of vertical wells within the refuse and pumping out the leachate, but compacted domestic waste is surprisingly impermeable. Many systems based on the drilling of pumping wells as a means of dewatering the site, after refuse deposition has been completed, have found serious limitations on how much fluid can be extracted.

10.9 REFERENCES

Bass, J. M., Ehrenfeld, J. R., and Valentine, J. N., 1984. *Potential Clogging of Landfill Drainage Systems.* Municipal Environmental Research Laboratory. U.S. Environmental Protection Agency Report U.S. EPA-600/52-83-109. Cincinnati, Ohio, February.

Cedergren, H. R. 1967. *Seepage Drainage and Flow Nets.* John Wiley, New York.

Childs, E.C. 1971. Drainage of Groundwater Resting on a Sloping Bed. *Water Resources Research*. 7, no.5 (October):1256–1263.

Demetracopoulos, A., and Korfiatis, G. 1984. Design Considerations for Landfill Bottom Collection Systems. *Civil Engineering for Practicing and Design Engineers*. 3(10): 967–984.

Donald, S., and McBean, E. 1994. Statistical Analyses of Compacted Clay Landfill Liners. *Canadian Journal of Civil Engineering*. October .

Essig Kohlhoff, Laber, Limbach, SchickeL, 1981. Untersuchung von Wirkung and Langzeitverhalten von Basisabdichtungen aus tonigem Erd-meterial und von Sickerwasserdranagen zur Sohlentwasserung bei Hausmulldeponien. Teilbericht 2: Sickerwasserdranagen Umweltforschungsplan des BMI, Forschungsbericht 1 03 02 210.

Gordon, M. E., Huebbner, P. M., and Miagga, T. J. 1989. Hydraulic Conductivity of Three Landfill Clay Liners. *ASCE Journal of Geotech Engineering Division* 115(8): 1148–1160.

Ham, R., Reinhardt, J., and Sevick, G. 1978. Density of Milled and Unprocessed Refuse. *ASCE-Journal of the Environmental Engineering Division* 104 (EE1).

Harr, M. 1962. *Groundwater and Seepage,* McGraw-Hill, New York.

Kmet, P., Quinn, K., and Slavik, C. 1984. Analysis of Design Parameters Affecting the Collection Efficiency of Clay-Lined Landfills. *Proceedings of the Fourth Annual Madison Waste Conference*. University of Wisconsin at Madison, pp. 250–265.

Korfiatis, G., and Demetracopoulos, A. 1986. Flow Characteristics of Landfill Leachate Collection Systems and Liners. *ASCE Journal of the Environmental Engineers*. 112(3): 538–550.

Matrecon Inc. 1980. *Lining of Waste Impoundment and Disposal Facilities*. Municipal Environmental Research and Development Laboratory. U.S. Environmental Protection Agency, September.

McBean, E. A., Mosher, F., and Rovers, F., 1993. "Reliability-Based Design for Leachate Collection Systems." Fourth International Landfill Symposium, Cagliari, Sardinia, Italy. October 11–15.

McBean, E. A., Poland, R., Rovers, F., and Crutcher, A. 1982. Leachate Collection Design for Containment Landfills. *ASCE Journal of the Environmental Engineering Division*, 108 (EE1):204–209.

McEnroe, B. M. 1989a. "Hydraulics of Leachate Collection and Cover Drainage." 2d International Landfill Symposium, Sardinia, Italy, pp. XIII-1–XIII-8.

McEnroe, B. M. 1989b. Drainage of Landfill Covers and Bottom Liners: Unsteady Case. *ASCE Journal of the Environmental Engineering Division*. 115, no. 6 (December):1103–1113.

McEnroe, B. M. 1990. Steady Drainage of Landfill Covers and Bottom Liners. *ASCE Journal of the Environmental Engineering Division*. 115,(6):1114–1122.

McEnroe, B., and Schroeder, P. 1988. Leachate Collection in Landfills: Steady Case. *ASCE Journal of the Environmental Engineering Division*. 114(5):1052–10(2.

Moore, C. 1983. *Landfill and Surface Impoundment Performance Evaluation*. Office of Solid Waste and Emergency Response. U.S. Environmental Protection Agency, EPA/530/SW-869. Cincinnati, Ohio.

Petyon, R. L., and Schroeder, P. R. 1990. Evaluation of Landfill-Liner Designs. *ASCE Journal of Environmental Engineering* 116(3).

Ramke, H. G. 1987. "Leachate Collection Systems of Sanitary Landfills." *International Symposium on Process, Technology, and Environmental Impact of Sanitary Landfill*. Cagliari, Sardinia, Italy, October19–23.

Richardson, G., and Koerner, R. 1987. *Geosynthetic Design Guidance for Hazardous Waste Landfill Cells and Surface Impoundments*. Contract no. 68-03-3338. Hazardous Waste Engineering Research Laboratory. Office of Research and Development. U.S. Environmental Protection Agency, Cincinnati, Ohio.

Spigolon, S., and Kelly, M. F. 1984. *GeoTech. Quality Assurance of Construction of Disposal Facilities*. U.S. Environmental Protection Agency, EPA-600/2-84-040, p. 49.

Stecker, P. 1992. Lecture notes from Sanitary Landfill Gas and Leachate Management. University of Wisconsin at Madison.

Towner, G. D. 1975. Drainage of Groundwater Resting on a Sloping Bed with Uniform Rainfall.

Water Resources Research 11(1):144–147.

Wong, J. 1977. The Design of a System for Collecting Leachate from a Lined Landfill Site. *Water Resources Research* 13(2): 404–410.

10.10   PROBLEMS

10.1 Demonstrate that equations for the continuous-slope and sawtooth configurations become equivalent when the slopes becomes zero.

10.2 Describe how you could mathematically model the height of mounding with a surface cover layer in a sawtooth configuration with a 1:6 slope, a 2 ft thick vegetative layer, and a 12 in. thick drainage layer of 10^{-3} cm/s sand.

10.3 Determine the maximum spacing of drains such that no impingement in the refuse occurs. The configuration of clay liner, sand layer, and refuse contains 12 in. of 10^{-3} cm/s sand. Assume an infiltration rate of 30 cm/yr. Select other coefficients as needed.

10.4 Utilizing the quantitative information in Problem 10.3, determine the maximum spacing of drains in a landfill that is 1000 m by 1000 m.

    (a)   Develop a collection piping layout for the landfill, including an estimate of the amount of piping required. Specify the liner slope to be utilized in your design.

    (b)   Assuming a maximum height of mounding (so as to prevent impingement into the refuse) of 30 cm, determine the volume of material required for the drainage layer.

    (c)   Develop sensitivity estimates in response to changes in liner slope, infiltration rate, and adjustments in the layout design.

    (d)   Estimate the total flow to be collected in the leachate collection system.

10.5 For Problem 10.2, check the Reynolds number assumption in the vicinity of the drain to ensure that Darcy's law is valid.

10.6 With leachate collection pipes at 200 ft intervals, percolation of 2 in./month, and a drainage layer with hydraulic conductivity of 10^{-3} cm/s, find the maximum head acting on the liner, assuming a bottom slope of 0.02.

10.7 Using the data in Problem 10.6, compare the maximum height of mounding estimated by the different models.

10.8 The landfill liner at a specific site is required to perform to the equivalent of a 1.2 m clayey silt liner having an hydraulic conductivity of 1×10^{-8} cm/s.

    (a)   During the construction of the liner in the downslope area of the landfill it was confirmed that clayey silt was a lesser quality material having a hydraulic conductivity of only 1×10^{-8} cm/s as compared with 5×10^{-9} cm/s for the previously constructed liner. Because the engineers are conservative and wish to maintain the same level of safety for the new area of liner, how far apart should they put the main leachate collectors to maintain the same advective flux through both liners? Assume that the design head (leachate mound) for the existing liner, based on the detection capabilities of the underlying aquifer, is 0.3 m above the liner.

    (b)   What required increase in thickness of the new liner area will achieve the same advective flux through the liner?

    (c)   If there are adequate silt resources to make the liner only 0.6 m thick, what thickness of PVC liner will be required with a hydraulic conductivity of 1×10^{-12} cm/s to achieve the same advective flux through the liner? Assume that the equivalent of Darcy flow through the PVC liner has been confirmed at 1×10^{-12} cm/s.

10.9 The leachate collection system and low-permeability barrier layer can be either continuous slope or sawtooth. Discuss the relative advantages and disadvantages of each configuration with respect to

    (a) mounding depths,

    (b) modifications of pipe failure, and

    (c) slope stability considerations.

TREATMENT OF LEACHATE

11.1 INTRODUCTION

Numerous chemical and biological reactions occur as infiltrative water percolates through the waste materials. As a result, organic and inorganic compounds leach out from the waste. A simple schematic (Figure 11.1) depicts some of these interactions. The products of the complex combination of reactions are potentially transported further by the percolating leachate and by the gases produced. During subsequent movement, physical processes such as sorption and diffusion take place in addition to the chemical and biological reactions.

In previous chapters we focused on procedures for estimating the quantity of leachate that will be produced. Now we turn to characterization of the leachate as it collects at the base of the landfill and design of leachate treatment facilities. The quantity of leachate is obviously an important consideration, but the quality of the leachate is equally important, since it greatly affects the options that must be considered during the design of the treatment system.

Leachate contains many constituents, and its quality is multidimensional. Much can be learned about the status or age of refuse within a landfill by monitoring leachate quality. The basic processes of waste decomposition affect the characteristics of the landfill gas and the quality of the leachate. This information is very important in designing the leachate treatment system for the present situation and for projecting likely changes that will occur.

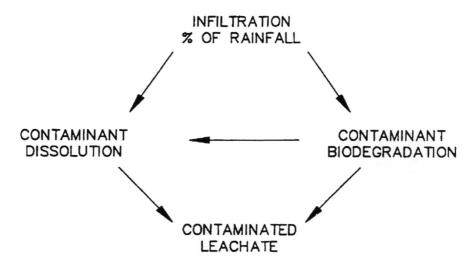

Fig. 11.1 Leaching: contaminant release from solid wastes into solution.

There are three major phases in the decomposition of solid waste:

Phase I-   Aerobic decomposition occurs rapidly, typically for a duration of less than one month. Once the available oxygen within the waste is used up (except in the vicinity of the surface), this phase of decomposition dies out.

Phase II- Anaerobic and facultative organisms (acetogenic bacteria) hydrolyze and ferment cellulose and other putrescible materials, producing simpler, soluble compounds such as volatile fatty acids (which produce a high BOD value) and ammonia.

Phase III- Slower-growing methanogenic bacteria gradually become established and start to consume simple organic compounds, producing the mixture of carbon dioxide and methane (plus various trace constituents) that constitute landfill gas. This phase is more sensitive than phase II.

Phase I is brief, lasting perhaps only a few days or weeks. However, it may persist for longer periods and produce significant quantities of carbon dioxide in shallow (less than 3 m) deposits of waste where air can readily enter the waste or be drawn into the waste by landfill gas pumping. Significant quantities of hydrogen (up to about 20 percent by volume) can be produced, particularly if the site is dry.

Phase II can last for years, or even decades. Leachates produced during this stage are characterized by high BOD values (commonly greater than 10,000 mg/L) and high ratios of BOD to COD (commonly greater than 0.7), indicating that a high proportion of soluble organic materials are readily biodegradable. Other typical characteristics of Phase II leachates are acidic pH levels (typically 5 to 6), strong, unpleasant smells, and high concentrations of ammonia (often 500–1000 mg/L). The aggressive chemical nature of this leachate assists in dissolution of other components of the waste, which typically translates into high levels of iron, manganese, zinc, calcium, and magnesium in the leachate. Gas production consists mainly of carbon dioxide with lesser quantities of methane and hydrogen.

The transition from Phase II to Phase III can take many years, and may not be completed for decades (and is sometimes never completed). However, wastes have been known to reach Phase III in a few months. In Phase III, bacteria gradually become established that are able to remove the soluble organic compounds (mainly fatty acids) which are largely responsible for the characteristics of phase II leachates. These bacteria thrive in the absence of oxygen and convert the soluble organic compounds into methane and carbon dioxide, which are then emitted as landfill gas.

Leachates generated during Phase III are often referred to as "stabilized," but at this stage the landfill is biologically at its most active level. A dynamic equilibrium is eventually established between acetogenic and methanogenic bacteria, and wastes continue to actively decompose. This active production of gas can last several years at a relatively high rate. It may then continue at a gradually decreasing rate over a period of many decades before the landfilled wastes are largely decomposed, and the gas production slows to such an extent that atmospheric oxygen can once more diffuse into the fill.

Leachates produced during Phase III are characterized by relatively low BOD values and low ratios of BOD to COD. However, ammonia nitrogen continues to be released by the first-stage acetogenic process and is present at high levels in the leachate. Inorganic substances such as iron, sodium, potassium, sulfate, and chloride may continue to dissolve and leach from the landfill for many years.

It should be evident from the preceding paragraphs that the leachate coming from a single location is highly variable over time. However, consider now the variability in the refuse from location to location. Some locations will be at one phase of decomposition, while others will be at a very different stage. The leachate at the bottom of the refuse is to some degree a result of the processes that have occurred in the refuse above it. As a result, much of the focus on research has been on the use of lysimeters; dealing with landfills as a whole is more difficult because of the spatial variability of the refuse in terms of the stages of decomposition.

 Factors affecting the composition of leachates include:

- Solid waste composition
- Age of the refuse
- Operation of the landfill
- Climate
- Hydrogeologic conditions in the vicinity of the landfill site
- Conditions within the landfill such as chemical and biolog.
  moisture content, temperature, pH, and the degree of stabiliz.

Because these factors vary considerabl𝑦 from landfill to landfill, significant variability exists in leachate composition. Examples of leachate strength variability over time are provided in Table 11.1. However, as noted, there can be significant degrees of variability from site to site. Examples of ranges of concentrations of various constituents for young leachates are listed in Table 11.2.

Table 11.1 INDICATION OF LEACHATE CONTAMINATION CONCENTRATION TRENDS FOR VARIOUS CONSTITUENTS

Constituent	Concentration (mg/L)		
	1 year	*5 years*	*15 years*
BOD	20,000	2,000	50
TKN	2,000	400	70
Ammonia-N	1,500	350	60
TDS	20,000	5,000	2,000
Chloride	2,000	1,500	500
Sulfate	1,000	400	50
Phosphate	150	50	–
Calcium	2,500	900	300
Sodium, potassium	2,000	700	100
Iron, magnesium	700	600	100
Aluminum, zinc	150	50	–

From Chapters 4 and 5, the methodologies to estimate total mass of leachate are available. However, it is very difficult to define the time dimension (to develop rate coefficients). Landfill leachate may be producing leachate at one location while refuse is still being placed at another point in the landfill. Due to the problems in trying to model the leachate concentrations, investigations at many sites have generally utilized only field monitoring observations, with minimal attempts to predict future quality of the leachate.

11.2 FACTORS CONTRIBUTING TO LEACHATE COMPOSITION

Historical efforts to unravel the processes controlling leachate contaminant quality and variability have generally focused on lysimeter studies. Some researchrs have characterized leachate strength by providing typical concentration

Table 11.2 RANGES OF CONTAMINANT CONCENTRATIONS IN YOUNG LEACHATE

(a) Alkaline Earth

	Concentration (mg/L)		
Metal	*1–2 Years*	*4–5 Years*	*Typical Drinking Water Standard*
Calcium (Ca)	1000-3000	100-1000	500
Sodium (Na)	1000-3000	100-1000	20
Magnesium (Mg)	500-1000	100-1000	–
Potassium (K)	500-1000	100-1000	–

(b) Heavy Metals

	Concentration (mg/L)		
Metal	*1–2 Years*	*4–5 Years*	*Typical Drinking Water Standard*
Iron (Fe)	500-1000	100-300	0.03
Aluminum (Al)	100-200	10-50	0.1
Zinc (Zn)	100-200	10-50	5
Copper (Cu)	<10		1
Lead (pb)	<10		0.05
Cadmium (Cd)	<1.0		0.005
Mercury (Hg)	<1.0		0.001

(c) Anions

	Concentration (mg/L)		
Anion	*1–2 Years*	*4–5 Years*	*Typical Drinking Water Standard*
Chloride (Cl)	1000-3000	500-2000	250
Bicarbonate $(HCO_3)^-$	1000-3000	1000-2000	—
Sulfate $(SO_4)^{2-}$	500-1000	50-500	500
Phosphate $(PO_4)^{3-}$	50-150	10-50	—

rates or relating leachate composition to the refuse landfill age (e.g., McGinley and Kmet, 1984; Lu et al., 1981; Robinson and Maris, 1979). Other investigators have used selected lysimeter studies in an effort to control or eliminate various external factors affecting leachate composition (e.g., Wigh, 1984). As early as 1954, Merz constructed several lysimeters to examine the degradation processes of MSW. Numerous lysimeter studies and several field experiments followed.

These studies identified the most important components of landfill design and operation that significantly affect leachate composition.

Refuse Composition   The mass of refuse stored in an MSW landfill represents a finite source of pollutants. The mass of pollutants available for leaching is largely a function of the physico-chemical nature of the waste, the extent of waste stabilization, and the volume of infiltration into the landfill (Lu et al., 1984). Household and commercial wastes are composed of a wide variety of components but can be surprisingly similar in overall composition, as noted in Chapter 2 (McGinley and Kmet, 1984).

In the early stages of decomposition, the more readily metabolized materials such as sugars, starches, fats, and proteins are degraded, resulting in the depletion of essential nutrients, especially phosphorus. The remaining more complex organic matter such as cellulose materials is slow to decompose and may be hindered further by the lack of nutrients (Rovers, 1972). Pesticides and herbicides, although generally only found in trace amounts, may also be present. Some of the more common trace pesticides are aldrin, BHC, and diazinon (McGinley and Kmet, 1984).

Moisture Addition   The rate of water addition to landfill refuse influences the quality of the leachate. Solute dissolution, microbial decay, and particle entrainment are all affected. At low infiltration rates, anaerobic microbial activity is thought to be a significant factor in governing leachate organic strength, but at high flow rates, soluble organic and even microbial cells may be flushed from the refuse. In such cases, microbial activity appears to play a lesser role in determining leachate quality (Straub and Lynch, 1982).

Depth of Refuse   Refuse depth influences leachate composition. Increases in refuse depth allow the percolate to approach its solubility limit, thereby reducing its leaching potential for the lower depths of refuse. Thus, to the solubility limit, increased depths result in a stronger percolate as the contact time between the solid and liquid phase increases. This can increase leachate strength but also increases the time for refuse stabilization (Lu et al, 1984; McGinley and Kmet, 1984; Qasim and Burchinal, 1970).

Refuse Temperature   Temperature affects bacterial growth and refuse decomposition. Bacteria operate within specific temperature ranges, and optimum efficiency occurs over a relatively narrow range. Optimum mesophyllic temperature for methanogenesis is between 30 and 40 °C (Ham et al., 1979). Elevated temperatures also favor the solubility of most salts and the kinetics of most chemically-driven reactions.

Sorption and complexation of metals in the leachate are important processes influencing the attenuation of trace metals. In general, complexation acts to increase metal solubility (Snoeyink and Jenkins, 1980). Ligands such as chloride, ammonia, phosphate, and sulfide, as well as an array of organic compounds, provide conditions ideal for the complexing of metal ions. Sulfides effectively compete with most complexing agents, however heavier metals will precipitate as sulfides.

11.3   LEACHATE CHARACTERIZATION

Attempts at MSW leachate characterization have generally focused on two approaches. The first and simplest approach has been to empirically fit leachate concentration histories to equations that describe the shape of the contaminant curve (e.g., Revah and Avnimeleih, 1979, Lu et al., 1981, and Wigh, 1979, 1984). The contaminant curve is generally developed as either contaminant concentration versus time, or the cumulative leachate volume per mass of refuse. Because concentration histories depend on the operation and design parameters of the landfill or lysimeter, these curves tended to be site-specific.

A second approach has been to quantitatively describe the physical, chemical, and biological processes that are believed to occur during leaching. Because of the dynamic nature of MSW landfills, these models have tended to be complex and to contain many simplifying assumptions. Examples of this approach include Qasim and Burchinal (1970) and Straub and Lynch (1982).

Qasim and Burchinal (1970) assumed that sorption of solute by the leachate was proportional to the concentration of solute in the liquid phase, and to the difference between the actual and maximum concentration of solute sorbed on the solid particles. Straub and Lynch (1982) developed models that described the leaching processes for organic and inorganic contaminants. The models consisted of a simple mixed reactor and an unsaturated vertical flow model with microbial digestion for the organics. The model assumed no biodegradation of the solid-phase organic compounds.

The leachate characterizations just described focused on prediction of leachate strength from a single lysimeter study. To reflect leachate quality dependence on operational and design variables, Lu et al. (1981) developed curves showing changes in concentration with increasing landfill age, with data from various landfills of different ages. They showed that the concentrations of various leachate constituents varied considerably (up to four orders of magnitude) from landfill to landfill. Landfill age was found to be the most relevant factor affecting leachate composition.

Major environmental problems have arisen at landfills because of the migration of leachate from the landfill site and the subsequent contamination of surrounding land and water. Whether leachate is to be collected and treated or be allowed to discharge to the soil (by natural attenuation, addressed in Chapter 12), it is essential to have estimates of leachate flow and strength, and the variation of these with time, as the site develops through closure and after closure. Although these estimates are essential, their preparation is a difficult and uncertain process.

In recent years, priority organics have become a primary concern among environmental regulators. Organic compounds can occur naturally but are largely associated with the increase in chemical manufacturing over the last 50 years. These manufactured organics, generally found in trace concentrations in MSW leachates (including pharmaceuticals, solvents, cleaners, lawn and garden herbicides, and pesticides), are of concern because of their potential toxic, carcinogenic, or mutagenic properties and their occasional persistence in the environment (McGinley and Kmet, 1984). The typical ranges of concentrations of priority organics found in leachate are presented in Table 11.3.

Table 11.3   TYPICAL PRIORITY ORGANIC POLLUTANTS IN LEACHATES

Organic	Range (ppb)	Average	Source
Absorbable organic halides	320-3500	2000	1
	<10-45,000	2500	2
Phenols	1-4000	1210	1
	<10-11,300	1750	2
		120	3
Polycyclic aromatic hydrocarbons	<1-3		1
	<10-100		2
Benzene	100-600	230	2
Toluene	<10-3200	720	2
Ethylbenzene	<10-4900	400	2
Chlorobenzene	<1-7	1	2

Source:
1) Stegmann and Ehrig, 1989
2) McGinley and Kmet, 1984
3) Conestoga-Rovers & Associates, 1987.
Municipal solid waste landfill leachate data from 83 landfills is reported in Subtitle D of RCRA (U.S. EPA, 1989), and for hazardous sites in U.S. EPA, 1987.

11.3.1 The Temporal Variability of Leachate

The types, amounts, and production rates of contaminants appearing in the leachate at a landfill site are influenced by numerous factors, including refuse type and composition, refuse density, placement sequence, depth, moisture loading to refuse, temperature, time, and pretreatment.

Accurate quantification of these factors and their impact is difficult due to the heterogeneity of the refuse found in a landfill. The mechanisms and extent

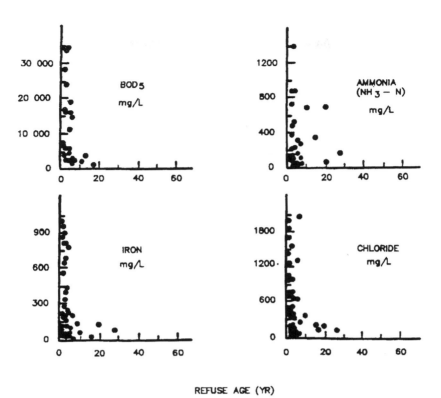

Fig. 11.2   Variation of leachate contaminant concentration with refuse age.
(From Lu, Eichenberger and Stearns, 1985)

to which they influence contaminant release and contaminant levels over time
cannot be easily quantified. It is therefore necessary to rely on data and experience from other landfill investigations and to apply them to landfills under study.

Nevertheless, there are certain trends in the leachate concentration levels
over time. Lu et al. (1985) produced an extensive review of investigations reporting leachate production and contaminant concentrations. They combined the
data obtained from those studies to produce contaminant production curves similar to those shown in Figure 11.2, where plots for BOD_5, iron (Fe), chloride (Cl),
and ammonia nitrogen (NH_3-N) are presented. In particular, some studies used
atypical lysimeters and generally produced the higher concentrations shown.
Consequently, the plots and models produced by Lu et al. (1985) represent upper
limits for leachate contaminant concentrations at field installations.

Some investigators (Fungaroli and Steiner, 1979, Ham, 1980, Wigh and
Brunner, 1981, and McGinley and Kmet, 1984) have produced data that appear

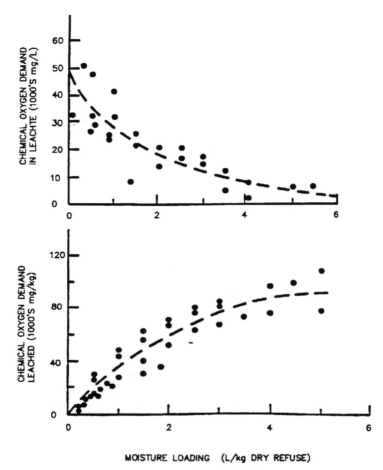

Fig. 11.3 Leachate COD production curves.
(Adapted from McGinley and Kmet, 1984)

to reflect field conditions more closely. Their work was also conducted and reported in such a way that the impact of some of the important factors including compacted density, moisture addition, depth, and refuse age could be evaluated.

McGinley and Kmet (1984) attempted to combine the data from these more realistic studies to produce leachate contaminant production curves. Examples of their plots of several data sets are shown in Figure 11.3. The graphs show leachate chemical oxygen demand (COD) as a function of moisture loading to the refuse in units of liters of leachate per kilogram dry refuse (as opposed to time) in an attempt to normalize the data. The upper graph expresses COD as milligrams per liter in the leachate, whereas the lower one uses milligrams COD leached per kilogram of dry refuse. Although the data are scattered, there is a reasonably good trend shown in each case. Young leachates exhibit CODs in the range of 30,000 to 50,000 mg/L, whereas leachates from old, extensively-leached refuse have CODs of generally less than 2000 mg/L.

Nevertheless, concentrations of the individual constituents do not change at the same rate at different sites, because time is only one of the important factors—another is infiltration. An alternative is to utilize moisture loading flux as the explanatory variable (as opposed to time). With knowledge of the infiltration, as can be computed using the water balance model, it is possible to estimate the concentrations of contaminants. This is not an unreasonable means of characterizing the concentration as a function of time, since there are finite amounts of chemicals within the refuse. Once they are leached out by the successive volumes of water, then their concentrations diminish over time. However, in the overall assessment of a landfill, concern is not with the water quality from a single location but over the entire site, which requires a means of extrapolating from individual lysimeter results.

On this basis Reitzel et al. (1992) examined lysimeter studies to develop regression curves for chloride, COD, total phosphates, ammonia, iron, cadmium, and lead. For each of these constituents, they normalized the time scale for the regression curves with regard to moisture addition and refuse mass and presented the data as cumulative liters of leachate per kilogram of refuse, as illustrated in Figure 11.4. The equations for the individual constituents are listed in Table 11.4, for the leaching of single MSW cells. Because a landfill is composed of multiple cells at different stages of stabilization, use of the curves requires the landfill to be analyzed as cells identified with respect to their time of placement.

Changes Expected Over Time Increases in the size of individual sites, improvements in the compaction of wastes, and use of cover materials to exclude air and minimize water infiltration alter the manner in which deposited wastes decompose. Rapidly established anaerobic conditions generate very high concentrations of BOD compounds such as volatile fatty acids, which, as decomposition proceeds, become efficiently converted to large quantities of landfill gas. In addition, there are changes in waste composition over time of input to the landfill as a result of 3Rs initiatives as discussed in Chapter 2.

11.4   CONSIDERATIONS IN CHOOSING LEACHATE MANAGEMENT OPTIONS

In this section we examine the factors that affect leachate treatment choices. Although considerable experience dealing with the treatment of domestic and industrial wastewaters has been transferred to the application of landfill leachate treatment, there are specific features pertinent to the treatment of leachate that must be considered before the domestic/industrial wastewater treatment information can be successfully utilized. Numerous good reference texts exist for those not familiar with municipal wastewater treatment (e.g., Metcalf and Eddy, 1991).

A further complication in designing leachate treatment systems arises because much of the published technical literature has been developed from lab-

CHLORIDE DATA

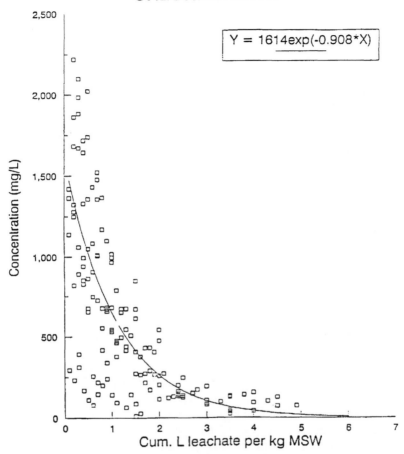

Fig. 11.4   Chloride prediction curve.

oratory- and pilot-scale plants. Few full-scale systems have been constructed, and from those that have, the results have seldom been published. As a consequence, there is a considerable opportunity for research and development in this field. Our intent is to organize the existing information and to provide guidelines as the data and experiences afford.

11.4.1 Sources of Variability

Landfill leachates from municipal solid wastes generally contain high concentrations of organic and inorganic chemicals. The inorganic ions include chlorides and sulfates, and metals such as iron, sodium, potassium, calcium, manganese, and zinc. Evaluation of alternatives for treatment of leachate must

Table 11.4   EQUATIONS DESCRIBING CONSTITUENT CONCENTRATIONS IN LEACHATE

Constituent	Equation
Chloride	$C = 1614 \exp(-0.908x)$
COD	$C = 40{,}000 \exp(-0.49x)$
Ammonia	$C = 655 \exp(-0.521x)$
Total Phosphate	$C = 23 \exp(-0.11x)$
Iron	$C = 375 \exp(-6.198x)$
Cadmium	$C = 13 \exp(-9.346x)$
Lead	$C = 10 \exp(-2.062x)$

Note:
x = cumulative liters of leachate per kg of MSW
C = concentration of constituent in leachate in mg/L

Source: After Reitzel et al., 1992.

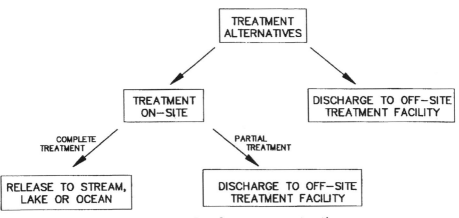

Fig. 11.5   Leachate treatment options.

consider the large temporal fluctuations in both the quantity and composition of leachate at a particular site (e.g., see Johansen and Carlson, 1976; Chian and deWalle, 1977a; Robinson and Maris, 1979). An important characteristic relevant to its treatability is then the change over time of certain components of leachate as the biological conditions change within the landfill.

Design of treatment systems for landfill leachate must consider the variability in the flow rates and the complex and temporally varying composition. To respond to the variabilities, the options for treatment of leachate may include, as depicted in Figure 11.5, (1) full treatment on site, (2) partial treatment on site and disposal to a publicly owned treatment works (POTW), or (3) transport off-site to a POTW directly. A part of the consideration of discharge possibilities is related to whether the leachate will have a detrimental impact on the POTW. The limited data that have been published on co-treatment indicate that generally 2 percent leachate by volume will produce acceptable results (Robinson and Maris, 1979). Due to the leachate volume contribution, the impacts on the POTW will probably include higher aeration requirements, a phosphorus deficiency for good biological treatment, higher sludge production due to the increased biomass and metal precipitates, elevated levels of metals in the sludge, increased foaming problems, and odors.

The option selected to treat leachate is a function of numerous factors, which are primarily a function of economics. The considerations include the relevant water quality standards or criteria that must be met prior to discharge, the extent of the variability in leachate flow and contaminant concentrations, the costs of exceeding the contaminant discharge criteria established by the POTW, and the physical proximity of the POTW. Thus, before decisions are made on whether to treat the leachate partially or fully on-site or off-site, it is first essential to understand the sources of both periodic and long-term variability.

The features that must be considered in the treatment options therefore include:

- The ability to recirculate leachate through the landfill mass
- The proximity of a sanitary sewer
- Trucking haul distance
- Distance and cost to construct a pumping system and sewer forcemain to transport leachate to POTW
- Wastewater treatment plant capacity and ability to treat the leachate
- Leachate strength
- Local sewer-use bylaw regulations
- Sewer surcharges
- Surface water discharge standards for on-site treatment.

Temporal Variations in Flow   The temporal variability in flow can be very substantial, as indicated in Figure 11.6(a) (after Robinson and Grantham, 1988). The variability in the initial stages occurs in part as various portions of the site attain field capacity. Thus, in the initial stages, there may be little or no leachate produced, although infiltration through the landfill surface continues to occur. In addition, there is the variability in response to changes in precipitation inputs from year to year, which, in turn, enter the refuse, percolate down (which takes time), and are collected in the leachate collection system to be treated. Unfortunately, the highest rates of leachate generation usually occur during the winter and spring months, when temperatures are lowest. Thus, for some locations, the low temperatures that inhibit biological treatment facilities, in general affect the utility of several types of treatment for the leachate (e.g., Robinson and Grantham (1988) reported that the majority of leachate is generated during the winter and very little during the summer due to extensive evapotranspiration).

In response to the variability in leachate flow, the general appro[...] leachate treatment typically include the following:

1. Equalize the flow by temporarily storing leachate in storage tanks or ponds.
2. Treat at a constant rate of flow and recycle the excess leachate flow back to the landfill during periods of high flow.
3. Utilize a relatively flexible treatment system in which components of the treatment system are scheduled to be utilized as the flow over the long-term grows and/or during periods of high flow.

Temporal Variations in Concentration Major concerns with potential pollution from raw or partially treated leachate include the organic content, the ammonia nitrogen (NH_3-N) content, and the heavy metals content. The most important quality parameters monitored are therefore BOD_5, COD, volatile fatty acids, NH_3-N, and metals.

Figure 11.6 provides an example of the variations in flow, temperature, and concentrations of contaminants over a 2 year time period. The types of variations depicted in Figure 11.7 for BOD and ammonia occur over a sizable time frame, measured in years. The BOD concentrations in the early years are measured in the thousands. Conversely, after a duration of perhaps 10 years, the concentrations of BOD are measured in hundreds. Thus, a leachate treatment system constructed to treat a young leachate will require a very different set of treatment processes than one constructed to treat an old leachate. Some of the differences between young and old leachates are summarized in Table 11.5.

Table 11.5 EXAMPLES OF DIFFERENCES BETWEEN YOUNG AND OLD LEACHATES

Young Leachates	Old Leachates
BOD measured in 1000s (mg/L) consists of volatile fatty acids (acetic acid, propionic acid, and butyric acid)	• BOD in 100s (mg/L) consists of humic and fulvic acids • COD in 1000s (mg/L)
BOD_5/COD ratio in the 0.4 to 0.8 range	• Many contain some priority organics such as toluene, benzene, and methylethylketone
BOD_5/NH_3–N ratio>>1.0 (After Forgie, 1988a)	• BOD_5/COD ratio approaches 0.1 • NH_3-N increases • BOD_5 decreases

As a result, knowledge of the age of the refuse that is generating the leachate is of great importance when designing a leachate treatment system. As apparent from Figure 11.7, some constituents such as BOD peak relatively quickly and decline over time. Other constituents increase to a plateau and tend to remain at a relatively constant level for a lengthy period of time. Therefore, the alternatives for treatment of landfill leachate differ depending on whether the leachate is derived from waste that has been relatively recently placed or

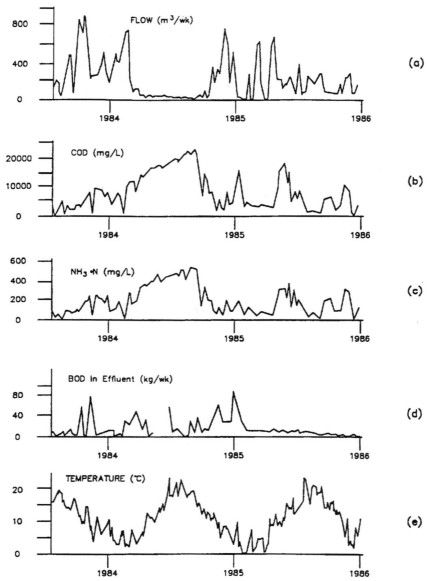

Fig. 11.6   Leachate production at Bryn Posteg Landfill.

from refuse that has been buried for a number of years. Young leachate is amenable to a number of types of biological treatment for reduction of its high but easily biodegradable organic content. If the leachate does not meet the characteristics of a young leachate, it is likely that there has been at least some conversion of the volatile fatty acids to methane within the landfill, and treatment will require a different set of processes.

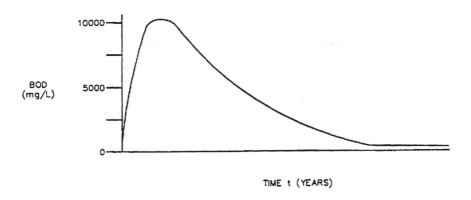

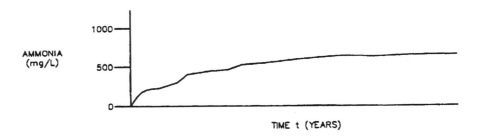

Fig. 11.7   Example of variations in concentrations vs. time.

Leachates are many times stronger than municipal sewage and thus leachates produce much higher quantities of organic sludge in the form of micro-organisms. In addition, leachates also produce high quantities of calcium, iron, and manganese, comprising greater than 50 percent of the sludge dry weight (as opposed to less than 10 percent in sewage sludge) (after Robinson and Grantham, 1988).

11.5   OVERVIEW OF ALTERNATIVE TREATMENT PROCESSES

With the exception of a short period of time after the refuse is placed, anaerobic conditions exist within the landfill during leachate production. Consequently, concentrations of nitrate and nitrite are generally low, since the oxygen has been utilized in the anaerobic decomposition. Leachates also contain high concentrations of soluble organic matter and inorganic ions. Before we examine how the variabilities affect the treatment processes and which treatment processes have the greatest potential for being cost-effective, we first briefly examine some of the attributes of the constituents to be removed in terms of treatability.

11.5.1 Chemical Composition of Leachate

Organics   Organics are characterized by the carbon–carbon bond.   One means of classifying the organics is into the three groupings indicated in Table 11.6.   There is increasing stability with respect to biodegradation from Group A through C.

Table 11.6   ORGANIC COMPOUND GROUPINGS

Group A Fatty Acids	*Group B Humic Acids*	*Group C Fulvic Acid–Like Substances*
Low molecular weight Examples: •acetic •propionic •butyric	**High molecular weight** Carbohydrate-like substances •carboxyl and aromatic hydroxyl groups	**Intermediate molecular weight**

Chian and deWalle (1975) conducted in-depth analyses to establish the composition of the organic fraction of the leachates. Their research revealed that "the volatile free fatty acids constituted from 20 to 70 percent of the total organic carbon (TOC) in leachate, depending on the age of the landfills." The percentages of these volatile fatty acids tend to decrease as the age of the landfill increases.   The remainder of the organics was distributed approximately in equal amounts between the groups of compounds of fulvic acids, tannic acids, lignins, celluloselike material, and other organic matter.

Studies by Qasim and Burchinal (1970) showed that concentrations of organic substances (TOC, COD, and BOD) and the ratio of BOD/COD are generally highest during the active stages of decomposition and gradually decrease as the landfill stabilizes.

Analyses of leachate samples collected by Chian and deWalle (1975) from more stabilized landfills confirmed a general decrease in volatile fatty acids in leachate over time and showed that most of the organic material was present as fulvic-like refractory molecules. Only 0.5 percent of the TOC was present as humic material of high molecular weight. These authors suggested that such old or stabilized leachates might be best treated by physical and chemical processes as opposed to biological treatment processes.

Nitrogen   Ammonia (NH_3-N) and organic-N, collectively referred to as total Kjeldahl nitrogen (TKN), represent a high percentage of the total soluble nitrogen compounds in leachate. Combined, they are typically in the hundreds of milligrams per liter and may be considerably higher.   Owing to the anaerobic conditions within landfills, concentrations of nitrite and nitrate are typically low.

If the resulting NH_3-N content is too high (e.g., greater than 1000 mg/L), nitrification may be inhibited. Nitrifying bacteria are also relatively sensitive to low temperatures. As a result, it may be necessary to partially reduce the level to a more acceptable concentration by physical-chemical methods prior to use of biological treatment methods.

Phosphates   Landfill leachate is frequently deficient in phosphorus for effective biological treatment. The highest concentration of soluble phosphorus reported in Robinson and Maris, 1979, was 0.41 mg/L, and concentrations below 0.1 mg/L were common. Ratios of BOD:P of greater than 7000:1 have been found for leachates from recently emplaced wastes (Robinson and Maris, 1979). Since an optimum ratio of BOD:P of 100:1 is widely recommended for biological wastewater treatment processes (Metcalf and Eddy, 1991), biological treatment of these leachates will be inhibited owing to phosphorus deficiency unless additions of phosphorus are made.

pH   Particularly during the acid-phase stage of decomposition, pH levels are acidic, so they will require adjustments to allow biological treatment. In addition, if removal of metals is required, pH will have to be adjusted to form precipitates.

Heavy Metals   In general, removal of metals will occur if either aerobic or anaerobic biological treatment is used (Forgie, 1988a) because the metals will precipitate out. If anaerobic treatment is used, the metals will tend to precipitate out as metal sulfides, whereas if aerobic treatment is selected, the metals will tend to oxidize and precipitate out as metal hydroxides. However, there may be specific instances when high concentrations of certain metals such as copper, zinc, and nickel may cause biological inhibition. In such cases, chemical precipitation may be needed.

On the negative side, however, removal of certain metallic ions such as Zn, Fe, Mn, and Pb in conjunction with sludge formation, either biologically or chemically, may lead to problems with sludge disposal. Because the sludges tend to concentrate the metals to two and three orders of magnitude greater than their influent concentrations, disposal problems may arise.

Dissolved Solids   Leachate typically has high levels of total dissolved solids (e.g., chlorides, sulfates, sodium). These constituents are not very reactive and therefore not easily removed. In the event of the need for removal, physical/chemical processes will be required.

11.5.2 Treatment Alternatives

Given the varying treatability of the different chemicals within leachate, it is not surprising that a number of different treatment alternatives exist. The alternatives include aerobic or anaerobic biological processes, and physical and chemical treatment methods. In addition, it is possible to recycle the leachate through the landfill as a partial means of treatment and then spray the leachate onto land as a disposal method. The focus of an individual treatment alternative is typically toward an individual chemical, with the entire system being sequenced to collectively treat the array of contaminants within the leachate. Table 11.7 summarizes some of the relationships.

Table 11.7 SUMMARY OF TREATMENT PROCESSES FOR THE MAJOR CONSTITUENTS IN LEACHATE

Chemical	Commentary	Likely Treatment Process
Organic strength	Young leachate—BOD in 10,000s (mg/L) -in the form of volatile fatty acids -amenable to biological treatment processes, barring any toxic inhibition	-Biological treatment
	Old leachate -BOD in 100s and COD in 1000s -in the form of humic and fulvic acids	-Carbon adsorption
Ammonia	-in the 1000s -some will be removed in bio-uptake	-biological nitrification/denitrification -Air stripping -Breakpoint chlorination -Ion exchange
Heavy metals	-in the 10s and 100s -iron (Fe) mainly, Zn, Pb, Cu also	-Chemical precipitation -Biochemical treatment
Phosphorus	-in 10s (mg/L)	-Supplementary additions needed for biological activity
pH	-leachate usually acidic	-Neutralization using lime or caustic
Conservative ions	Cl^- and SO_4^{2-} in 1000s (mg/l), K^+ and Na^+ in 1000s (mg/l), leachates typically have high total dissolved solids (eg., chloride, sulfate, sodium)	-Reverse osmosis and ultrafiltration

The treatment alternatives to be considered for application in a specific situation are a function of the necessary quality of the effluent at discharge. On-site treatment to allow discharge of the treated water to a nearby water body will necessarily involve much more extensive treatment than the alternative of partial treatment prior to discharge to a POTW.

11.6   BIOLOGICAL TREATMENT

The focus in biological treatment is to change the form of the organic constituents. The downside of biological treatment is that it may produce relatively large quantities of biomass sludge requiring subsequent disposal.

For a leachate with a high BOD:COD ratio (>0.4) the only major treatment options are aerobic or anaerobic biological treatment. Table 11.8 lists some of the alternative approaches. The first column differentiates between those processes in which the microorganisms are (1) suspended or (2) attached (e.g., via a film) to some type of fixed surface. Figure 11.8 illustrates the different types of aerobic and anaerobic approaches.

If the BOD is greater than 50 mg/L, treatment by biological treatment is essential. Biological treatment is very effective for reducing the biodegradable organics as characterized by BOD_5 (if BOD_5 represents the main part of COD).

Considerable bench-scale research has been conducted to determine whether concentrated leachates can be treated in biological treatment systems. Results from bench-scale "fill-and-draw" activated sludge studies conducted by Boyle and Ham (1974), Cook and Foree (1974), Uloth and Mavinic (1977), and Chian and deWalle (1977b) indicated that BOD_5 removals were excellent. Effluent COD values were high; however, Chian and deWalle (1977b) established that in many cases these may be primarily humic, fulvic, and tannic acids, and lignins, which are biologically inert.

Along with BOD_5 removal, biological treatment will assist with removal of suspended solids by sedimentation, NH_3-N and organic-N by bio-uptake (and to nitrification if aerobic treatment is employed), and of metals by biosorption and precipitation as oxides and carbonates.

As the pH and redox potential (Eh) increase, metals change from a lower oxidation state and higher solubility to a higher oxidation state and lower solubility (e.g., Fe^{2+} is much more soluble than Fe^{3+}). If anaerobic treatment is employed, the metals will precipitate out as sulfides, whereas if aerobic treatment is employed, the metals will oxidize and precipitate out as metal hydroxides and carbonates.

The presence or absence of oxygen differentiates aerobic and anaerobic decomposition. An example of aerobic decomposition is the oxidation of carbohydrates to carbon dioxide and water.

$$C_6H_{12}O_6 + 6O_2 \rightarrow 6CO_2 + 6H_2O \tag{11.1}$$

One volume of CO_2 is produced for each volume of oxygen utilized. Fatty materials, such as stearic acid, are similarly oxidized.

$$C_{18}H_{36}O_2 + 26O_2 \rightarrow 18CO_2 + 18H_2O \tag{11.2}$$

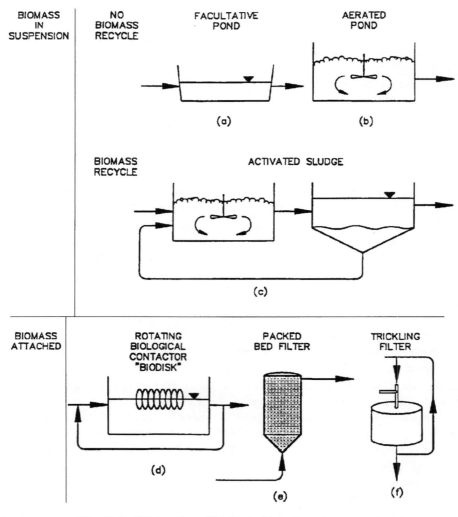

Fig. 11.8   Schematics of biological treatment process types.

Conversely, anaerobic microbial decomposition for the same initial materials is

$$C_6H_{12}O_6 \rightarrow 3CO_2 + 3CH_4 \qquad\qquad (11.3)$$

and

$$C_{18}H_{36}O_2 + 8H_2O \rightarrow 5CO_2 + 13CH_4 \qquad\qquad (11.4)$$

Table 11.8 SUMMARY OF BIOLOGICAL TREATMENT PROCESSES

Biomass	Treatment	Recycle	Comments
In suspension	Facultative ponds	no	–Involves both aerobic and anaerobic decomposition. Odors will result during process. –Care must be taken to avoid biomass washout
	Aerated ponds	no	–Effluent will be high in suspended solids –Aeration requires substantial energy –Care must be taken to avoid biomass washout
	Activated sludge	yes	–Less sensitive to shock loading –Care must be taken to avoid biomass washout
Attached	Rotating biological contactors	yes	–Less sensitive to biomass washout
	Packed filter bed	yes	–Less sensitive to biomass washout
	Trickling filter bed	yes	–Less sensitive to biomass washout

As apparent from Equations (11.3) and (11.4), oxygen is not necessary for the microbial decomposition.

Many treatment systems include both aerobic and anaerobic components as part of the overall treatment system, taking advantage of the beneficial features of each type of biological treatment.

11.6.1 Aerobic Treatment

Fatty acids are a product of the anaerobic decomposition of organic materials within the landfill and thus are in the leachate. However, these fatty acids are easily biodegradable using aerobic processes, barring any toxic inhibition. They do, however, require very large quantities of oxygen, and the processes generate huge quantities of biomass that must be disposed of.

Aerobic biological treatment processes include lagoons, activated sludge facilities, rotating biological contactors (RBCs), and trickling filters. All the aerobic processes work on the same principle—microorganisms acting on organic matter in the presence of oxygen. The processes are differentiated by whether the microorganisms are in suspension or fixed to a medium. Note that to keep the biological action in the aerobic phase, it is essential to supply large quantities of oxygen, particularly when treating young leachates with their associated high concentrations of organics. Thus, knowledge of the organic load of the leachate is essential for the design of an aerobic system. For example, organic matter consisting mainly of volatile fatty acids can be degraded readily by biological means. However, in leachates from older, more stabilized fills, a greater proportion of the organic fraction consists of refractory material; therefore, such leachates will be less amenable to biological treatment.

Because the organic load of the leachate varies with time, it is necessary for the system to adapt by varying the air supply. This can be accomplished by installing a diffused aeration system, where the air is supplied by several blowers, or by use of mechanical aerators. The use of diffused air is also advantageous during periods of very low temperatures, to minimize freezing. However, the diffused air must be supplied by large bubble aerators to avoid clogging as a result of precipitates. Some typical design guidelines for aerobic treatment are summarized in Table 11.9.

Table 11.9 TYPICAL DESIGN FEATURES FOR AEROBIC TREATMENT SYSTEMS

Consideration	Comments
Organic loading	Food: maintain the microorganism ratio at<0.3{(kg BOD)/(d kg VSS)} where VSS refers to volatile suspended solids concentrations.
Solids retention time	Design for large solids retention times, for example: >10 days at (20° C) and, >20 days at (10° C).
Sludge production rates	Upwards of 1 kg/(kg BOD removed). Plan for appropriate disposal of significant quantities of residual biomass. One kilogram of residual biomass per kilogram of BOD removed will be produced, which is approximately twice what a normal municipal wastewater treatment will produce. Approximately one-half of the residential biomass arises from the BOD removal and one-half from precipitates.
Oxygen supply	Provide for the high oxygen requirement of young leachates.
PO_4–P supplement	Usually required due to a phosphorus deficiency in the leachate. Biomass uptake of nutrients will occur in the approximate ratio of BOD:N:P = 100:5:1
NH_3–N conversion	Biomass uptake at ratio of BOD:N:P = 100:5:1 Nitrification may dictate design for old leachates Ammonia may be inhibitory to biomass growth at high concentrations
Recycle of biomass	Not required for high-strength leachate but is required for low strength leachate to retain the biomass
Precipitate formation	$CaCO_3$ and Fe_2O_3 precipitates can coat impellers and aeration components Consider use of equalization tanks to allow pre-precipitation
Sequencing batch reactors	Use for flows less than 100 m³ day.
Foam Control	Necessary inclusion in design.
Lagoons	Maintain loading at less than 6kg BOD_5/100m³/d.

Table 11.9 TYPICAL DESIGN FEATURES FOR AEROBIC TREATMENT SYSTEMS

(Table Continued)

Consideration	Comments
Activated sludge	Maintain temperature of >10 °C for activated sludge systems. Maintain a loading rate of less than 0.01kg BOD_5/kg mixed liquor volatile suspended solids or MLVSS/d, where MLVSS refers to mixed liquor suspended solids for effective treatment using activated sludge. Note: Good nitrification and good metals removal will be obtained within the specified conditions. [e.g., Uloth and Mavinic (1977) reported high percent removal of some metals in the settled biological floc, especially iron (>98), zinc (>99), calcium (>93), manganese (>95), cadmium (>96), lead (>79), and magnesium (>54)]

In addition to degrading organic carbon, biological treatment is also an effective way to incorporate ammonium into the biomass or to oxidize it to nitrate during aerobic decomposition followed by denitrification to gaseous nitrogen during anaerobic decomposition. As a result, aerobic biological treatment is frequently an important component of a leachate treatment sequence.

Aerated Lagoons Aerated lagoons or ponds are aerated by a series of aerators (mechanical or diffuser) as indicated in schematic form in Figure 11.8(b). The diffused air can be used to create mixing as well as to supply the oxygen. Note that aerobic lagoons have a limited ability to effectively handle increasing leachate strength.

Chian and deWalle (1977b) showed that, in general, most of the organic matter in effluents from aerated lagoons consists of stable refractory materials, often with a high molecular weight. These effluents were similar to leachates from older, relatively stabilized landfills.

Table 11.10 gives data obtained from various studies showing the treatment effectiveness of aerated lagoons. Another case study by Robinson and Luo (1991), using pilot-scale treatment facilities for a Hong Kong landfill, reported that NH_3-N was present at levels up to 5000 mg/L, and high treatment standards were met by aerobic systems.

Robinson (1990) reported effective treatment of high ammonia concentrations in the influent from leachate in southwest England, using a 15-day retention period and low-speed floating surface aerators. Conventional activated sludge systems, with much shorter hydraulic retention times (hours instead of days,) lack the substantial dilution and buffering that an extended aeration system provides against short-term variations in influent quality, flow, and ambient temperature.

Safferman and Bhattacharya (1990) reported that an extended aeration facility effectively removes low concentrations of Resource Conservation Recovery Act(RCRA)-specified organic compounds without interfering with BOD_5,

Table 11.10 SUMMARY OF RESULTS OF LEACHATE TREATMENT IN PILOT AERATED LAGOONS

Study	BOD (mg/l)		T (·C)	Retention Time, t_0 (days)	Comment
	In	Out			
Boyle & Ham (1974)	2900	200	24	5	Experienced foam problems
Cook & Foree (1974)	7100	26	22	10	Lime + phosphates, P, added
Uloth & Mavinic (1977)	36,000	32	20	20	N + P added Foam problems experienced
Chian & deWalle 1977(b)	35,200 (COD)	1030 (COD)	24	7	P added COD measured, effluent contained fulvic acid and tannic acid
Zapf-Gilje (1979)	13,600	26	25	6	

ammonia, and suspended solids removal. They used a pilot-scale extended aeration system with dual-media secondary effluent filters in which the hydraulic retention time of each system was 16.8 hours, and the average solids retention time was 21 days.

Activated Sludge The flow diagram of Figure 11.8(c) indicates the recirculation characteristic of the activated sludge process. The detention time in activated sludge plants can be considerably shorter than in aerated lagoons because the bacteria levels are three to five times higher and can be controlled to a considerable degree through recirculation. The recirculation is achieved by using a settling tank following the aeration tank and recirculating the sludge back into the aeration tank. The treatment effectiveness of the activated sludge facility as reported in Ehrig, 1989, is summarized in Table 11.11.

Table 11.11 EXAMPLE OF LEACHATE TREATMENT IN FULL SCALE ACTIVATED SLUDGE PLANTS

	Influent	Effluent
BOD_5 (mg/L)	5294	254
COD (mg/L)	12,359	1566

Source: After Ehrig,1989

Besides BOD_5 reduction, the nitrification of ammonium is an important aspect of the treatment in many activated sludge plants. Nitrogen elimination

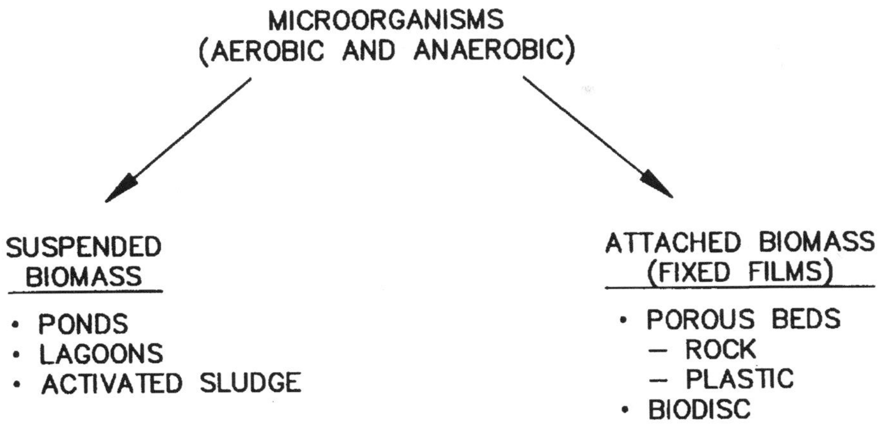

MICROORGANISMS
(AEROBIC AND ANAEROBIC)

SUSPENDED
BIOMASS

ATTACHED BIOMASS
(FIXED FILMS)

- PONDS
- LAGOONS
- ACTIVATED SLUDGE

- POROUS BEDS
  - ROCK
  - PLASTIC
- BIODISC

Fig. 11.9   Alternative biological treatment methodologies.

becomes more and more important with the aging of a landfill due to the increase in nitrogen levels. During aeration, the pH increases, with the result that equilibrium shifts from ammonium to free ammonia. Concentrations of free ammonia may have inhibiting effects on nitrifying bacteria.

Overloading during nitrogen degradation results in the nitrification of an increasing part only to nitrite, as opposed to nitrate. To reduce the high nitrate content in leachate effluent and to stabilize pH conditions in activated sludge plants, a denitrification step may be necessary.

Rotating Biological Contactors (RBCs) and Trickling Filters (Fixed-Film Reactors)   The RBC unit is a fixed-film biological process in which the biomass accumulates on rotating disks, as depicted in Figure 11.8(d). A fixed-film biomass system tolerates hydraulic or chemical shock loadings, which are common in landfill leachate, better than do suspended growth processes. However, RBC units tend to clog with calcium deposits that prevent the growth of the biomass. As a result, many treatment systems utilize NaOH preceding the RBC system to raise the pH and precipitate out the metals.

Alternatively, trickling filters, as depicted in Figure 11.8(f), use a fixed surface, such as rocks, as the medium on which the biomass grows, and provide a discontinuous trickling of the water being treated over the biomass.

The RBC and trickling filter processes (attached or fixed biomass to the surface of the material of the contactor or filter) differ from the activated sludge process (suspended biomass). A summary is schematically illustrated in Figure 11.9. Air is supplied naturally for both RBCs and trickling filters; for example, the rotating contactor is partly in the air and partly in the water while rotating. The air vents through a trickling filter from the bottom to the top [as depicted in Figure 11.8(f)]. These methods consume low amounts of energy. However, there are limitations to the treatment capability for leachates high in organics, since the precipitates and/or biomass will clog the system and/or may not supply oxygen at the rate needed to keep the system aerobic.

Nitrification processes are more effective in fixed-film reactors. Knox (1985) investigated trickling filters and demonstrated the importance of temperature (due to the sensitivity of biological treatment to temperature). He obtained good removal until loading reached more than $2 \text{ g N/m}^2/\text{d}$, at which point ammonium oxidation was incomplete, and increasing nitrite concentrations were produced.

Detrimental Aspects of Aerobic Treatment   The following are negative characteristics of aerobic treatment:

1. Aerobic treatment does not function well in the presence of toxic metals (e.g., Cu, Zn, and Ni all inhibit nitrification).
2. Typically, the BOD_5:P ratio in leachate is well above 100:1, requiring the addition of supplemental phosphorus to allow effective aerobic treatment.
3. It is difficult to maintain effective treatment when low ambient temperatures are encountered. This is particularly critical during periods of low water temperature, due to the decreasing elimination rates of nitrogen.
4. Leachate foaming problems frequently occur during aeration due to increased surface tension, requiring the use of a defoaming mechanism.
5. The potential exists for $CaCO_3$ and/or iron precipitation, which causes operational difficulties with aeration equipment. As the redox potential increases, the oxidation state increases, and calcium and iron become less soluble.
6. There is the potential for very high NH_3-N concentrations. If the NH_3-N content is too high (e.g., greater than 1000 mg/L), nitrification may be inhibited. As a result, the ammonia concentration must be partially reduced to a lower level by physical-chemical means (e.g., air stripping) before nitrification can proceed (Keenan et al., 1984). Specific inhibitory metals may also require leachate pre-treatment to raise the pH by, for example, NaOH addition.
7. Energy to maintain oxygenation is expensive. Over the lifetime of a treatment plant, the operation and maintenance costs can easily exceed initial capital costs.

Ammonia concentrations in the effluent from aerobic treatment facilities can be very low when nitrification has occurred; nitrate may be correspondingly high. If there is a nitrate concentration limitation for the effluent, biological denitrification (using anaerobic decomposition) may have to be included in the treatment process train.

Summary of Concerns Regarding Aerobic Treatment Processes   Aerobic treatment systems are effective for young leaches when the BOD:COD ratio is greater than 5. Leachates from recently placed refuse may be treated satisfactorily by aerobic processes on a small scale. However, aerobic biological treatment of leachates

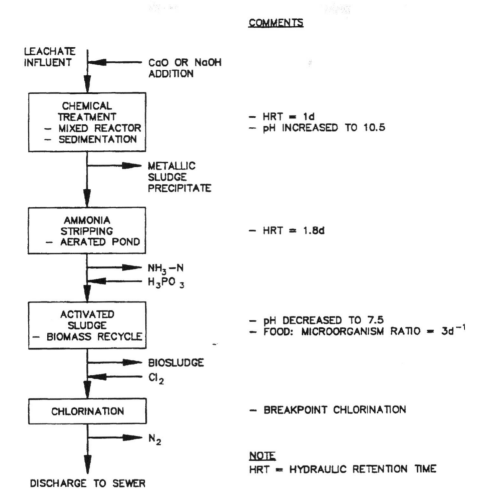

COMMENTS

LEACHATE
INFLUENT ◄──── CaO OR NaOH
ADDITION

CHEMICAL
TREATMENT
— MIXED REACTOR
— SEDIMENTATION

— HRT = 1d
— pH INCREASED TO 10.5

──► METALLIC
SLUDGE
PRECIPITATE

AMMONIA
STRIPPING
— AERATED POND

— HRT = 1.8d

──► NH_3–N
◄── H_3PO_3

ACTIVATED
SLUDGE
— BIOMASS RECYCLE

— pH DECREASED TO 7.5
— FOOD: MICROORGANISM RATIO = $3d^{-1}$

──► BIOSLUDGE
◄── Cl_2

CHLORINATION

— BREAKPOINT CHLORINATION

──► N_2

NOTE
HRT = HYDRAULIC RETENTION TIME

DISCHARGE TO SEWER

Fig. 11.10   Example of aerobic leachate treatment system for leachate prior to discharge to sanitary sewer.

will not be successful at high organic loadings and low retention periods without the addition of nutrients and large aeration rates. Most of the organic matter in effluents from aerated lagoons consists of stable refractory materials, often with a high molecular weight. Figure 11.10 summarizes an example (after Fungaroli and Steiner, 1979) of a leachate treatment that includes aerobic treatment as an essential component that would be effective prior to discharge to a sanitary sewer.

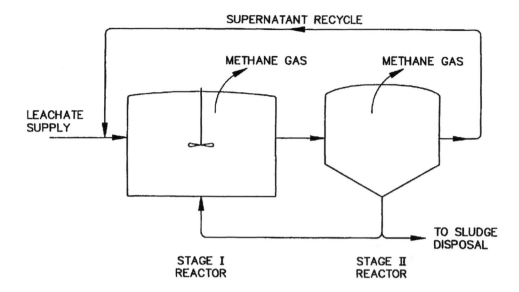

Fig. 11.11   Schematic of anaerobic treatment in a two-stage reactor.

11.6.2 Anaerobic Treatment

Anaerobic biological activity (i.e., in the absence of dissolved oxygen) is the natural degradation process that occurs within a sanitary landfill. This can be followed by further anaerobic treatment in the leachate treatment facility.

In an anaerobic wastewater treatment system, complex organic molecules in the influent wastewater are fermented by bacteria to volatile fatty acids, mainly acetic, propionic, and butyric. These in turn are converted by methanogenic bacteria to methane and carbon dioxide, resulting in a low production of biological solids requiring disposal. (A similar sequence of conversions occurs during the anaerobic decomposition of refuse within the landfill).

Figure 11.11 depicts an example of anaerobic leachate treatment using a two-stage system. Note the enclosed chambers to prevent oxygen entry and to allow capture of the resulting gases. The Stage I reactor is stirred to ensure good contact between the microorganisms and the organic substrate. Heat is sometimes applied to assist with the speed of biodegradation. The second vessel is quiescent to allow improved separation of the leachate into a supernatant (which is recycled back to Stage I for further treatment) and a sludge. Methane gas is produced in both vessels. This system is seldom used; most new systems are now designed as upflow systems, described below.

Many other configurations for anaerobic treatment are also employed—the primary key is to avoid oxygen access, thereby keeping the biological action anaerobic. A disadvantage of anaerobic treatment processes is that the microorganisms (methanogenic bacteria in particular) are easily inhibited by acidic pH values and are also sensitive to the presence of some metals (Mosey and Hughes,

1975). These inhibitions may cause reduced growth rates and lead to a net washing-out of microbial cells from a completely-mixed reactor system. Such problems may be overcome by the addition of a buffer solution to the influent leachate and by the use of an anaerobic filter.

Anaerobic filters, illustrated in Figure 11.8(e), slowly pass the leachate through a medium. The filter is kept submerged, and a film of anaerobic bacteria builds up on the surface of the filter material. Because of the low yield of the anaerobic process, bacteria are retained in the filter for a lengthy period of time. Filters are non-labor-intensive and are widely utilized because they represent a simple treatment system.

Use of anaerobic fixed-film reactors is a significant trend. The films have been shown to be more effective than digesters. The potential for biomass washout is reduced, with the result that higher loadings are possible, (e.g., 20 versus 2 kg COD/m^3). Table 11.12 gives examples of treatment efficiencies of fixed-film reactors.

Table 11.12   EXAMPLES OF ANAEROBIC TREATMENT EFFICIENCY USING FIXED FILM REACTORS

Source	Leachate Strength (mg/L)	COD Removal (%)	Support Medium
Chian and deWalle (1977a)	54,000	95	Granular carbon
Henry et al., (1985)	2000	91	Plastic film
Soyupak (1979)	5800	90	Sand
Wright et al., (1985)	22,800	97	Plastic rings

Anaerobic biological processes for the treatment of leachates have several potential advantages over aerobic biological processes. The advantages include the generation of methane gas as a byproduct and the much lower production of biological solids in the form of sludges or suspensions. In addition, the systems have no need of aeration equipment and its considerable energy requirements.

Anaerobic processes need adequate temperatures, and they must be designed carefully, otherwise the filters may clog (e.g., greater than 90 percent BOD removal is obtained with a solids retention time greater than 10 days when temperatures are greater than 15°C). Depending on the local meteorologic conditions, some form of heating above ambient winter conditions may be necessary.

Disadvantages of anaerobic treatment include:

- Higher temperatures, for example, 15 to 35°C are necessary for optimal treatment
- Relatively long retention times
- Incomplete organic removals
- Limited NH$_3$-N reduction.

If BOD$_5$ concentrations are very high, then the effectiveness of the anaerobic treatment processes is a great advantage, due to its low energy requirement.

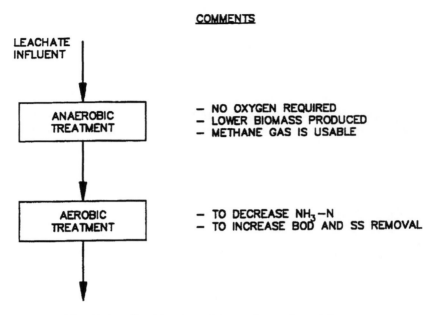

Fig. 11.12   Combination of anaerobic and aerobic treatments.

However, a combination of anaerobic treatment followed by aerobic treatment is frequently the most effective approach, as summarized in schematic form in Figure 11.12.

Effluents from anaerobic biological leachate treatment are typically similar to older natural leachates, that is, a BOD_5:COD ratio less than 0.3, COD in the 1000 to 3000 mg/L range, and relatively high NH_3-N content (e.g., 500 to 1000 mg/L). As a result, it is unlikely that anaerobic leachate treatment alone will be sufficient to permit discharge.

The greater efficiency of filters when compared with corresponding completely mixed digesters is explained by the fact that microorganisms are largely retained within the filter, whereas they may be lost in the effluent from a digester. For example, Foree and Reed (1973) found the suspended solids concentrations in the filter effluent to be approximately 40 mg/L, whereas they reached levels of 2754 mg/L in the effluent from the digester.

Chian and deWalle (1977(c)) examined removal of toxic metals from leachate in an anaerobic filter. They concluded that the percent removal of iron, zinc, nickel, cadmium, lead, and chromium increased with increasing concentrations of metals in the leachate, and also with increasing hydraulic retention time. Metals were precipitated as sulfides, carbonates, and hydroxides.

Wright and Austin (1988) used a facility involving upward flow through a sludge blanket and media layer. The influent flow was mixed with recirculated reactor contents and proceeded through a heat exchanger into the distribution piping at the bottom of the reactor. Flow moved upward through a 2 m anaerobic sludge blanket and then through a 1 m transition area before entering the media. The media was a cross-flow block type installed in a 3 m layer. An overflow area above the media provided further clarification and degasification of the liquid before it was discharged.

11.6.3 Summary of Biological Treatment

Anaerobic leachate treatment is an effective process, but the effluents still have high COD values of 1000 to 4000 mg/L and a BOD_5:COD ratio greater than 0.3 (Ehrig, 1989). Thus, after the anaerobic treatment, the leachate usually has to be treated aerobically to final effluent standards. Both aerobic and anaerobic treatment have limits with respect to the removal of organics. As the easily degraded organics are removed, those that remain are increasingly more difficult to degrade. As a result, the BOD_5 content decreases much more rapidly than the COD content, causing the BOD_5:COD ratio to decrease.

11.7   PHYSICAL AND CHEMICAL TREATMENT

As the landfill stabilizes, there is a decrease in the proportion of readily biodegradable organic compounds contained in the leachate. The effectiveness of biological leachate-treatment processes therefore decreases as the landfill waste stabilizes, and other forms of treatment such as physical and chemical techniques may become more appropriate.

Therefore, physical and chemical treatment alternatives represent an addition to, or replacement for, the aerobic and/or anaerobic biological treatment of landfill leachate. A standalone physical/chemical system is applicable only for very old leachates. Physical and chemical methods that have been used to treat leachate (or to 'polish' biologically treated leachate) include the addition of chemicals to precipitate, coagulate, or oxidize inorganic and organic fractions; adsorption by activated carbon and ion exchange resins; and treatment by reverse osmosis membrane techniques. Precipitants and coagulants (lime, ferric chloride, and alum) have little effect on high concentrations of organic matter. There are numerous possible applications of physical/chemical treatment.

An example of a possible sequence of physical/chemical processes is depicted in Figure 11.13. This sequence is frequently utilized in treating groundwater contaminated with solvents and heavy metals.

Granular Filtration   Granular filtration removes suspended solids and is typically employed prior to the use of activated carbon to prevent clogging of the carbon by the suspended solids. Granular filtration of biologically treated landfill leachate may also be needed to meet stream quality discharge standards for suspended solids.

Carbon Adsorption   Treatment through sorption onto activated carbon generally occurs in a column configuration, although dispersed carbon treatment systems also exist. Columns may be downflow or upflow, with the latter having

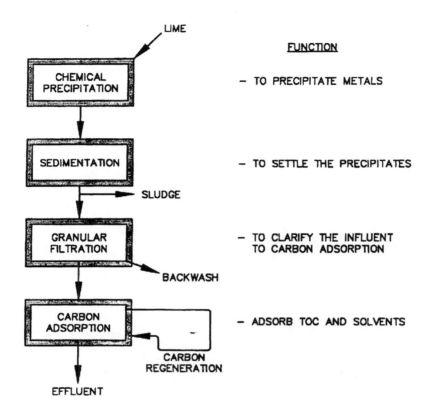

Fig. 11.13 Physical/chemical process sequence.

either packed or suspended carbon. Carbon treatment of raw young leachates generally yields poor TOC removal due to the poor affinity of carbon for free volatile fatty acids. Use of granular activated carbon or powdered activated carbon has been shown to be very effective for treatment of poorly biodegradable organics, solvents, pesticides, humic acids, and the like. Carbon adsorption can thus be used to reduce COD in old leachates or to remove color and refractory organics contributing to residual COD. Effluent CODs in the range from 0 to 100 mg/L are obtainable.

Chemical Precipitation If the raw leachate is treated biologically, either anaerobically or aerobically, heavy metals will be removed either as metal sulfides in anaerobic treatment or as metal hydroxides in aerobic treatment. If no biological treatment is involved in the treatment process sequence, or if the metal content is still too high, the treatment options include chemical precipitation with lime or caustic, aeration, or chemical oxidants such as chlorine, hydrogen peroxide, or potassium permanganate. Experiments have shown that dosages of chemicals on the order of 1000 mg/L will remove some high molecular weight organic species.

Table 11.13   OPERATION PARAMETERS AND TREATMENT RESULTS OF THREE REVERSE OSMOSIS PLANTS

	(a)	(b)	(c)
No. of Stages	1	2	1
Operation pressure (kPa)	360	–	400
CF	4.9	~4	5
Permeate Flow Rate(L/m²/hr)	12.6	≈ 12	14.5
Permeate Concentration			
BOD_5 (mg/l)	112	<5	–
COD (mg/l)	456	<10	–
NH_4 (mg/l)	564	<10	–
Cl (mg/l)	1020	<30	–
AOX (μg/l)	<110	–	–
Elimination			
BOD_5 (%)	89	>95	98
COD (%)	89	>99	99
$NH_4{}^+$(%)	58	>98	94
Cl^-(%)	43	>98	95
AOX (%)	93	–	–

Source: Grosse, 1988; Logemann, 1987; Ryser, 1985

Notes:

a: Rastatt (FRG)
b: VAM (Netherlands)
c: Uttigen (Switzerland)
CF: concentration factor = (permeate flow)/(concentrate flow)

Ultrafiltration   Ultrafiltration is an effective means of removing high molecular weight material from leachate. However, smaller molecular weight species frequently escape through the filter. Pretreatment with a biological process improves the effectiveness of ultrafiltration by removing particulate matter that would otherwise foul the filter.

Reverse Osmosis   Reverse osmosis is a separation process that may be appropriate in specific leachate treatment applications. With new developments in membrane material, it is possible to get very good quality effluent with the exception of acetic-phase leachate. The acetic-phase leachates have very small organic molecules that tend to pass through ultrafiltration and are not retained in reverse osmosis (Forgie, 1988b). Conventional cellulose acetate membranes, despite their excellent salt preclusion, tend to be highly pervious to acetic acid.

In these cases, biological pretreatment is required. Table 11.13 presents technical and operational data and elimination rates of several reverse osmosis plants (after Ehrig, 1989). deWalle and Chian (1977) found reverse osmosis to be the most effective method of removing COD, but fouling of the membrane was a continuing problem. Additionally, the highly concentrated rejection liquids produced by reverse osmosis require disposal.

Breakpoint Chlorination The use of breakpoint chlorination provides a removal mechanism for nitrogen. Specifically, as chlorine is added, readily oxidizable substances (e.g., Fe^{2+}) and organic matter react with the chlorine. After meeting this immediate demand, the chlorine continues to react with ammonia to form chloroamines. With continued chlorine additions, some of the nitrogen is released as N_2 gas. Formulas that indicate possible reactions for this are presented in Metcalf and Eddy (1991, p. 335).

Air Stripping Air stripping can be used in conjunction with anaerobic treatment to remove ammonia from the effluent. However, the air stripping towers can be prone to freezing in the winter and carbonate encrustation of the tower packing.

Air stripping can also be effective prior to biological treatment to remove ammoniacal nitrogen concentrations. An advantage of air stripping is that it can be adjusted relatively easily to changes in volume and strength of the leachate. A disadvantage of air stripping is that the efficiency drops off dramatically at low temperatures (e.g., 0–5°C). Also, the air discharge often has to be treated with activated carbon.

Ion Exchange The success of removal of organic matter by ion exchange is strongly dependent on organic matter type and the ion exchange resin being used. Most investigators have found relatively poor removal efficiencies, particularly with the low molecular weight organic compounds typical of raw young leachates. In contrast, ion exchange has achieved excellent COD removal from effluents from aerobic biological treatment systems.

In general, the most difficult ions to remove are the less reactive ones such as those of the alkaline earth metals: K^+, Na^+, Ca^{2+}, and Mg^{2+}. Being less reactive, they are also less responsive to treatment. Under batch equilibrium conditions, up to 95 percent removal of these ions has been achieved through the use of ion exchange processes. For feed liquids of less than 1000 mg/L, ion exchange is usually less expensive to use than reverse osmosis. For concentrations in excess of 1500 mg/L, reverse osmosis is usually less expensive.

11.8 TREATMENT SEQUENCES

11.8.1 Overview of Treatment Process

Due to the array of constituents within leachates, no single treatment alternative is suitable. Instead, a combination of processes is used, with each process playing a specific role in treating the leachate.

To determine what treatment system components are needed, a series of essential steps have been depicted in Table 11.14. However, typically there is a series of water quality constraints; thus, the most effective treatment is actually a sequence of treatment processes. Which components are necessary for a specific application are very much a function of disposal opportunities: whether the treated leachate is to be discharged to a POTW, whether pretreatment is necessary prior to discharge to the POTW, or whether the leachate is fully treated on site to the extent that it can be acceptably discharged to surface water or soil. The determination of which option to follow must be based on economics. If the leachate is to be treated in an existing POTW, total costs will be ery much influenced by costs for transportation and delivery of the leachate to t e POTW.

Table 11.14   STEPS IN PLANNING THE TREATMENT AND DISPOSAL OF LEACHATE

1. Estimate leachate flow, Q, using the HELP model or other water balance method.

2. Estimate the leachate contaminant concentrations for each constituent given the age of the site.

3. Identify the treatment and disposal options given the leachate quality discharge constraints and costs.

4. Select the treatment and disposal system to reflect uncertainty and flexibility.

Forgie (1988a) suggested criteria for treatment-type selection decisions. When a leachate has a high COD (e.g., 10,000 to 30,000 mg/L), a relatively low NH_3-N (e.g., 200 mg/L), a BOD_5:COD ratio in the 0.4 to 0.3 range, and a significant concentration of low molecular weight volatile fatty acids, the young leachate is amenable to both aerobic and anaerobic biological treatment. Physical/chemical treatment (e.g., coagulation and flocculation) of this type of leachate is generally not appropriate because of the predominant low molecular weight nature of the organic material.

If the leachate does not meet the characteristics of a young leachate, it is likely that there has been at least some conversion of the volatile fatty acids to methane within the landfill. As a result of this conversion, the COD will be in the 1500 to 3000 mg/L range, and the BOD_5:COD ratio will be less than 0.4, reflecting a significant decrease in the biodegradable organic fraction in the leachate. In addition, such old(er) leachates will likely have a significant NH_3-N concentration due to the anaerobic decomposition of organic matter within the landfill.

Both aerobic and anaerobic treatment systems have limits with respect to the removal of organics (Forgie, 1988a). As the easily degraded organics are removed, those that remain are increasingly more difficult to degrade. If the BOD_5/COD ratio is in the 0.1 to 0.4 range and the NH_3-N content is high, aerobic biological treatment is appropriate because it can remove NH_3-N through nitrification in addition to reducing BOD_5 and COD. If the BOD_5:COD ratio is less than 0.1, it indicates that the remaining organics are becoming increasingly more difficult to degrade biologically. Therefore, the decision to use aerobic

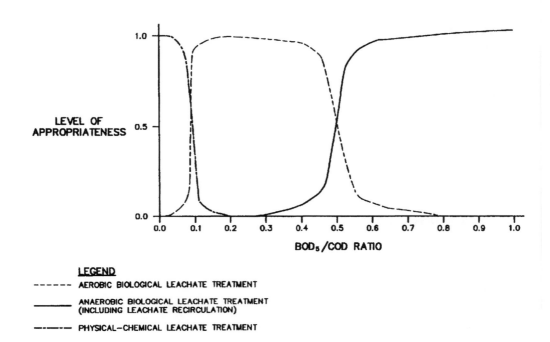

(AFTER FORGIE, 1988c)

Fig. 11.14 Proposed appropriateness levels for biological and physical-chemical leachate treatement.

biological treatment would be made only on the basis of NH_3-N removal requirements. At a BOD_5:COD ratio of 0.1 or less, the level of volatile fatty acids is lower, and physical/chemical treatment becomes the preferred alternative.

On this basis, it is possible to develop graphic interpretations of the degree of appropriateness of aerobic biological, anaerobic biological, and physical/chemical treatment over the range of likely BOD_5:COD ratios, as shown in Figure 11.14. Although the exact shape of the curves in this figure cannot at present be firmly fixed, their general shapes and threshold values are reasonable.

On a more general scale, it may be appropriate to utilize a sequence of treatment schemes. A possible treatment sequence, with an indication of different entry points into the train of processes, is illustrated in Figure 11.15.

The following is a description of the role of each component.

1. (If needed) pretreatment with caustic soda (NaOH) is used to adjust the pH and to precipitate the heavy metals. High calcium concentrations in the raw leachate discourage the use of lime due to scaling problems. The addition of NaOH has two purposes:

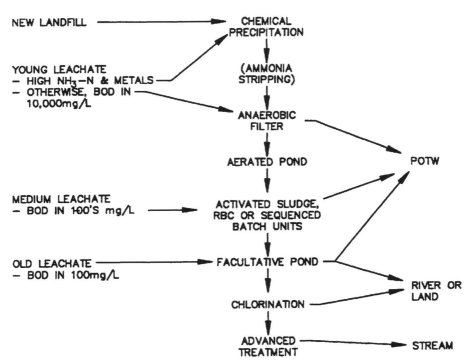

Fig. 11.15 Biological treatment process selection.

- At pH levels less than 8, most ammonia is in an anionic form and does not readily volatilize. When the pH is raised to between 9 and 10, volatilization of ammonia occurs.
- The addition of NaOH promotes the precipitation of calcium, iron, and manganese out of the raw leachate.

2. Anaerobic reactors are to reduce organic strength.
3. Polishing lagoons are used to improve the removal of BOD, ammonia, and suspended material.
4. Equalization tanks are used to equalize the influent flow and plant recycle streams for a constant discharge through succeeding units of treatment.
5. Acid is added to bring the pH of the leachate back to a neutral level (i.e., following the NaOH addition). This process is required to bring the pH back to the range of 6–8, where the biomass operates most effectively.

11.8.2 Treatment of Leachate at Municipal Sewage Works

As an alternative to treatment on site, the treatment of leachates at municipal sewage works has been reviewed by Robinson and Maris (1979). Treatment of leachates in combination with domestic sewage at a POTW is a potential disposal opportunity where access to the sewer system is available near the landfill site. In particular cases, the use of tankers for transporting leachate to sewage disposal facilities may also be justified for short periods of time, but the expense is likely to make this uneconomical as a permanent means of removal. The key that allows treatment within the POTW is that the volume must be a small proportion of the total sewage flow.

Leachate treatment provided in the POTW is BOD_5 and suspended solids removal, disinfection, conversion from NH_3 to NO_3, and adsorption/precipitation of some metals and organics. There is little impact on Cl, SO_4, Na, K, and Mg other than that which occurs by dilution. Palit and Qasim (1977) showed that leachate could be treated biologically using a conventional activated sludge process, although occasional problems with sludge bulking and poor solid/liquid separation were encountered. The addition of phosphate was also necessary.

A municipal treatment system or POTW is obviously going to have a greater resiliency in treatment by virtue of the large dilution of leachate flows. However, there is a degree to which the influent leachate can be accepted without disrupting the ongoing treatment of the municipal wastewater.

A number of authors have determined experimentally the proportion of leachate that can be tolerated in a POTW influent without causing a deterioration in effluent quality. Comparison of their results is made difficult by differences in composition of both leachate and sewage and by differing experimental procedures, but there were impacts.

Boyle and Ham (1974) investigated the treatment of various proportions of leachate with domestic sewage using leachate with a BOD of 8800 mg/L. Addition of 2 percent leachate by volume to sewage had little effect on effluent quality or on the performance of the POTW. Alternatively, 5 percent leachate in the sewage resulted in poor treatment, with low dissolved oxygen and poor settling characteristics.

Chian and deWalle (1977b), with leachate of 24,700 mg/L and 2 percent leachate in sewage, obtained good treatment. However, 4 percent leachate in sewage resulted in poor treatment, as reflected in high effluent BOD and deteriorating sludge characteristics. This was attributed to increases in the ratio of BOD:P to values above 130:1 in the effluent (a ratio of 100:1 being commonly recommended).

Based on the results of studies by Palit and Qasim (1977), Boyle and Ham (1974), and Chian and deWalle (1977b), Robinson and Maris (1979) reported that "the experimental evidence does ... indicate the apparent ability of municipal sewage works to treat sewage containing up to 2 percent by volume of high-strength leachate with no adverse effects." There is still some concern with using the 2 percent guideline, since the continuous-flow bench-scale studies of Chian and deWalle (1977b) showed that the presence of 0.5 percent leachate in sewage influent caused an impairment of the sludge settling characteristics.

Additional considerations in treating leachate at municipal sewage works include the following:

- The resulting metals levels in the POTW sludge may be of concern.
- Leachates from young landfills tend to encrust trickling filters, mostly by precipitating carbonates.
- The ability to transfer the leachate to a POTW is a function of physical proximity of the landfill to the POTW. If the leachate quantities are too great or the sewer surcharge is too high, then on-site treatment (at least in part), is appropriate.

If trucking to a sanitary sewer is seriously being considered, care must be taken to choose a remote location on the sewage collection system for discharge of the leachate. Public opposition can rapidly develop when the discharge point is in a populated area.

Table 11.15 TREATMENT EFFICIENCY OF HALIFAX, NOVA SCOTIA, PLANT

Parameter	Concentration (mg/l) at			
	A	B	C	D
COD	22,800	NM	693	325
BOD_5	16,100	NM	254	15
TOC	8,100	NM	224	120
Humic acid	416	NM	368	381
NH_3-N	406	NM	382	37
Fe	937	155	12	1.4
Zn	68	4	0.5	0.1
Ca	1,740	2,020	108	27
$Cl-$	1,110	NM	1,015	1,080
SO_4^{2-}	831	NM	28	54
Alkalinity	3,850	4,200	2,563	1,800
Total dissolved solids	15,300	19,200	4,220	4,215
pH	5.6	7.2	NM	NM

Note: NM—not measured

11.8.3 Case Studies of Leachate Treatment

Halifax, Nova Scotia   An on-site treatment sequence is successfully being used in Halifax, Nova Scotia, for leachate from a 700 t/d landfill and treating 80 m^3/d of leachate. The treatment sequence employed is illustrated in Figure 11.16.

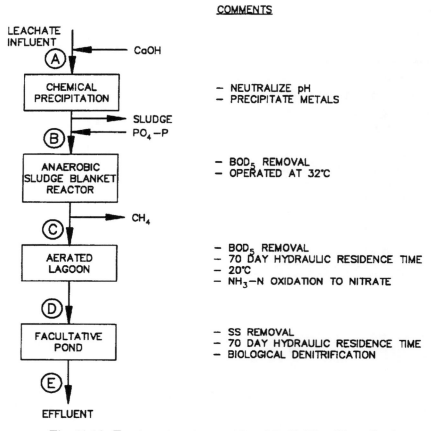

Fig. 11.16   Treatment system employed in Halifax, Nova Scotia.

Steps utilized in the treatment include:

1. Lime is used for pretreatment to adjust pH and precipitate inhibitory metals.
2. Anaerobic sludge blanket reactors are used for reduction of organic strength. The anaerobic reactors pass flow upward through a sludge blanket (2 m thick) and then a fixed-film media layer before overflow to discharge in the effluent and/or recirculation. Phosphates are added for nutrients and methane gas is recovered.
3. An equalization tank equalizes the influent flow and plant recycle to allow a constant discharge to the succeeding units of the treatment system.
4. Indications of the treatment efficiency from a pilot-scale at various loca-

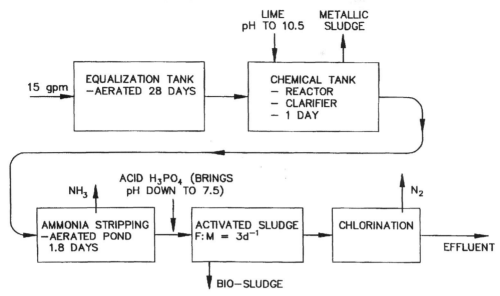

Fig. 11.17   Schematic of Grows Landfill treatment sequence.

tions in the process stream are provided in Table 11.15.

5. Removal of color, suspended solids, and metals (particularly iron) was good, although high sludge yields resulted. The addition of lime can remove three-quarters of the suspended solids, one-half of the soluble organic material, and some of the metals but has little impact on chloride concentrations.

6. Capital costs were $3 million (1988), and operating and maintenance costs were $155,000 per year.

Grows Landfill, Bucks Co., Pennsylvania   The Grows Treatment Facility was an EPA-supported demonstration project for a landfill of 50 acres, 800 t/d, 85 percent MSW (Fungaroli and Steiner, 1979). Treatment was required to meet sewer standards. The flow, as expected, was variable, but the average was in the range of 10 to 15 gpm. The treatment system is characterized in Figure 11.17. Leachate initially enters an equalization tank, since the leachate concentrations were found to be highly variable with time. Chemical pretreatment by flash mixing a lime slurry with the leachate is required to decrease the heavy metals and organics fraction. Mixing, flocculation, and clarification all occur within one unit, with a hydraulic retention time of 1.7 hours at a flow rate of 380 L/min. After the chemical pretreatment, air stripping of ammonia is accomplished in a lagoon. Prior to the biological process, sulfuric and phosphoric acids are added to reduce the pH and to supply nutrients (orthophosphate) that were previously precipitated out by the lime addition.

Biological treatment consists of an aeration tank and a secondary clarifier utilizing a conventional activated sludge process. After the secondary clarifier, effluent is either directed to a chlorine contact and then discharged to the Delaware River or recirculated by pumping the leachate back into the landfill when precipitation is minimal, to maintain optimal use of the landfill in treating the leachate in situ. Treatment effectiveness obtained is summarized in Table 11.16.

Table 11.16   REMOVAL EFFICIENCIES AT GROWS LANDFILL

| | Concentration (mg/L) | | | |
| | Water Quality | | | |
Parameter	Influent	Effluent	Standard	% Removal
BOD_5	8480	66	100	99
NH_3–N	695	30	35	96
SS	585	84	—	86
Fe	9.7	0.7	7	93
Cl	3172	2925	—	—

The treatment problems encountered at the Grows Landfill in Pennsylvania included:

- Variability in leachate concentrations—5000<COD<50,000 mg/L and a flow of 15 gpm on average but variable
- NH_3-N toxicity—1900 mg/L, which required the addition of air stripping. pH adjusted up to 10.5 for air stripping of ammonia
- Phosphorus deficiency—addition of phosphorus was required
- Freeze-up—biomass was poorly acclimatized when temperatures dropped below 2°C;
- Slow start-up—biomass poorly acclimatized
- pH adjustment to get the activated sludge to work
- Chlorination during the winter, mainly to enhance NH_3-N removal.

Omega Hills, Milwaukee, Wisconsin   An anaerobic upflow filter is being used at the Omega Hills Landfill in Milwaukee. The facility was designed for discharge of treated effluent into the Metropolitan Milwaukee Sanitary sewer system. The plant schematic and pertinent design features of the facility are depicted in Figure 11.18. Noteworthy is that the design conditions were not the same as the field conditions encountered. The differences between design and field conditions are listed in Table 11.17. Nevertheless, excellent treatment is being obtained. The system is underloaded, so various possibilities for treating other

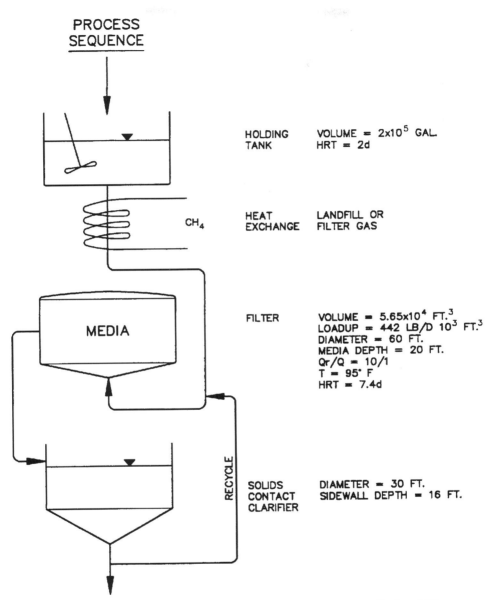

PROCESS
SEQUENCE

HOLDING   VOLUME = 2×10^5 GAL.
TANK   HRT = 2d

CH$_4$ HEAT   LANDFILL OR
            EXCHANGE   FILTER GAS

MEDIA

FILTER   VOLUME = 5.65×10^4 FT.^{3}
            LOADUP = 442 LB/D 10^3 FT.^{3}
            DIAMETER = 60 FT.
            MEDIA DEPTH = 20 FT.
            Q_r/Q = 10/1
            T = 95° F
            HRT = 7.4d

RECYCLE

SOLIDS   DIAMETER = 30 FT.
CONTACT   SIDEWALL DEPTH = 16 FT.
CLARIFIER

Fig. 11.18   Anaerobic upflow filter in use at Omega Hills Landfill.

wastewater are being examined to more fully utilize the available treatment capacity.

Bryn Posteg Landfill, Wales, United Kingdom   As described in Robinson and Grantham (1988), the Bryn Posteg Landfill has been designed as a containment landfill. The treatment system in place is depicted in Figure 11.19. The high

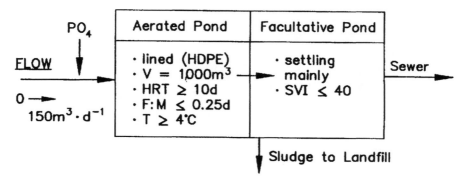

Fig. 11.19   Leachate treatment system in use at Posteg Landfill, Wales,
United Kingdom.

Table 11.17   DESCRIPTION OF OMEGA HILLS LEACHATE TREATMENT SYSTEM

I.

Design Conditions	Field Conditions Encountered
105 gpd (flow)	2-3 x 10^4 gpd
38,000 mg/L (BOD)	7,000 mg/L (average BOD)
900 lb/d 10^3 ft^3 (loading)	67 lb/d 10^3 ft^3

II.   *Treatment Efficiency (as measured September 1985 to May 1986)*

Contaminant (mg/L)	Influent	Effluent
BOD	3700–24,000	350–700
TSS	2500–15,000	100–500
Cd	0.1	0.08
Cn	1.84	0.21
Pb	3.19	0.51
Ni	1.39	0.68
Zn	9.47	1.61

quality of the nearby River Severn made it infeasible to treat and discharge
locally.   The on-site treatment plant reduces contaminant levels prior to discharge to a small municipal treatment system.

Phosphoric acid is added due to nutrient deficiency. The plant was commissioned during May and June of 1983 and has run continuously since start-up.

The lagoon is aerated continuously. High-quality effluent has been maintained at all times, with substantial removal of COD, BOD_5, ammonia, and metals. Excellent removal of organic compounds and of ammonia has been maintained even when COD and ammonia levels in leachate have risen to as high as 23,000 and 600 mg/L, respectively (Robinson, 1991). Efficiency of treatment for selected contaminant levels and associated costs of removal are listed in Table 11.18. Excellent removals of lead and zinc have also been maintained.

Table 11.18   EFFICIENCY OF TREATMENT AT BRYN POSTEG LANDFILL

	Concentration (mg/L) Influent		
Parameter	Peak	Average	Effluent
BOD_5	+10,000	3,700	24
NH_4–N	+1,000	129	12
Fe	+500	254	3.8
pH (units)	5.8	—	8.0
COST:			
Capital:	$120,000 annual (U.S., 1985)		
O & M:	$1/1000 U.S. gal		
Sewer surcharge:	$9/1000 U.S. gal		
w/o treatment savings	$68,000/yr		

Love Canal (Southern Sector), Niagara Falls, New York   The leachate treatment system in place at Love Canal, Niagara Falls, New York, is described in McDougall (1980). The collection of leachate and contaminated groundwater follows the schematic of Figure 11.20. The designers have utilized a simple but highly effective system employing a reactivated bituminous granular activated carbon to remove dissolved organic compounds. The system has been operational since December 1979. The treatment efficiency attained is quantified in Table 11.19.

Sarnia Landfill, Ontario, Canada   The Sarnia Landfill has been continuously active since 1971 with an area of 21 ha accepting domestic and commercial waste. The leachate currently consists of approximately 75 percent old leachate (i.e., at least 5 years old) and the remaining factor new leachate (i.e., less than 5 years old). The average annual precipitation is 780 mm/year. The average depth of refuse is 18 m.

On-site treatment of leachate for surface water discharge was completed in 1990. The treatment sequence consisting of pretreatment, biological suspended

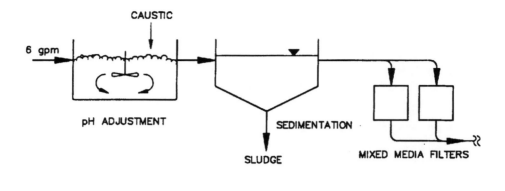

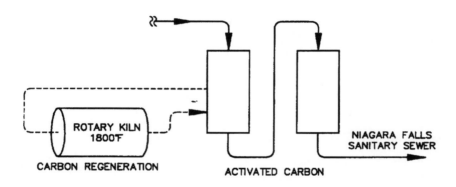

Fig. 11.20   Leachate treatment system inplace at Love Canal in
Niagara Falls, N.Y.

growth treatment, sand filtration, and postaeration in lagoons and wetlands is
depicted in Figure 11.21. The criteria for direct discharge to the wetland that
indirectly discharges into the St. Clair River are BOD_5 ≤15 mg/L, SS ≤15 mg/L,
NH_3 ≤10 mg/L, and P ≤5 mg/L.

General leachate characteristics are listed in Table 11.20. In general, metals concentrations in the leachate are relatively low, except for relatively high
levels of calcium, magnesium, and iron. The concentrations of organics, in general, are moderate to low, with the majority of organics being below detectable
levels. Table 11.21 lists the organic concentrations in the leachate.

Variability of the leachate in response to seasonal changes is minimal, since
the completed portion of the landfill has a low-permeability cap that greatly
reduces infiltration. Since start-up of the full-scale plant, flow rates have ranged
from 41 m^3/d to 111 m^3/d, with an average flow rate of 68 m^3/d. The plant design
flow rate is 91 m^3/d.

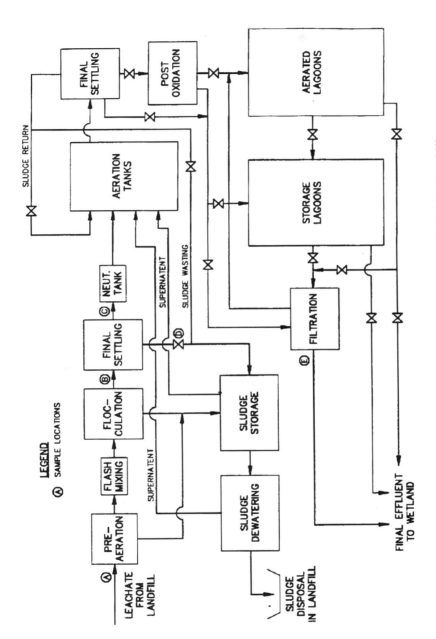

Fig. 11.21 Leachate treatment flow diagram for Sarina Landfill.

Table 11.19   TREATMENT EFFICIENCY ATTAINED AT LOVE CANAL FOR PERIOD NOVEMBER 1978–JUNE 1979

Parameter	Influent	Effluent
TOC	300–1,200	50–7,180
Priority Pollutants		
13 in mg/L range		
Carbon tetrachloride	61.0 (highest)	<0.01
Chlorobenzene	50.0	0.012
2,4–Dichlorophenol	25.0	ND
82 in μg/L range		
Methylene chloride	140	46 (highest)
Ethylbenzene	590	<10
Hexachlorobenzene	110	ND

Table 11.20   GENERAL PARAMETER LEACHATE CHARACTERIZATION FOR THE SARNIA LANDFILL

Parameter	Raw Leachate Concentration (mg/L)
BOD_5	50-2,540
COD	281-2,800
TOC	250-900
TKN	46-810
NH_3–N	9.2-689
NO_3^-	<0.01-69
VSS	8-240
TSS	12-685
pH (units)	6.2-7.5
Phosphorus (total)	0.11-12.2
Fe	9-65
Ca	92.2-337
Mg	294-398
SO_4^{2-}	<1-19.6

Notes:
   Concentrations are expressed as mg/L unless otherwise specified.

During pilot-scale studies calcium was determined to be inhibiting the biological processes. Reduced substrate removal rates were observed as a result of

the calcium encrustation of the biological floc, which drastically inhibited the oxygen transfer to the microbial biomass. Chemical pretreatment of the leachate with air oxidation and sodium hydroxide was employed to increase the flocculation of calcium, followed by sulfuric acid addition to decrease the pH to approximately 7.0 (otherwise the high pH would kill the majority of the biomass).

With the chemical pretreatment, calcium and iron concentrations are removed at rates exceeding 60 and 95 percent, respectively, prior to the biological degradation process. The leachate is generally phosphorus-deficient, so phosphoric acid is added prior to biological treatment on an as-required basis.

Aeration basins are working in a plug-flow mode with a designed hydraulic retention time of 5 to 10 days. Organic loadings in terms of BOD_5 and NH_3 range from 2 to 80 kg BOD_5/d and 3 to 33 kg NH_3/d. The F:M ratios for BOD_5 have varied from 0.002 to 0.09 kg BOD_5/kg VSS/d. Similarly, in terms of ammonia, F:M ratios have ranged from 0.002 to 0.145 kg NH_3/kg VSS/d.

Suspended solids levels leaving the secondary clarifier prior to the sand filtration exceeded 15 mg/L approximately 15 percent of the time. However, the addition of aluminum prior to the secondary clarifier produced an effluent from the secondary clarifier consistently below 5 mg/L.

Approximately 12.9 kg/d of primary sludge is produced, and approximately 18.6 kg/d of the secondary sludge is produced, with calculated average sludge solids concentrations of 40 percent.

Effluent from the mechanical portion of the treatment system is discharged into lagoons for final polishing, postaeration, and storage prior to discharge into the wetlands. Table 11.22 presents results of analysis of the leachate at various stages of the treatment process. Utilization of the wetlands was designed as a part of the treatment process, but this discharge can occur only during the growing season (March to November). Natural treatment is provided within the wetland by plants that take up phosphorus and nitrates for growth, as well as by the soil, which adsorbs phosphorus.

Table 11.21   LEACHATE INFLUENT ANALYTICAL RESULTS (OPEN CHARACTERIZATION—VOLATILE AND EXTRACTABLE ORGANICS) SARNIA LANDFILL LEACHATE TREATMENT FACILITY

Open Characterization: Volatile Organics		Open Characterization: Extractable Organics	
Parameter	Concentrations (ppb)	Parameter	Concentrations (ppb)
1,2,3-trimethylbenzene	8	2(3H)-benzothiazolone	30
1,2,4-trimethylbenzene	10	2-butoxyethanol phosphate	10
1,3,5-trimethylbenzene	4	2-chloroethanol phosphate	10
1-ethyl-2-methylbenzene	4	2-methylhexanoic acid	300
1-ethyl-3-methylbenzene and/		2-piperidinone	50
or1-ethyl-4-methylbenzene	8	alcohol	20
2-butanol	2		
2-hexanone	7	alkene (molecular weight 210)	20

Table 11.21   LEACHATE INFLUENT ANALYTICAL RESULTS (OPEN CHARACTERIZATION—VOLATILE AND EXTRACTABLE ORGANICS) SARNIA LANDFILL LEACHATE TREATMENT FACILITY

Open Characterization: Volatile Organics		Open Characterization: Extractable Organics	
Parameter	Concentrations (ppb)	Parameter	Concentrations (ppb)
acetone	9	benzeneacetic acid	100
benzene	6	benzenepropanoic acid	600
$C_{10}H_{14}$-alkylbenzene	5	benzoic acid	50
cis-1,2-dichloroethene	10	beta-eudesmol	10
dichloromethane	1	bisphenol A	10
diethyl ether	7	C9 alkyl benzoic acid (molecular weight 150)	10
ethylbenzene	50		
fenchone	4	cis-terpin hydrate	10
indan	3	cycloalkene	30
isopropylbenzene	2	cycloalkene (molecular weight 208)	30
ketone (molecular weight 142)	2		
ketone (molecular weight 154)	2	cycloalkene (molecular weight 210)	20
m-xylene and/or p-xylene	70		
methyl ethyl ketone	20	cyclohexanecarboxylic acid	500
methyl isobutyl ketone	3	decanoic acid	80
methyl isopropyl ketone	20	deet	40
o-xylene	40	diethyl phthalate	20
propylbenzene	2	heptanoic acid	900
tetrahydrofuran	3	hexadecanoic acid	20
toluene	50	hexanoic acid	3000
		ketone	40
		m-cresol & p-cresol	100
		molecular sulfur	300
		nonanoic acid	100
		octanoic acid	1000
		tetraglyme	10
		tolmetin	20

A baseline assessment of the wetlands was performed prior to discharging effluent there. From this baseline it was subsequently determined that hydraulic loading was the major impact on the wetlands. Hydraulic impacts were alleviated by discharging to the wetlands during periods that coincide with the naturally wet periods of the year.

Hyde Park, Niagara Falls, New York   The Hyde Park Landfill is located in the northwest portion of the town of Niagara Falls, New York. The site is approximately 6.1 ha in size, and operated between 1953 and 1975. Industrial and chemical wastes were disposed of at the site during this period.

A compacted clay cover was constructed over the landfill in 1978, and a perimeter subsurface leachate collection system was constructed in 1979. Leachate ($33 \text{ m}^3/\text{d}$) is drained to a sump and then pumped into a two-component lagoon. In the first lagoon, the leachate is allowed to settle and supernatant is drawn off and is stored in the second lagoon before being trucked to the nearby privately owned Niagara Treatment Plant, where it is pretreated and discharged to the sanitary sewer. Figure 11.22 is a schematic of the treatment facility.

At the privately owned treatment facility, the wastewater is initially separated from the nonaqueous liquid (NAPL), and the NAPL is trucked off-site to be incinerated. The wastewater is stored and its pH is adjusted before being transferred to an aeration tank and an air stripper to remove volatiles. Solids are removed in a settling tank, and the wastewater is then passed through a preliminary liquid-phase carbon treatment followed by biodegradation in a sequencing batch reactor (SBR). The SBR allows the biological reaction and subsequent settling processes to occur in one unit. Finally, the supernatant is passed through a series of activated carbon filters before being discharged into the sanitary sewer.

11.9   LEACHATE RECIRCULATION

In some cases, leachate is collected and returned to the top of the landfill. This approach has the benefit of accelerating the stabilization of organic materials present in the waste. The use of recirculation does not eliminate the ultimate need for treatment, since eventually the excess leachate will have to be treated. Both spray irrigation of the leachate onto the top of the landfill and well injection are employed. Recirculation systems need to operate in such a manner that the amount of recycled leachate does not exceed the storage capacity of the upper layers of the landfill.

Leachate recirculation can be utilized during the early stages of landfill development, when leachate production quantities are low. In addition, recirculation can be utilized in later stages of development to eliminate problems of off-site transport during peak production periods or during downtimes of the transport devices. Recirculation reduces the hydraulic peaks and can serve to even out the chemical and biological concentration variations of the liquid wastes.

Table 11.22   ANALYTICAL RESULTS FOR SAMPLES COLLECTED MARCH 24, 1993

Parameter	Raw Leachate Influent Sampling Port	Inflow to Primary Clarifier	Primary Clarifier Effluent (Overflow Weir)	Primary Clarifier Sludge Wasting Line	Final Effluent Filter Clear Well
Parameter	Location A	Location B	Location C	Location D	Location E
TSS	42	436	25.4	1,458	4.8
BOD	570	NA	NA	NA	7.05
COD	NA	NA	1,554	NA	390
Ammonia	237	NA	273	NA	0.39
Nitrate	7.5	NA	22.5	NA	340
TKN	347	NA	358	NA	25.0
TOC	419	NA	451	NA	111
DOC	382	NA	421	NA	109
Iron	25.0	NA	4.05	NA	NA
Calcium	170	NA	26.3	NA	NA
Magnesium	217	NA	193	NA	NA

Notes:
NA—not analyzed
All results in mg/L
Sample locations A-E are shown on the leachate treatment flow diagram for the
Sarnia Landfill, Figure 11.21.

Pohland (1973, 1975) and Pohland and Kang (1975) reported on the influence of recirculation on refuse stabilization rates. Over a 2- to 3-year period, they observed markedly reduced concentrations of BOD, TOC, and COD in their pilot-scale investigations. Control of pH and initial seeding with sludge further enhanced efficiency.

The approach essentially uses the landfill itself as an uncontrolled anaerobic percolating filter to treat the leachate. Considerable volumes of liquid can also be lost by evaporation during recirculation, and spray equipment may be selected to encourage this loss.

Apparent advantages of using leachate recirculation include the following:

- It delays disposal of leachates
- It provides treatment for BOD and speeds up decomposition
- It enhances CH_4 production rates
- It lowers treatment cost
- It allows buffers and nutrients to be added if needed to accelerate anaerobic decomposition.

Disadvantages of leachate recirculation include field plumbing problems—settling, clogging, and freeze-up; odors; and the necessity to design the leachate collection system to handle higher hydraulic loadings.

The effect of leachate recirculation on the enhancement of biological degradation processes in landfills is not clear. Under certain circumstances (e.g., low precipitation rate) it may be an important factor (Stegmann and Ehrig, 1989a). Leachate neutralization and recirculation may be used to promote decomposition.

11.10   TREATMENT OF LEACHATE BY SPRAYING ONTO LAND

The spraying of leachate onto land (as opposed to onto the landfill) may result in a significant volume reduction due to evapotranspiration. Also, as the leachate percolates through, it provides an opportunity for microbial degradation of organic compounds, removal of inorganic ions by precipitation or ion exchange, and uptake of ammonia by plants (Robinson and Maris, 1979). Transpiration by plants accounts for a substantial proportion of total evaporative loss. Lawlor (1976) and Nordstedt et al. (1975) reported that Pb, Cr, Cd, and Ni are effectively adsorbed by soils (although these metals may still be available for uptake by plants). Nevertheless, operational problems with this treatment approach may develop over the long term.

11.11   SUMMARY

Different methods for leachate treatment exist, and there is no universal preferred alternative. Instead, the best approach frequently is a combination of alternatives.

- The evaluation of treatment processes should include not only possible effluent values and maintenance costs but also the production of residuals with new pollution potential.
- Generally, a biotreatment system with a long solids retention time is needed for leachate treatment (e.g., an extended aeration activated sludge process

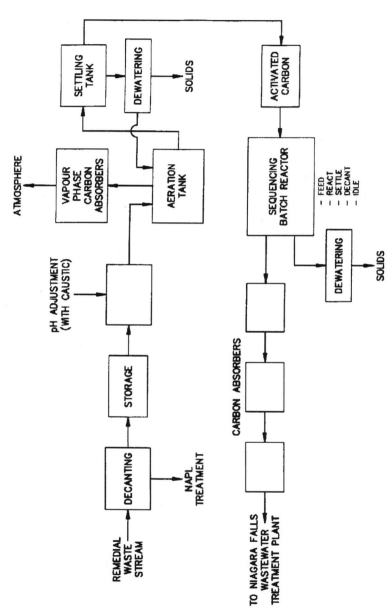

Fig. 11.22   Leachate treatment flow diagram for Hyde Park Landfill.

with secondary clarification and sludge recycle).

- If the ambient temperatures are low in the geographic region of concern, aerated lagoon systems will not work efficiently.
- Flow equalization will usually be beneficial.
- Nutrient additions, particularly phosphates, will probably be required.
- There will be higher concentrations of heavy metals and potentially toxic trace organics in the sludges.
- The composition will vary in the long term, as the nature of the leachate being produced changes, and the volume of leachate will fluctuate in the short term, in response to variations in infiltration. Due to the variability in flow and concentration levels, the best philosophy is to maintain as much operational flexibility as possible.
- Treatment costs are very much a function of leachate strength and quantities and available disposal options.
- In general, removal of metals will not be a problem if either aerobic or anaerobic biological leachate treatment is used. With anaerobic treatment, the metals will tend to precipitate out as sulfides; with aerobic treatment the metals will tend to oxidize and precipitate out as hydroxides and carbonates.
- Biological sludges should be sent to a POTW digester. Metallic sludges must be sent to a hazardous waste site.
- Anaerobic decomposition gases can be sent to a burner or be treated by air stripping or activated carbon.
- Most facilities are focused on concentrating the contaminants rather than destroying them.
- Experimental evidence indicates the ability of municipal sewage works to treat sewage containing up to 2 percent by volume of high-strength leachate with few adverse effects.
- Since treatment plants have to be planned before the leachate can generally be tested, it should be noted that 100% removal is rare, and priority organics are problems.

Because wastes are emplaced over a period of time at most landfill sites, it is difficult to arrive at a meaningful figure for the age of any particular site.

Leachate quality will change with further implementation of the 3Rs. As the composition of the waste at landfills is altered, the leachate will also change. If the majority of the organic and paper component is removed from the site, the leachate should have a reduced organic loading. Future leachate will then potentially consist of a higher mass loading of conservative species such as metals and higher-order organics.

References

Boyle, W., and Ham R. 1974. Biological Treatment of Landfill Leachate. *Journal of Water Pollution Control Federation* 46:860–872.

Chian, E. S. K., and deWalle, F. B., 1975. Characterization and Treatment of Leachate Generated from Landfills. Symp. ASChE Symposium Series. 71:319–327.

Chian, E. S. K., and deWalle, F. B., 1977a. *Evaluation of Leachate Treatment, Volume I: Characterization of Leachate*. U.S. Environmental Protection Agency EPA-600/2-77-186a, September.

Chian, E. S. K., and deWalle, F. B. 1977b. *Evaluation of Leachate Treatment. Volume 2: Biological and Physical - Chemical Processes.* U.S. Environmental Protection Agency EPA-600/2-77-186b, November.

Chian, E. S. K., and deWalle, F. B. 1977c. "Removal of Heavy Metals from a Fatty Acid Wastewater with a Completely Mixed Anaerobic Filter." In *Proceedings of the 32nd Annual Industrial Waste Conference.* Purdue University, Lafayette, Indiana, pp. 920–928.

Conestoga-Rovers & Associates. 1987. *Sarnia Landfill Leachate Treatment Options.* Report no. 792-50, February.

Cook, E. N., and Foree, E. G., 1974. Aerobic Biostabilization of Sanitary Landfill Leachate. *Journal of Water Pollution Control Federation* 46(2):380–392.

de Walle, F., and Chian, E. S. 1977. "Leachate Treatment by Biological and Physical-Chemical Methods - Summary of Laboratory Experiments" *Management of Gas and Leachate in Landfills*, ed. S.K. Banerji. *Proceedings of the 3rd Annual Solid Waste Research Symposium*. St. Louis, Mo., March.

Ehrig, H. J. 1989. "Leachate Treatment Overview." *2nd International Landfill Symposium*. Sardinia, Italy.

Foree, E. G., and Reid, V. M. 1973. *Anaerobic Biological Stabilization of Sanitary Landfill Leachate*. University of Kentucky, Office of Research and Engineering Services. Report no. VKY TR65-73-CE17, January, p. 51.

Forgie, D. 1988a. Selection of the Most Appropriate Leachate Treatment Methods. Part 1 A Review of Potential Biological Leachate Treatment Methods. *Water Pollution Research in Canada* 23:308–328.

Forgie, D. 1988b. Selection of the Most Appropriate Leachate Treatment Methods Part 3 A Decision Model for the Treatment Train Selection. *Water Pollution Research in Canada* 23:341–355.

Fungaroli, A. A., and Steiner, R. 1979. *Investigation of Sanitary Landfill Behavior.* Vols. 1 and 2. U.S. Environmental Protection Agency EPA-600/2-79-053a. July

Grosse, G., 1988. Betriebserfahrungen durch Einsatz des Umkehrosmuseverfahrens zur Aufbereitung von Sickerwasser einer Hausmülldeponie. *Tagungsunterlagen Technische Akademie. Esslingen 21/22.3.*

Ham, R. 1980. *Decomposition of Residential Solid Waste in Test Lysimeters.* U.S. Environmental Protection Agency SW-190C.

Ham, R., et al. 1979. *Recovery, Processing and Utilization of Gas from Sanitary Landfills.* U.S. Environmental Protection Agency EPA-600/2-79-00, February.

Henry, J. G., Pradad, D., and Young, H. 1985. "Studies in Anaerobic Filter Treatment of Leachates." *Proceedings Conference on New Directions and Research in Waste Treatment and Residuals Management.* International Conference, University of British Columbia. Vancouver, pp. 297–306, June.

Johansen, O. J., and Carlson, D. A. 1976. Characterization of Sanitary Landfill Leachates. *Water Research* 10:1129–1134.

Keenan, J., Steiner, R., and Fungaroli, A. 1984. Landfill Leachate Treatment. *Journal of Water Pollution Control Federation* 56(1):27–33.

Knox, K. 1985. Leachate Treatment with Nitrification of Ammonia. *Water Research.* 19(S):895–904.

Lawlor, M. 1976. Dealing with Leachate at Solid Waste Disposal Sites. *Public Works* 107(5): 74–76.

Logeman, F. 1987. *Behandlung von Sickerwasser mittels Umkehrosmose*, Firmenunterlagen Storck Friesland B.V.

Lu, J. C. S., Morrison, R. D., and Stearns, R. J. 1981. "Leachate Production and Management from Municipal Landfills: Summary and Assessment, Land Disposal: Municipal Solid Waste." *Proceedings of the Seventh Annual Symposium*. EPA-600/-9-81-002a, pp. 1–17.

Lu, J. C. S., Eichenberger, B., and Stearns, R. J. 1984. *Production and Management of Leachate from Municipal Landfills: Summary and Assessments*. U.S. Environmental Protection Agency EPA-600/2-84-092, May.

Lu, J., Eichenberger, B., and Stearns, R. J. 1985. *Leachate from Municipal Landfills Production and Management*. Noyes, Park Ridge, N.J. pp. 109–121.

McDougall, W. J. 1980. Containment and Treatment of the Love Canal Landfill Leachate. *Journal of the Water Pollution Control Federation* 52(12): 2914–2924.

McGinley, P. M., and Kmet, P. 1984. *Formation, Characteristics, Treatment and Disposal of Leachate from Municipal Solid Waste Landfills*. Wisconsin Dept. of National Resources, Special Report, August.

Merz, R. C. 1954. "Final Report on the Investigation of Leaching of a Sanitary Landfill." Publication no. 10, California State Water Pollution Control Board, Sacramento, 91 p.

Metcalf & Eddy Inc. 1991. *Wastewater Engineering: Collection, Treatment, and Disposal*, 3d ed. McGraw-Hill, New York.

Mosey, F., and Hughes, D. 1975. The Toxicity of Heavy Metal Ions to Anaerobic Digestion. *Water Pollution Control* 74:18–39.

Nordstedt, R., Baldwin, L., and Rhodes, L. 1975. Land Disposal of Effluent from a Sanitary Landfill. *Journal of the Water Pollution Control Federation* 47:1961–1970.

Palit, T., and Qasim, S. R. 1977. Biological Treatment Kinetics of Landfill Leachates. *ASCE Journal of the Environmental Engineering Division* 103 (EE2):353–366.

Pohland, F. 1973. *Sanitary Landfill Stabilization with Leachate Recycle and Residual Treatment*. Environmental Protection Technology Series. School of Civil Engineering, Georgia Institute of Technology. EPA 600/2/75-043. Atlanta, Georgia. October, p. 106.

Pohland, F. 1975. Accelerated Solid Waste Stabilization and Leachate Treatment by Leachate Recycle Through Sanitary Landfills. *Progress in Water Technology* 7(3/4): 753–765.

Pohland, F., and Kang, S. 1975. "Sanitary Landfill Stabilization with Leachate Recycle and Residual Treatment." Water: 2. Municipal Waste Treatment, American Institute of Chemical Engineers Symposium Series 71 (145):308.

Qasim, S. R., and Burchinal, J. C. 1970. Leaching from Simulated Landfills. *Journal of the Water Pollution Control Federation* 42:371–379.

Reitzel, S., Farquhar, G., and McBean, E. 1992. Temporal Characterization of Municipal Solid Waste Leachate. *Canadian Journal of Civil Engineering*, August.

Revah, A., and Avnimeleih, Y. 1979. Leaching of Pollutants from Sanitary Landfill Models. *Journal of Water Pollution Control Federation* 51 (11):2705–2716.

Robinson, H. D., 1989. "Development of Methanogenic Conditions within Landfills." Presented at Second International Landfill Symposium, Porto Conte, Sardinia, Italy, October 9–13.

Robinson, H. D. 1990. On-Site Treatment of Leachates from Landfilled Wastes. *Journal of Institute of Water and Environmental Management* 4, no. 1 (February): 78–89.

Robinson, H. D. 1991a. "Leachate Collection, Treatment, and Disposal." Presented at Engineering of Landfill, Treatment, and Disposal, Institution of Water and Environmental Management, Shrewsbury, October 30.

Robinson, H. D. 1991b. "Groundwater Pollution and Aquifer Protection in Europe" Presented at 1991 Annual Symposium, Palais des Congres, Paris, October.

Robinson, H. D., and Grantham, G. 1988. The Treatment of Landfill Leachates in On-Site Aerated Lagoon Plants: Experience in Britain and Ireland. *Water Research* 22 (6):733–747.

Robinson, H. D., and Luo, M. 1991. Characterization and Treatment of Leachates from Hong Kong Landfill Sites. *Journal of Institute of Water and Environmental Management* 5, no. 3 (June):326–335 (a).

Robinson, H.D., and Maris, P. J. 1979. *Leachate from Domestic Waste: Generation, Composition, and Treatment: A Review*. Water Research Centre Technical Report TR108. Medmenham Laboratory, England, March.

Rovers, F., 1972. "Seasonal Effects on Landfill Leachate and Gas Production." Master's thesis, Dept. of Civil Engineering, University of Waterloo, Waterloo, Ontario.

Ryser, W., and Ritz, W. K. 1985. Behandlung von Sickerwassser durch Membrant rennverfahren auf der Mülldeponie Uttigen. *Fachtagung: Sickerwasser aus Mülldeponien - Einflusse und Behandlung.* 21/22.3, Braunschweig.

Safferman, S. I., and Bhattacharya S. K. 1990. *Treatability of RCRA Compounds in a BOD/Nitrification Wastewater Treatment System with Dual Media Filtration.* U.S. Environmental Protection Agency EPA-600/52-90/013, Cincinnati, Ohio, August.

Snoeyink, V. D., and Jenkins, D. 1980. *Water Chemistry.* John Wiley, Toronto.

Soyupak, S. 1979. "Modifications to Sanitary Landfill Leachate Organic Matter Migration through Soil." Ph.D. thesis, Dept. of Civil Engineering, University of Waterloo, Waterloo.

Stegmann, R., and Ehrig, H. J. 1989a. "Leachate Production and Quality: Results of Landfill Processes and Operation. Second International Landfill Symposium, Sardinia, Italy, October 9–13.

Stegmann, R., and Ehrig, H. J. 1989b. Operations and Design of Biological Leachate Treatment Plants. *Progress in Water Technology* 12, IAWPR/Pergamon Press, pp. 919–947.

Straub, N. A., and Lynch, D. R. 1982. Models of Landfill Leaching: Organic Strength. ASCE *Journal of Environmental Engineering Division* 108 (EE2): 251–268.

Uloth, V. C., and Mavinic, D. S. 1977. Aerobic Bio-Treatment of a High-Strength Leachate. *ASCE Journal of the Environmental Engineering Division,*103 (EE4): 647–661.

US EPA, 1987. Composition of Leachates from Actual Hazardous Waste Sites. EPA/600/2-87/043

US EPA, 1989. Draft Background Document. Summary of Data on Municipal Solid Waste Characteristics.

Wigh, R. J. 1979. *Boone County Field Site Interim Report.* EPA-600/2-79-058.

Wigh, R. J. 1984. *Comparison of Leachate Characteristics from Selected Municipal Solid Waste Test Cells.* EPA-600/2-84-124, July.

Wigh, R. J., and Brunner, D. 1981. "Summary of Landfill Research, Boone County Field Site." Presented at Seventh Annual Research Symposium, Land Disposal: Municipal Solid Waste. EPA-600/9-81-002a, pp. 209–242.

Wright, P., Kennedy, K., and Robson, D. 1985. "Utilization of an Anaerobic Reactor Pilot Plant to Assess Methane Gas Production and Treatability of a Landfill Leachate." *Proceedings Conference on New Directions and Research.* University of British Columbia, Vancouver, B.C., pp. 262–279.

Wright, P. J., and Austin, T. P. 1988. Nova Scotia Landfill Leachate Treatment Facility First of its Kind in Canada. *Environmental Science and Engineering* October:12–14.

Ying, W., Bonk, R., Lloyd, V., and Sojka, S. 1986. Biological Treatment of a Landfill leachate in Sequencing Batch Reactors. *Environmental Progress* 5 (1):41–50.

Zapf-Gilje, R. 1979. "Effects of Temperature on Two-Stage Biostabilization of Landfill Leachate." Master's thesis, Dept. of Civil Engineering, University of British Columbia, Vancouver.

11.12 PROBLEMS

11.1 Identify and justify the type and sequence of processes that you would use to provide the treatment described below. Deal with the contaminants of major concern (no more than five).

   (a)   leachate from a landfill (average age = 2 years) treated to the quality of domestic sewage;

(b)   groundwater contaminated by a spill of mixed organic solvents treated to background groundwater quality;

(c)   leachate from a 20-year-old landfill treated for discharge to a nearby stream providing 50:1 dilution;

(d)   groundwater contaminated by a municipal landfill 150 m upgradient, that must be treated to drinking water quality.

11.2   What potential problems should be considered when leachate from a young municipal landfill is added to the influent of a municipal secondary sewage treatment system? What can be done to overcome these problems?

11.3   The landfill leachate system in Lingen, West Germany (Stegmann and Ehrig, 1989) consists of five aerated lagoons in series. Estimate the effluent BOD from the system for the following conditions:

• influent BOD of 1500 mg/L, with a BOD/COD ratio of 0.4,

• average lagoon temperature of 10°C,

• average hydraulic residence time of 20 days.

11.4   Contrast an old and a young landfill leachate in terms of the chemical contaminants listed in the table below. Calculate the loading of each contaminant influent for each leachate type to a municipal treatment plant with an influent flow of 22.5×10^3 m^3/d of which 2.5percent by volume is leachate. Describe the impact of each on the municipal treatment plant. What steps could you take to remedy the problems caused?

Contaminant	BOD	COD	NH_3-N	PO_4-P	Cl	Fe	Pb
Typical municipal sewage concentration (mg/L)	250	400	25	12	55	5	<1

11.5   Compare the use of an aerated lagoon and an upflow, anaerobic biofilter for on-site treatment of landfill leachate. Consider operation and effluent quality.

11.6   The objective is to calculate the average concentrations of iron (Fe) and ammonia (NH_3) in a landfill leachate over a 4-month period. The section of the landfill to be analyzed has surface area dimensions of 35 m x 50 m and receives 48 cm of percolation during the 4-month period. The section of the landfill was placed in three distinct layers at three different times with the following characteristics:

Layer	Depth (m)	MSW Compacted Density (kg/m^3)	Previous Total Amount of Percolate/Fixed
Top	3.5	485	53
Middle	3.0	485	87
Bottom	3.6	485	187

Leaching data from a lysimeter study are to be used to calculate the Fe and NH₃ concentrations. A total of 540 kg of MSW was packed into the lysimeter, and water was allowed to percolate through it for several months. The leachate was collected and analyzed on six occasions. The following data were obtained:

Collected Leachate Sample Number	Volume of Leachate Collected (L)	Contaminant Concentrations	
		Fe (mg/L)	NH3 (mg/L)
1	54	1203	3000
2	108	852	1148
3	270	459	541
4	432	213	119
5	378	85	58
6	378	29	29

Total leachate collected: 1620 L

11.7 A new landfill is being designed with an expected life of 20 years. Leachate will be collected on a liner and transported to a municipal pollution control plant. It is necessary therefore to estimate the leachate flow and contaminant composition during the period for which treatment is required. Describe the steps and important factors that you would use to make these estimates. Which contaminants do you think should be chosen for this analysis? Give reasons for your selections.

Draw schematic, qualitative diagrams of total leachate flow and contaminant concentrations (for two representative contaminants) versus time.

11.8 Leachate collected at a municipal landfill is analyzed and found to have the following concentrations:

COD	=	30,000 mg/L
BOD	=	15,000 mg/L
VFA	=	15,000 mg/L

(a) Estimate the approximate age of the refuse from which the leachate is being produced.

(b) Design a treatment process to treat the leachate. Indicate why you selected each of the processes.

(c) Could ammonia concentrations be utilized to indicate the age of the refuse from which the leachate is being produced?

DESIGN OF NATURAL ATTENUATION SITES

12.1 INTRODUCTION

Attenuation is defined as the reduction of contaminant concentrations during transport through the soil environment. A number of factors associated with the soil environment provide this natural attenuation capability. Thus, the environment has the ability to assimilate wastes, but the assimilative capacity is limited. The processes that influence attenuation can be as simple as dilution by uncontaminated, infiltrating water or as complicated as physico-chemical interactions that fix, or retard, contaminant movement through the soil materials.

Historically, use of the natural attenuation capacity of a soil environment was common. Thus, the natural attenuation landfill design concept has been in use for many years. Despite this, it would be considered a relatively high risk method of controlling leachate migration today. The method is high risk because there are two key areas of uncertainty: prediction of contaminant loading and quantification of leachate attenuation mechanisms. The regulatory trend is definitely toward maximizing containment and removal of the leachate before release to the environment.

A containment site is one that is capable on the basis of its geology, hydrogeology, and appropriate engineering measures to contain indefinitely both deposited materials and site-generated leachate. A natural attenuation site permits the slow migration of liquids, allowing the natural processes of attenuation and dispersion in and beyond the site to reduce concentrations of the pollutants to safe levels. Although current designs do not utilize natural attenuation capacity as a design philosophy, the natural attenuation capacity is still considered to be an important feature as a backup to the leachate collection facility.

In this chapter we examine some of the mechanisms by which natural attenuation occurs. The future of landfill disposal is clear: acceptable disposal sites will be difficult to find, and their location will be approved only after certain geologic and hydrogeologic criteria are met. In large measure, these criteria will be met if there is a natural attenuation capacity that, in the event of a failure of the containment system, will provide a safety factor to mitigate detrimental effects to the environment.

Table 12.1 THE PRIMARY DIFFERENCE BETWEEN A CONTAINMENT SITE AND A NATURAL ATTENUATION SITE

Type	
A containment Site	is one that is capable on the basis of its geology, hydrology, and appropriate engineering measures to contain indefinitely, both deposited materials and site-generated leachate.
A natural attenuation site	permits the slow migration of liquids, allowing the natural processes of attenuation and dispersion in and beyond the site to reduce concentrations of the pollutants to safe levels.

12.2 THE IMPORTANCE OF A NATURAL ATTENUATION CAPABILITY

Historically, landfills more than approximately 20 years old servicing major municipalities were managed as natural attenuation sites. Since the early 1970s more attention has been directed to controlling the impacts of municipal landfills on the environment through the use of engineering controls. These controls include, but are not limited to, site location evaluation, surface water management, low-permeability landfill bases and caps, and leachate collection and treatment. However, even to this day, some smaller municipalities (e.g., towns and villages) currently use nonengineered sites that are basically open dumps with only limited operations to control litter and vermin.

At engineered sites the quantity of leachate discharged to the natural environment is generally some value larger than the theoretical value of zero. Even for those sites designed to achieve zero discharge, some discharge may still occur. Thus, for both engineered and non-engineered sites, natural attenuation processes still play an important role.

Some other situations in which natural attenuation can play an important role are, as follows:

- Spills
- Manufacturing facilities, both operational and nonoperational
- Land treatment systems
- Areas where chemicals are introduced to the ground by humans.

Due to their methods of operation, most natural attenuation landfills generate considerable volumes of leachate with elevated concentrations of an assortment of chemical compounds. Typical chemical compounds and concentrations were discussed in Chapter 11.

As a direct result of the generation of leachate within a landfill, and the considerable levels of contamination that may be associated with the leachate, the potential for creating polluted groundwater is certainly evident. Often, these landfills are located either close to or within aquifers that supply groundwater to the local community. Nearly one-half of the U.S. population use groundwater from wells or springs as their primary source of drinking water (Water Resources Council, 1978).

In fact, the potential pollution of groundwater is a serious consideration in terms of the siting of new landfills. Horror stories exist about the impact that historical disposal practices have had upon the groundwater. Fried (1975) and Garland and Mosher (1975) cited several examples of pollution, and Apgar and Satterthwaite (1975) gave an example of severe economic damage incurred when a drinking water aquifer was polluted by leachate from a landfill in New Castle, Delaware. Leachate from the landfill migrated more than 800 ft and contaminated the Potomac aquifer four years after the landfill site had been closed.

Obviously, the likelihood of leakage is much greater at old sites, since many of them do not have a leachate collection system in place. However, even for new systems, the possibility exists for failure of the leachate collection system, so the potential for release of contaminants to the natural environment must be considered.

12.3 COMPONENTS OF A NATURAL ATTENUATION SYSTEM

A natural attenuation landfill is designed to utilize the natural, in-place soils and the groundwater flow system beneath the site to attenuate or cleanse contaminants leached from the solid wastes. This typically occurs primarily in the unsaturated zone beneath the base of the landfill. As the leachate migrates toward the water table, various physical, chemical, and biological reactions occur. Upon reaching the water table, the chemical concentrations are further reduced by dilution by the groundwater flow system underlying the site and continuation of the physical, chemical and biological reactions. In addition, when the mixed leachate/groundwater migrates beyond the area impacted by the verti-

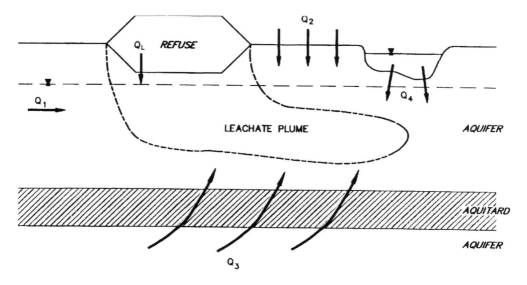

Fig. 12.1   Dilution schematic.

cal migration of leachate from the landfill (Q_L in Figure 12.1), the chemical concentrations are further reduced by infiltration of clean recharge waters (Q_2 in
Figure 12.1).   Although dilution reduces the chemical concentrations of the
leachate, the actual mass of chemicals is unaffected.

Attenuation, then, includes numerous processes, as indicated in
Figure 12.1.   The effectiveness of the natural environment to attenuate solute
levels depends on numerous factors, including hydraulic, geologic, geochemical,
and microbiological phenomena, and those listed in Table 12.2.   Although each of
the individual processes has been investigated, the combined effect in various
hydrogeologic environments is difficult to assess.

In some cases, natural processes are sufficiently effective to prevent problems of groundwater pollution.   For example, clay soils have a greater capacity
for physicochemical attenuation of contaminants than do coarse sands or fis-
sured rock.   In permeable geologic deposits, contaminants not affected by chemical attenuation can be transported great distances.   Most of the processes have
limited and/or time-dependent capacities that must be considered in design.

12.4   QUANTIFICATION OF NATURAL ATTENUATION CAPABILITY

The following sections describe in detail the features required for the natural
attenuation site.

Table 12.2   MECHANISMS OF NATURAL ATTENUATION

Physical Processes

filtration

dilution

hydraulic dispersion

volatilization

physical sorption

Chemical Processes

molecular diffusion

adsorption/absorption/desorption

precipitation

co-precipitation

ion exchange

oxidation/reduction

Biological Processes

microbial biodegradation

12.4.1 Physical Attenuation Processes

Physical attenuation processes include filtration, dilution, hydraulic dispersion, volatilization, sorption, and diffusion.

Filtration   Filtration is the physical process of removing suspended solids from the landfill leachate as it migrates through the soils underlying, and downgradient of, the landfill. Thus, any precipitates formed are removed by natural filtration in the soil profile near the point of formation. The degree of filtration is initially a function of the soil characteristics, but if there is sufficient removal, the filtration may become a function of the material due to clogging.

The degree of filtration provided depends on (1) the grain size of the geologic materials through which the landfill leachate migrates and (2) the particle size of the suspended solids contained in the leachate. These components may consist of microorganisms (e.g., bacteria and viruses) and metal oxides and hydroxides.

In general, the finer the geologic material, the smaller the particle size that will be filtered. The process and effects of filtration are often observed in artificial infiltration basins. For example, even at a suspended solids loading as low as 10 mg/L (Rice, 1974) clogging occurred in the coarse-textured soils used as the basin floor.

It has historically been accepted that colloid-sized particles (10^{-6} m) do not move significant distances through the subsurface. However, a recent paper by Mills et al. (1991) indicated that "colloidal particles can move through the subsurface thereby providing a mobile solid phase for contaminant migration" and that "if contaminants do associate with colloid particles, the net rate of migration of chemicals would be significantly enhanced." The paper continues with a brief description of a model (COMET) whose development was supported by the EPA to simulate colloid-facilitated transport of metals.

Dilution   Dilution is the physical process of decreasing leachate concentrations by adding non-leachate water to the leachate. At a landfill, the addition of such "clean" water occurs primarily by two methods: (1) via upgradient groundwater, Q_1 in Figure 12.1, and (2) infiltration downgradient from the landfill, Q_2 in Figure 12.1. Other potential sources of clean water are upwelling (Q_3) and contributions from streambeds, where the stream acts as a source of recharge water (Q_4).

Hydraulic Dispersion   As a plume migrates, it tends to spread out in both the longitudinal and the lateral (vertical/horizontal) direction. Hydraulic dispersion (or mechanical dispersion) is one of the two processes that affect this spreading of the plume. The other, molecular diffusion, will be discussed in a subsequent section. Hydraulic dispersion is caused by mechanical mixing during fluid transport and occurs on both the microscopic and macroscopic scale. A schematic of the effect of dispersion is shown in Figure 6.1.

According to Freeze and Cherry (1979), on the microscopic scale, dispersion is caused by three mechanisms. The first occurs in individual pore channels because molecules travel at different velocities at different points across the pore channel due to the drag exerted on the fluid by the roughness of the pore surfaces. The second process is caused by the difference in pore sizes along the flow paths followed by the water molecules. Due to the differences in surface area and roughness relative to the volume of water in individual pore channels, different pore channels have different bulk fluid velocities. The third dispersive process is related to the tortuosity, branching, and interfingering of pore channels. The spreading of the solute in the direction of bulk flow is known as *longitudinal dispersion*. Spreading in directions perpendicular to the flow is called *transverse* or *lateral dispersion*. Longitudinal dispersion is normally much larger than lateral dispersion.

Dispersion is a mixing process. Qualitatively, it has an effect similar to turbulence in surface water regimes. For porous media, the concepts of average linear velocity and longitudinal dispersion are closely related. Longitudinal dispersion results when some of the water molecules and solid molecules travel more rapidly than the average linear velocity and some travel more slowly. The solute therefore spreads out in the direction of flow, resulting in declines in the peak concentration.

Macroscopic-scale dispersion can be subdivided into two categories, large scale and small scale. Large scale refers to variations in materials (i.e., heterogeneities) that can be identified and mapped by careful drilling, sampling, and subsurface geophysical logging, such as a thin layer of coarse-grained sand within a matrix of fine-grained sand. Small-scale variations usually cannot be identified by conventional techniques.

According to Freeze and Cherry (1979), "in granular aquifers, heterogeneities of this type are ubiquitous. Hydraulic conductivity contrasts as large as an order of magnitude or more can occur as a result of almost unrecognizable variations in grain-size characteristics." These variations can significantly affect the pathway, speed, and spread of the plume. As a direct result of the nature of the processes (spreading, rather than removing), dispersion becomes less effective as the length of the distribution increases.

For conservative and/or poorly reactive contaminants such as chloride, sodium, potassium, sulfate, and certain refractory organic species, dilution and dispersion represent the only means of peak concentration reduction. Dispersion affects all contaminant concentrations and is the major driving force, when compared with diffusion, in high-permeability soils such as sands and gravels.

Volatilization   Volatilization is the process by which a chemical evaporates to the vapor phase from another environmental phase. It occurs primarily in the unsaturated zone and at the water table surface. Volatilization of organic compounds can be an important migration pathway of organic chemicals with a high Henry's law constant (e.g., vinyl chloride). Lyman et al. (1990) elaborated on several estimation methods for evaluating this migration pathway.

Due to the current state of environmental awareness (e.g., separate household hazardous waste collections to avoid disposal in MSW sites), it is presumed that the volume of volatile organic compounds being deposited at natural attenuation landfills is currently very small. Historically, however, significant volumes of such chemicals were commonly disposed of in natural attenuation landfills.

Physical Sorption   In physical sorption the solute is held to the surface of the soil particles by van der Waals forces (Mortimer, 1968). It is currently believed that the organic content of a soil is the primary factor in determining the amount of sorption that can occur. Thus, sorption plays a greater role in soils that have a higher organic content, such as clays.

Diffusive Processes   Diffusive processes create mass spreading due to molecular diffusion in response to concentration gradients. Molecular diffusion is particularly important in the spreading of contaminants in directions perpendicular to the direction of fluid movement. However, because molecular diffusion processes are extremely slow, their effect is significant primarily in situations with slow-moving, small contaminant pulses. The diffusive process, as compared with hydraulic dispersion, is the major driving force in low-permeability soils such as clays.

12.4.2 Chemical Attenuation Processes

Precipitation   Precipitation usually involves the formation of insoluble salts of multivalent metallic ions. The major anions involved include carbonates (CO_3^{2-}), hydroxides (OH^-), silicates (H_3SiO_4,) and phosphates (PO_4^{3-}). The advantages of precipitation as a mechanism for contaminant removal are its high capacity and its low reversibility.

The two main types of precipitation reactions are (1) ion exchange phenomena in which, for example, Na^+ ions exchange for Ca^{2+} ions on colloid particles, causing the colloids to clog the pores, and (2) chemical precipitation of sulfides, hydroxides, and carbonates under anaerobic conditions.

Examples of the latter include the following:

Metal		Anion		Precipitate
Zn^{2+}	+	S^{2-}	$\rightarrow$	ZnS
Pb^{2+}	+	$2OH^-$	$\rightarrow$	$Pb(OH)_2$
Mn^{2+}	+	$CO_3{}^{2-}$	$\rightarrow$	$MnCO_3$

Griffin et al. (1976) showed that precipitation was operative in the removal of Pb^{2+}, Zn^{2+}, Hg^{2+}, Cd^{2+}, Cu^{2+}, and Cr(IV). In general, precipitation improves with increasing pH. For example, precipitation of the heavy metal cations in leachate is an important attenuation mechanism at pH values of 5 and above. The results of Fuller and Korte (1975, 1976) and Farquhar (1977) indicated that most multivalent ions except calcium and magnesium are kept in low concentrations through precipitate formation. The basic chemistry of metallic ion solubility is described in Stumm and Morgan (1970).

Precipitation removal tends to have a high capacity in comparison to sorption and most often occurs close to the base of the landfill. If precipitation is going to occur, it occurs quickly in that it is not affected by rates of formation, or kinetics.

Sorption and Ion Exchange   The processes of adsorption, absorption, and desorption, partition the contaminants between the groundwater and the mineral or organic solids in the soil. It is difficult to differentiate between sorption processes and ion exchange reactions (to be discussed below) in experiments; however, they are different (sorption may cause a lowering of total dissolved solids, whereas ion exchange does not.) Therefore, sorption may be considered to attenuate leachate, whereas ion exchange reactions simply change the type of ions present in the exfiltrate (Bagchi, 1987).

Sorption processes thus transfer contaminants from the liquid phase to the solid phase—the sorption is due to surface tension and affects both ions and organics. The degree of sorption improves with high contents of clay, organic carbon and/or hydrous oxides. Cation exchange, electrostatic forces, and hydrophobic forces are also involved.

The degree of adsorption is typically summarized in the form of equilibrium adsorption isotherms, such as depicted in Figure 12.2. For further examples, see

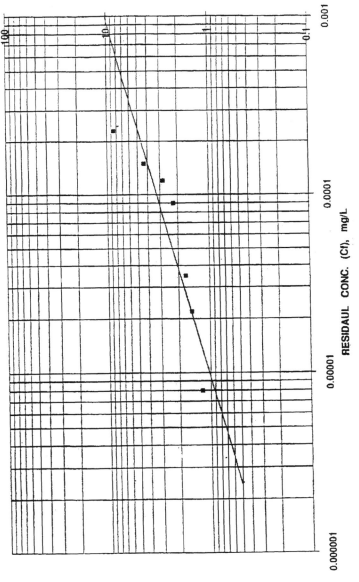

RESIDAUL CONC. (Cf), mg/L

X/M, mg ABSORBED/ gm CARBON

Fig. 12.2

Dominico and Schwartz (1990). The isotherm curves are generated from experiments that contact soil with contaminants and are specific to the contaminant and the soil. There are limited numbers of adsorption sites, so the contaminant loading is important. Mathematically, the Freundlich isotherm data assume the form

$$s = kC^n \tag{12.1}$$

where:

s　　　= (mg contaminant adsorbed)/(kg soil), which is approximately ppm

C　　　= equilibrium concentration,

k and n = constants.

Note that adsorption processes are often controlled by kinetics, not equilibrium.

Complications in the assessment of adsorption arise in that contaminants compete for the available adsorption sites. For example, chemical X may adsorb preferentially in comparison with chemical Y. Also, desorption may occur when the concentration of chemical X in the leachate is reduced. Adsorption represents a collection of processes that remove contaminants into or onto solid soil surfaces. Ion exchange, surficial adhesion, and internal diffusion are three of the processes involved (Weber, 1972). Some of the sorption processes are highly reversible and therefore can result in the return of the chemical to the fluid phase as conditions change. Ion exchange processes are poorly reversible. Capacity and reaction rate limitations affect most adsorption processes, depending on the conditions.

The taking up, and giving off, of positively charged ions by a soil is referred to as *cation* or *base exchange*. The total cation exchange capacity (CEC) of a soil is defined as the number of milliequivalents (meq) of cations that 100 g of soil can adsorb. The CEC of a soil depends on the quantity of mineral and organic colloidal matter present in the soil matrix.

Typical CEC values, at a pH of 7, are 100 to 200 meq/100 g for organic colloids, 40 to 80 meq/100 g for 2:1 clays (montmorillonite minerals), and 5 to 20 meq/100 g for 1:1 clays (kaolinite minerals). The reported CEC values are affected by the pH of the solution, dropping to about 10 percent of the given values at a pH of 4.

EXAMPLE

Assume that the CEC of a landfill liner material is 100 meq/100 g. If the density of the clay material used in the liner is 2195 kg/m^3 (137 lb/ft^3) or specific gravity of 2.2, then 105,970 meq of cations can be adsorbed per cubic meter of liner material. For a value of 20 mg/meq for the heavy metals, 2,120 g of metal can be adsorbed per cubic meter. If the concentration of heavy metals in the leachate is 100 mg/L, then 1 m^3 of the liner material can remove heavy metals from about 21,200 L of leachate.

In general, heavy metals are removed in natural attenuation sites by ion exchange reactions, whereas the trace organics are removed primarily by sorption. The ability of a soil to retain the heavy metals found in leachate is a function of the cation exchange capacity of the soil.

Anion exchange falls more in the category of specific ion sorption. The anion chloride is considered a conservative or noninteracting ion. It has a negative charge and is attracted only by the positively charged sites on the soil surface. At neutral pHs, these positively charged sites occur mostly at the edge of clay minerals. Oxygen and hydroxyl (OH^-) are the negative ions constituting the clay lattice, so to substitute for oxygen in the lattice, the anion must be approximately the same size. The chloride ion is approximately two and one-half times as large as the oxygen ion and hence is unable to replace or coordinate with oxygen or the hydroxyl ions. In contrast, the fluoride ion is about the same size and can replace and coordinate with the oxygen and hydroxyl ions (Griffin, 1982).

Table 12.3 summarizes the different types of constituents. Sorption capacity appears to increase with increased soil surface area, free lime content, pH, and organic content.

Table 12.3 EFFECTIVENESS OF SORPTION FOR DIFFERENT CONSTITUENTS

Type	Removal level	Constituent
Metals and other	good	Heavy metals, ammonia
Cations	poor	Sodium (Na^+), potassium (K^+)
Anions	good	Phosphate (PO_4^{3-})
		Chromate (CrO_4^{2-})
		Dichromate ($Cr_2O_7^{2-}$)
		Arsenate (AsO_4^{3-})
	poor	Chloride (Cl^-)
		Sulfate (SO_4^{2-})
		Nitrate (NO_3^-)
Organics	good	Hydrophobic, charged, e.g., pesticides, PCBs
	poor	Hydrophilic, neutral, e.g., most organic material

Sorption of heavy metals is strongly reduced in the presence of high amounts of organic acids, especially in soils with a low organic matter content (Hoeks et al., 1979). Griffin et al. (1976b) and Griffin and Shimp (1978) found that many of the heavy metals were actively sorbed onto clay minerals. The mechanism of removal was identified as ion exchange increasing in intensity with increased concentration and pH. Cation removal was accompanied by Ca^{2+} release to the liquid phase. For some metals, a limiting concentration was defined beyond which precipitation of the cation occurred. As an example, Pb^{2+} ion exchange prevails as a mechanism up to pH near 7.5, at which point precipitation begins to be dominant as the removal mechanism.

In assessing the suitability of a site for waste disposal, it is clear that the presence of rocks and soils with high attenuating capacities (i.e., high CEC values) is a positive attribute (Fuller, 1980).

Solubilization   The reaction of carbon dioxide with water results in the formation of carbonic acid:

$$CO_2 + H_2O \rightarrow H_2CO_3 \qquad (12.2)$$

If solid calcium carbonate is present in the soil structure, the carbonic acid will react with it and form calcium bicarbonate ions:

$$CaCO_3 + H_2CO_3 \rightarrow Ca^{2+} + 2HCO_3^- \qquad (12.3)$$

Similar reactions occur with magnesium carbonates. If a given free carbon dioxide concentration is present, the reaction shown in Equation (12.3) will proceed until equilibrium is reached:

$$H_2O + CO_2 \qquad (12.4)$$

$$\updownarrow$$

$$CaCO_3 + H_2CO_3 \rightarrow Ca^{2+} + 2HCO_3^-$$

Thus, any process that increases the free carbon dioxide available to the solution will cause more calcium carbonate to dissolve. The resulting increase in hardness is the principal effect of the presence of CO_2 in groundwater.

12.4.3 Microbial Degradation

Aerobic and anaerobic degradation processes occur as biodegradable organic materials pass through the soil. The biological degradation may initiate chemical reactions (e.g., oxygen depletion by microbial degradation processes, creating anaerobic conditions and the initiation of redox chemical reactions). The oxidation–reduction or redox potential is a critical factor that, along with pH, controls many chemical and biological reactions. The redox potential is a measure of the tendency of the system to oxidize or reduce. Thus, there is a great deal of interrelation among the various processes.

Of all the contaminants in landfill leachate, particularly young leachate, the highest concentrations are exhibited by organic matter, expressed as BOD. These may exist in the range of 10,000 mg/L and thus represent a significant potential source of groundwater contamination. Nevertheless, despite the concentrations of organic matter in leachate, incidences of groundwater contamination by these materials have been identified only rarely. As demonstrated by Soyupak (1979), the minimal incidence of BOD-contaminated aquifers results mainly from the active microbial decomposition in soils adjacent to the landfill. Even biological action that produces microbial slimes, which in turn cause soil clogging, creates particles that can be removed by filtration or adsorption.

The details of the biological mechanism were examined in Chapters 4 and 5 and thus will not be repeated here. Suffice it to indicate the general character of the equation as

$$\left\{ \text{fatty acids} \right\} \atop \left\{ \quad \right\} \rightarrow \atop \left\{ \text{cellulose} \right\}$$

$$\begin{array}{l} CO_2, NH_3, CH_4, H_2S, N_2, \text{ and } NO \\ + \\ H_2O \\ + \\ \text{cell mass} \end{array}$$

It is sufficient to acknowledge that the region immediately below a liner is frequently robust with biological action. The availability of nutrients and high BOD values have frequently combined to create a very effective removal process. Robinson and Blackley (1991) reported that anaerobic conditions rapidly became established within the full 6 m depth of the unsaturated zone beneath a landfill by diffusion of landfill gases well in advance of any leachate percolating from the waste at Stangate East landfill site. Hoeks and Borst (1982) indicated that leachates high in volatile fatty acids experienced extensive removal of these acids in soils by methane fermentation. Because adsorption of leachate organic matter onto soil surfaces has been found to be negligible (Griffin et al. 1977), biological degradation represents the primary attenuating process. Biodegradation is an irreversible process.

Microbial methylation is also an important process in the landfill environment. The Hg^{2+} ion is readily converted to the highly volatile methyl forms in both aerobic and anaerobic sediments. Arsenic and selenium have also been shown to be methylated by microorganisms (Wood, 1974).

12.4.4 Summary of Attenuation Mechanisms

As apparent from the number of processes briefly described, the subsurface migration of many organic and inorganic contaminants can be attenuated. However, several qualifying factors require careful consideration.

1. The degree of attenuation depends on the charge and physical and chemical character of the leaching chemical.
2. The attenuation capacity of materials is limited. The capacity limit is more likely to be exceeded in large landfills with large volumes of leachate, as opposed to smaller landfills.
3. The degree of attenuation depends on the charge, physico-chemical nature, and available migration pathways of the matrix through which migration occurs (e.g., in a bedrock unit in which fracture flow is the primary route of migration, the attenuation capacity of the geologic materials is low and thus will have little influence on contaminant migration).

As a result of these considerations, natural attenuation is generally recognized as unsuitable for all but the smallest MSW landfills.

Inward Gradient Landfills An additional type of attenuation landfill is an inward gradient (zone-of-saturation) landfill. This type of landfill is developed by excavating into a saturated clay soil environment. The excavation remains dry prior to refuse placement because the rate of evaporation exceeds groundwater seepage from tight clay deposits (Kmet, 1991). Once the base and sidewalls of the excavation are covered with refuse, inflowing groundwater and percolation from rainfall and refuse decomposition and compression have to be removed to maintain the inward gradients. Theoretically, as long as both vertical and horizontal gradients remain toward the landfill, the potential for groundwater contamination is eliminated.

A major concern with such a landfill configuration is the assurance that such a site is within a "homogeneous" clay environment. As a result, agencies typically require 1 ft thick granular drainage blankets to maintain dry base conditions (Gordon and Huebner, 1984). In addition, it is prudent to avoid disposal of liquid waste at zone-of-saturation sites and to avoid clay in daily cover material in order to minimize the washing of fines into the granular drainage blanket.

Impact of Liners Griffin et al. (1976a) investigated and evaluated the attenuating properties of clay minerals used as liners (natural or artificial) for sanitary landfills. Their studies indicated that chloride, sodium, and water-soluble organic compounds (COD) were relatively unattenuated by passage through the clay. Potassium, ammonium, magnesium, silicon, and iron were moderately attenuated. Metals such as lead, cadmium, mercury, and zinc were strongly attenuated, whereas concentrations of calcium, boron, and manganese were markedly higher in the effluents than in the original leachate.

12.5 ASSESSMENT FACTORS FOR A NATURAL ATTENUATION SITE

The unsaturated zone may be thought of as a potential buffer zone that offers protection to the underlying groundwater. The various physical, chemical, and biological processes alter the nature and quantities of the contaminants arriving at the water table as a function of travel time within the unsaturated zone. Not surprisingly, this region is frequently characterized by high microbial activity, which promotes biodegradation. Vapor-phase transport also may be responsible for migration losses of the volatile contaminants. Precipitates may be filtered out, changing the hydraulic conductivity of the soil regime. However, these are only some examples of the attenuation processes.

The ideal natural attenuation site has uniform, moderately textured soil (soils with moderate amounts of silt and clay) of considerable thickness between the bottom of the waste and the water table. These soils are desirable because they have a relatively high amount of surface area and a moderate to low rate of water movement through them. This provides more opportunity and time for attenuation mechanisms to occur as the leachate percolates through them prior to entering the groundwater flow system. Heavily textured soils such as tight clays should be avoided for this type of design because they do not allow passage of leachate at a rate sufficient to prevent the buildup of a head of leachate within the site unless provisions are made for the removal and treatment of such leachate. In some situations, leachate removal and treatment may be required only during the operating period of the site. For example, if the site is capped on closure with a cap that allows less infiltration to pass through it than will pass through the base of the landfill, then it may be possible to locate the landfill within a deposit of heavily textured soils.

In assessing existing landfills and in designing new ones, consideration of leachate strength must center initially on the contaminant type and concentration and its variation with respect to time. The problems of contamination tend to occur when (a) the extent of contaminant transport is high, (b) the point of compliance is close to the source, and (c) the quality standards imposed are stringent.

Obviously it is not going to be trivial to assess the importance of these various features in the design of a natural attenuation site. The best that can usually be accomplished in design is a reasonable characterization utilizing conservatively low assumptions (i.e., those that overpredict potential leachate volumes and chemical concentrations).

12.6 MATHEMATICAL MODELING

12.6.1 Introduction

The previous sections focused on the elements of natural attenuation mechanisms. Now we turn to the use of mathematical models as a means of evaluating the effectiveness of the various mechanisms. Thus, models are used to gain an understanding of the movement of landfill-generated leachates into the subsurface. This movement depends on the geologic and hydrologic character of the site as well as the physical and chemical interactions between the leachate and the soil.

A mathematical model in the context of natural attenuation is the solution of an equation or a set of equations. The equations range from "black box" models, such as linear regression, to those that describe the physics underlying the movement of a convecting fluid and its dissolved chemical constituents. Physics-based governing equations are based on the conservation principle, augmented by constitutive relationships and proper initial and boundary conditions.

The array of available mathematical models for examining natural attenuation concerns is considerable. The models range from the simple to the complex and are the foci of entire books themselves. Our intent is to provide only a very brief coverage and an indication of some of the technical literature.

12.6.2 Linear Regression Models

One form of linear regression model that has been developed by the EPA to estimate the leachate concentration for organic compounds migrating from the base of a landfill from waste concentration (Poppiti, 1985) is

$C'_{(1)}$	=	2.14×10^{-5} CS	for C ≤1 mg/kg
$C'_{(2)}$	=	$0.044\ C^{0.7}\ S^{0.31}$	for C > 10 mg/kg
C'	=	$C'_{(1)} + m\ (C - 1)$	for $1 < C < 10$ mg/kg, m = slope between $C'_{(1)}$ and $C'_{(2)}$ at 1 and 10 mg/kg respectively.
or C'	=	$2.11 \times 10^{-3}\ C^{0.678}\ S^{0.373}$	

where:

C'	=	leachate concentration (mg/L),
C	=	waste concentration (mg/kg),
S	=	water solubility of organic compound (mg/L).

This model does not take into account any of the natural attenuation processes that occur within the landfill base liner.

12.6.3 Simple Plume Delineation Models

The simplest type of leachate plume model, based on system physics, is one that mixes with groundwater and moves at the same rate and in the same direction as the groundwater. In this case, plume velocity (ignoring attenuation and other geochemical factors) can be expressed using Darcy's law (from Equation 6.2) as

$$\bar{v} = -\frac{Ki}{n} \qquad (12.5)$$

where:

$\bar{\upsilon}$ = average linear velocity of the plume (L/T),

K = hydraulic conductivity of the aquifer (L/T),

i = the hydraulic gradient or the slope of the water table or the potentiometric (pressure) surface (L/L),

n = the porosity of the aquifer.

The direction of movement of simple plumes is controlled by the gradient of the potentiometric surface. In unconfined aquifers, this gradient commonly follows the surface topography. The gradient, however, can be affected by a variety of hydrogeologic and geologic factors. These factors include:

- Aquifer geometry — the shape of the water-bearing zone through which the groundwater flows
- Aquifer uniformity — the consistency of the hydrogeologic properties of the aquifer
- Geologic discontinuities — faults, fractures, bedding planes, joints, and other planes of greatly different permeability
- Hydrogeologic barriers — areas of recharge or discharge that control potentiometric gradients.

12.6.4 More Sophisticated Modeling

The model described in the previous section is the simplest of model types based on system physics and will not demonstrate attenuation (but will denote the contaminant front for the conservative ions). More complicated models may be structured on the basis of the advection-dispersion equation. In a general form, the advection-diffusion equation may be written as

$$\frac{\partial c}{\partial t} = \frac{\partial}{\partial x}\left('D\frac{\partial c}{\partial x}\right)\bigg|_{y,z} - \upsilon\frac{\partial c}{\partial x}\bigg|_{y,z} - 'R(x,t)\ \big|_{y,z} \qquad (12.6)$$

where:

c = contaminant concentration at a point at some time,

　　　x, y, and z indicate coordinate directions,

y, z = evaluation at location y and z,

υ = interstitial velocity (L/T),

R = contaminant reaction,

D = dispersion coefficient (L^2/T), which can also be written as $D_d + D_m$.

D_d is the coefficient of mechanical dispersion and D_m is the coefficient of molecular diffusion. It is by means of the dispersion coefficient term, which peaks as a result of pulse loadings of leachate, that contaminants will be attenuated or dispersed. This is a potentially important attenuation mechanism for some problem types. Equation (12.6) may be written in one, two, or three dimensions depending on the problem being analyzed.

The assignment of R reflects the various attenuation mechanisms (i.e., sorption, precipitation, and biological decomposition). The magnitude of R is, of course, very contaminant-specific. Sykes et al. (1982) expanded the R term to two components, one for biochemical reaction and the second for adsorption. Additional attenuation by dilution is reflected by, for example, influx of infiltration water.

Although the equation can reflect the various attenuation mechanisms, the assignment of the magnitudes is, of course, a considerable challenge. There is also the problem of the numerical aspects of solving the resulting equation. For applications of nonisotropic and nonhomogeneous media, the solution of the resulting equations is a considerable challenge, and numerical solutions involving use of finite elements or finite difference procedures are essential. However, when simplifying assumptions can be made, it becomes possible to integrate the differential equation and thus to develop an analytic solution.

The number of analytic solutions is substantial, and the interested reader is referred to van Genuchten and Alves (1982) and Javandel et al. (1984) for further information. We limit our discussion to the Ogata-Banks solution, which is a useful model for the situations in which the necessary assumptions used in deriving the equation are appropriate.

EXAMPLE

The following is an outline of the procedure to estimate groundwater concentrations for organic compounds from waste or leachate concentrations.

1. If only waste characterizations are available, the model described in Section 12.6.2 can be used to estimate leachate concentrations migrating from the base of the landfill. If possible, it is preferable to collect and analyze a leachate sample to obtain laboratory results and forego the modeling step.

2. After obtaining the leachate concentration, C', it is necessary to use a linkage model for migration through the unsaturated zone to estimate the strength of the leachate when it contacts groundwater. Various models are available. A simple one is (Donigan et al., 1983)

$C_0 = (C' qL)/\upsilon\ b$

where:

C_0=concentration in the saturated zone (mg/L),

C'=previously defined,

q=recharge rate (cm/yr),

L=width of leachate plume at the water table (m),

υ =   Darcy velocity (cm/yr),

b   =   effective saturated aquifer thickness or zone of mixing, beneath the site (m).

Note that this method does not explicitly account for the thickness of the unsaturated zone, and the attenuation processes in the unsaturated zone, through which the leachate must in fact travel before reaching the water table.   Other one-dimensional models have been developed that take into account various unsaturated zone processes (e.g., Thomson, 1992).

3. After the leachate reaches the groundwater table, various models, as mentioned in Sections 12.6.2 and 12.6.4, can be used to estimate groundwater concentrations at the downgradient compliance point.

12.7   SUMMARY

The following generalizations regarding natural attenuation can be made:

- Some contaminants are poorly attenuated by natural processes.
- Some soils attenuate poorly, for example, sandy soils and acidic soils.
- Capacity limits exist for attenuation mechanisms.
- Desorption may occur.
- Accurate predictions at the design stage are very difficult because the data for soils and leachates are limited.

Therefore, to be effective, attenuation should be relied on only for small landfills or landfills with large buffer zones, and/or treatment layers (lime, carbon, peat, and the like) should be used below the landfill to remove specific contaminants.

References

Apgar, M., and Satterthwaite, W. B. 1975. "Groundwater Contamination Associated with the Llangollen Landfill, New Castle, Delaware." In *Proceedings of the Research Symposium on Gas and Leachate from Landfills: Formation, Collection and Treatment*, Rutgers University, New Brunswick, N.J., March 15–16.

Bagchi, A. 1987. Natural Attenuation Mechanisms of Landfill Leachate and Effects of Various Factors on The Mechanisms. *Waste Management and Research* 4: 453–464.

Dominico, P. A., and Schwartz, F. W. 1990. *Physical and Chemical Hydrogeology*. John Wiley, New York.

Donigan, A., Yo, T., and Shanahan, E. 1983. *Rapid Assessment of Potential Groundwater Contamination Under Emergency Response Conditions*. U.S. EPA, Washington, D.C.

Farquhar, G. 1977. *Leachate Treatment by Soil Methods: Management of Gas and Leachate in Landfills*. U.S. EPA Research Symposium Series, St. Louis, Mo.

Freeze, R. H., and Cherry, J. A. 1979. *Groundwater*. Prentice-Hall, Toronto.

Fried, O. 1975. *Groundwater Pollution*. Elsevier, Amsterdam, The Netherlands, 330 pp.

Fuller, W. 1980. Soil Modification to Minimize Movement of Pollutants for Solid Waste Operations. *CRC Critical Reviews in Environmental Controls*, 9 (3).

Fuller, W., and Korte, N. 1975. "Attenuation Mechanisms of Pollutants through Soils." In *Proceedings of the Research Symposium on Gas and Leachate from Landfills: Formation, Collection and Treatment*, Rutgers University, New Brunswick, N.J., March 15–16.

Fuller, W., and Korte, N. 1976. "Attenuation Mechanisms of Pollutants Through Soils." *Gas and Leachate from Landfills*. ed. E. J. Genetelle and J. Cirello. U.S. Environmental Protection Agency EPA-600/9-76-004, Cincinnati, Ohio, pp. 111–122

Garland, G., and Mosher, D. 1975. Leachate Effects from Improper Land Disposal, *Waste Age* 6:42–48.

Gordon, M. E. and Huebner, P. M. 1984. "An Evaluation of the Performance and Zone of Saturation Landfills in Wisconsin." Presented at the Seventh National Groundwater Quality Symposium, Las Vegas, Nevada, Sept. 26–28.

Griffin, R. A. 1982. "Mechanisms of Natural Leachate Attenuation." Presented at Sanitary Landfill Design Course, University of Wisconsin Extension at Madison, February.

Griffin, R., and Shimp, N. 1978. *Attenuation of Pollutants in Municipal Landfill Leachate by Clay Minerals*, EPA-600/9-80-010.

Griffin, R., Cartwright, K., and Shimp, N. 1976a. Attenuation of Pollutants in Municipal Landfill Leachate by Clay Minerals, *Environmental Geology Notes*, no. 78, *Illinois State Geological Survey,* November

Griffin, R., Frost, R., and Shimp, N. 1976. "Effect of pH on Removal of Heavy Metals from Leachate by Clay Minerals. Residual Management by Land Disposal." *Proceedings of Hazardous Waste Research Symposium*, Tucson, Arizona, February.

Griffin, R., Frost, R., An, A., Robinson, G., and Shimp, N. 1977. Attenuation of Pollutants in Municipal Landfill Leachate by Clay Materials. Part 2, Heavy Metal Adsorption. *Environmental Geology Notes*, No. 79. Illinois State Geological Survey, Urbana.

Hoeks, J., Beker, D., and Borst, R. 1979. *Soil Column Experiments with Leachate from a Waste Tip* 2. *Behavior of Leachate Components in Soil and Groundwater.* Institut voor Cultuur Technik & Watershurshouding. Wageningen, August.

Hoeks, J., and Borst, R. 1982. Anaerobic Digestion of Free VFA in Soils Below Waste Tips. *Water Air and Soil Pollution* 17:165–173.

Javendel, I., Doughty, C., and Tsang, C. F. 1984. *Groundwater Transport: Handbook of Mathematical Models*. AGU Water Resources Monograph 10., 240 pp.

Kmet, P., 1991. "Controlling Landfill Leachate Migration in Municipal Solid Waste Management." *Making Decisions in the Face of Uncertainty*, ed. M. Haight. University of Waterloo Press, Waterloo, Ontario.

Knutson, G. 1966. "Tracers for Groundwater Investigations." *Proceedings Interim, Symposium on Groundwater Problems*. Stockholm, Sweden.

Lyman, W. J., Reehl, W. F., and Rosenblatt, D. H. 1990. *Handbook of Chemical Property Estimation Methods*. American Chemical Society, Washington, D.C.

Mills, W. B., Liu, S., and Fong, F. K. 1991. Literature Review and Model (COMET) for Colloid/Metals Transport in Porous Media. *Groundwater* 29, no 2 (March/April):199–208.

Mortimer, C. E. 1968. *Chemistry: A Conceptual Approach*. Reinhold, New York.

Poppiti, J. 1985. Memo to Workshop Participants on the VHS Model and Method to Predict Organic Leachate Concentrations from Wastes. U.S. EPA, Washington, D.C.

Rice, R. C. 1974. Soil Clogging during Infiltration of Secondary Effluent. *Journal Water Pollution Control Federation* 46:708–716.

Robinson, H., and Blackley, S. G. 1991. *Attenuation of Contaminants: How Reliable a Philosophy*. Report of Aspinwall and Co., Shrewsbury, England.

Soyupak, S. 1979. "Modification to Sanitary Landfill Leachate Organic Matter Migration Through Soil." Ph.d. thesis, Dept. of Civil Engineering, University of Waterloo, Waterloo.

Stumm, W., and Morgan, J. 1970. *Aquatic Chemistry*. Wiley-Interscience, New York.

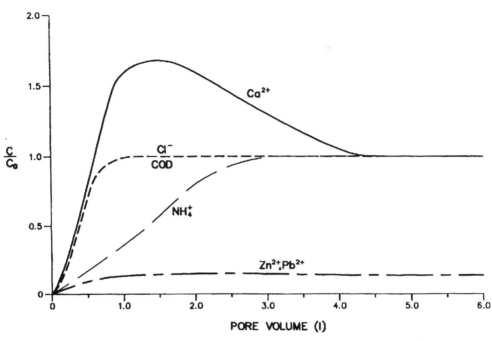

Fig. 12.3   Relative contaminant concentrations versus pore volumes.

Sykes, J., Soyupak, S., and Farquhar, G. 1982. Modeling of Leachate Organic Migration and Attenuation in Groundwaters below Sanitary Landfills. *Water Resources Research* 18 (1):135–145.

Thomson, N. R. 1992. *IDUSAT (One-Dimensional Unsaturated Flow and Transport) User's Guide*. Dept. of Civil Engineering, University of Waterloo, Waterloo, Ontario, Canada.

van Genuchten, M. Th., and Alves, W. J. 1982. *Analytical Solutions of the One-Dimensional Convective: Dispersive Solute Transport Equation*. U.S. Salinity Lab. Riverside, Calif.

Water Resources Council. 1978. *The Nation's Water Resources 1975–2000*, vol. 1. Water Resources Council, Washington D.C.

Wood, J. 1974. Biological Cycles for Toxic Elements in the Environment. *Science* 183:1049–1052.

Weber, W. J. 1972. *Physicochemical Processes for Water Quality Control*. Wiley-Interscience, New York.

12.8   PROBLEMS

12.1 Leachate passes from a landfill into the soil below, with contaminant concentrations in the leachate as listed. A section of soil below the landfill and along the line of flow has a volume of 10 m^3 and a porosity of 0.45. The data in Figure 12.3 show relative contaminant concentrations (C/C_0) as a function of the pore volumes of liquid passing out of the control volume section. Calculate the mass of contaminant removed in

the control volume and estimate the removal mechanism involved.
Leachate contaminant concentrations (in mg/L except, pH)

Ca^{2+}	Cl^-	COD	NH_4^+	Zn^{2+}	Pb^{2+}	pH
65.2	810	11500	960	12.1	9.7	6.3

12.2 In modeling contaminant transport in groundwater, the dispersion coefficient is usually obtained by fitting a numerically solved transport model to field data. The field
data generally consist of contaminant concentrations in samples collected in monitoring wells at various points in time and space. Explain the major sources of error
introduced by this process.

12.3 The following equation can be used to estimate contaminant flux through a clay
liner, due to diffusion. For a liner that has a leachate head on the liner of 0.8 m and
a leachate chloride concentration (Cl) of 1200 mg/L, calculate the leachate flux
through the liner with and without diffusion effects. Diffusion, D, for Cl is
10^{-6} cm^2/s. List the assumptions made.

$$'F = 'D\frac{dc}{dz}$$

where:

$\dfrac{dc}{dz}$ = concentration gradient (g/m^3)/m$^)$,

D = diffusion coefficient (m^2/s),

F = contaminant flux (g/m^2/s).

Permeameter Data

Compaction (kg/m^3)	Moisture Content (%)	Hydraulic Conductivity (cm/s)
1815	7.8	1×10^{-5}
1865	9.2	9×10^{-6}
1890	10.4	3×10^{-6}
1900	11.3	2×10^{-7}
1875	12.7	6×10^{-9}
1830	14.2	4×10^{-9}

12.4 If natural attenuation sites are left to themselves, their attenuation capacity is limited. Containment sites, on the other hand, have intensified certain natural attenuation mechanisms to make them more 'secure.' Identify those features of an engineered containment site that are an intensification of a natural attenuation process. Explain your answer.

12.5 A 12-year-old, unlined landfill has contaminated a limestone sand and gravel aquifer below the site, creating a contaminant plume extending 420 m downflow. The following contaminants and parameters are present in the plume to varying degrees. Assign to each a migration distance, either 20 m, 100 m, or 400 m, for 25% of the raw leachate concentrations. Describe the attenuating mechanism(s) that is likely to dominate for each

(a)	acetic acid		*(f)*	lead
(b)	ammonia		*(g)*	oxidation–reduction potential (negative)
(c)	chloride		*(h)*	pH (acidic)
(d)	conductivity		*(i)*	phenolics
(e)	iron		*(j)*	suspended solids
			(k)	trichloroethylene

LANDFILL GAS
MIGRATION

13.1 PATHWAYS OF MIGRATION

The microbiological processes and rates of generation of landfill gases were the subjects of Chapters 4 and 5. As indicated there, the production and accumulation of gases increases the gas pressures, and, under static conditions, internal landfill pressures, higher than atmospheric occur. The resulting gradient in pressure is one of the driving forces for these gases to travel beyond the confines of the refuse to exhaust to the atmosphere and/or into the surrounding soil environment. It is these migration pathways that are the subject of this chapter. Two mechanisms are responsible for the migration pathways of the gas, namely, pressure gradients and diffusion processes. It is important to understand how these mechanisms function in determining the extent of landfill gas migration within the subsurface environment around a landfill site.

The migration of landfill gas from disposal sites is an important consideration, since the landfill gases can result in serious safety and health hazards. Methane gas, one of two main landfill gas constituents, is explosive at concentrations above 5 to 15 percent by volume. Landfill gas migrates from the refuse through soils into adjacent structures and has produced explosions that have resulted in extensive property damage and, in some cases, loss of life (e.g., see Moore et al., 1979; MacFarland, 1970; and Zabetakis, 1962).

An explosion hazard develops when methane migrates from a landfill and mixes with air in a confined space, as demonstrated by the following examples:

- A garage in Toronto, Canada, situated near a former dump site exploded twice in 1966 and 1969, both times during the winter.
- Explosions took place in two houses in the vicinity of a 12-year-old dump site in Kitchener, Canada, in 1969. Concurrently, methane concentrations in excess of 5 percent of volume were measured in basements of other houses in the neighborhood, beneath the concrete floor of an ice arena 1/4 mi. from the dump site, and in the soil of a playground 1/4 mi. away.
- At the World's Fair in 1967 in Montreal, Quebec, Longeuil parking lot was paved over a former dump site. Workmen were found setting fire to gas escaping through cracks in the pavement to heat their lunch.

In addition to posing a safety and health risk, the migration of gases from landfills also poses a potentially serious problem by creating vegetation stress. Persistent high concentrations of combustible gases in the subsurface create anoxic conditions in the root zones of plants that lead to poor health and the eventual demise of the vegetation.

As an example, in the early 1970s, a peach farmer in Glassboro, New Jersey, lost a number of acres of peach trees following the filling of an adjacent 20 ft deep former sand and gravel pit with municipal domestic refuse. The pit belonged to the peach grower, who was paid by the municipality to have it filled with the refuse. He then planned to extend his peach orchard over the completed refuse landfill. However, he found that the trees planted adjacent to this landfill began to die a year or two after the refuse was placed against the bank nearest his orchard. Examination showed that the soil in areas where the trees had died contained high concentrations of the gases of anaerobic decomposition. Landfill gas concentration gradients indicated that the gases had migrated from the landfill. Although the landfill was only 7 m deep, the gases traveled 25 m from the pit edge, causing death and injury to the peach trees.

The odors caused by gaseous emissions from landfills have also necessitated the implementation of costly remedial procedures. The existence of these odors has resulted in the need to utilize flares to combust the gas due to complaints from nearby residents.

Other concerns with landfill include the following:

1. Toxicity — Hydrogen sulfide (H_2S) causes chronic conditions at concentrations greater than +20 ppm, and severe conditions at greater then +100 ppm; benzene and toluene sometimes cause problems and are examples of trace organics.
2. Vegetation stress — Landfill gas replaces oxygen within the root zone of the subsurface, creating anoxic conditions and vegetation stress.
3. Corrosion — CO_2 creates acidic conditions by the reaction:
   $$CO_2 + H_2O \rightarrow H_2CO_3$$
   Then, for example, if calcium carbonate is part of the soil structure,
   $$CaCO_3 + H_2CO_3 \rightarrow Ca^{2+} + 2HCO_3^-.$$

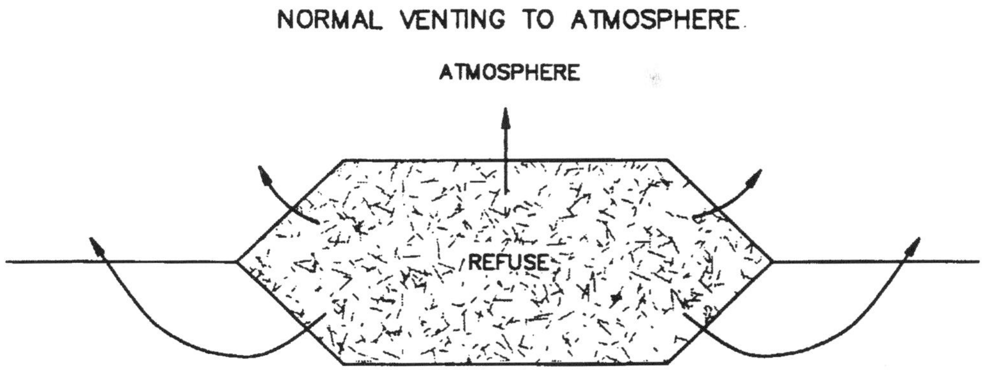

NORMAL VENTING TO ATMOSPHERE.

ATMOSPHERE

Fig. 13.1 Gas migration pathways.

Concern for these problems has stimulated extensive gas monitoring programs for existing sites and a growing trend toward the installation of gas control systems in, and adjacent to, landfill sites.

Gases generated in sanitary landfills flow in all directions from the refuse and can contaminate the soil atmosphere in areas adjacent to the landfill. Old sand and gravel pits are particularly noted for problems of lateral migration of the gases from landfills, due, in most cases, to the porosity of the adjacent soils and their lack of resistance to the movement of these gases. It has also been observed that when the adjacent soil stratigraphy consists of horizontal layers of sand and gravel interspersed with clay layers, the distance of migration is greatly enhanced, because the clay layers reduce the upward vertical movement and venting of the gases from the sand and gravel seams. Freezing of the ground surface in northern climates and extensive rainy weather also contribute to the increased lateral migration of the gases. Surface coverings such as parking lots and tennis courts also act to increase lateral migration.

Structures built on, or adjacent to, landfills usually have numerous pockets and openings in and around structures where landfill gases may accumulate. A number of such cases have been documented. The gases apparently seeped into and through the stone bedding of water or sewer pipes and into an enclosed space within the building.

Gas travels through porous media such as soil materials as a result of two processes: diffusion in response to a concentration gradient, and advection, due to a pressure gradient. It will be seen in Section 13.2 that both processes may contribute simultaneously to gas transport in soil.

Figure 13.1 shows idealized gas movement in soil away from waste disposal sites. The gas moves preferentially along paths of lowest resistance (high conductivity) with eventual discharge to the atmosphere unless otherwise prevented. If a low-permeability cover is in place (in comparison with adjacent

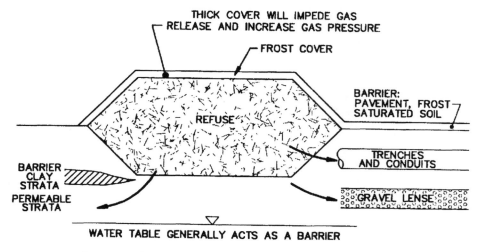

Fig. 13.2 Potential gas migration pathways and related causes.

soils), methane moves from within the fill, through the cover material and escapes into the atmosphere without causing problems to adjacent lands. This means that much of the gas will leave via the cover, with the amount of discharge increasing with cracking or fissuring of the cover material and decreasing with increasing cover thickness and decreased permeability. Some gas will move laterally into neighboring soils. Groundwater represents a lower boundary to landfill gas migration. Landfill gas will, however, move upward to discharge into the atmosphere a short distance from the landfill unless prevented by an impervious layer above, such as that created by frost, pavement, saturated soils, or clay strata or unless intercepted by highly permeable strata or inclusions. Thus, for containment sites, the extent of migration from the landfill is limited to 10 to 15 m and will cause few, if any, problems to adjacent land users.

In addition, any landfill gas generated in a saturated zone will remove rapidly to the surface of the saturated zone under buoyancy; there is virtually zero lateral movement of gas in saturated conditions. (Note that groundwater contaminated with high levels of organics will generate methane; thus, if there is significant groundwater movement, it may give the appearance of lateral movement).

Examples of situations that inhibit discharge of landfill gases to the atmosphere are schematically depicted in Figure 13.2. These activities/structures may potentially cause problems, as the landfill gas will follow paths of least resistance, in some cases migrating significant distances into neighboring soil environments. Preferential transport may occur under any of a number of conditions. In these situations, the extent of gas migration from the landfill may be much greater.

Predictions of methane migration in specific applications must include characterization of the available pathways. For existing situations, in situ monitoring programs can, of course, be utilized to identify the extent of existing migration. For example, extensive monitoring programs that were carried out on several landfills in the Los Angeles (1972) area showed that air contained in the soil pore space in a strip of 180 m (600 ft) maximum width contained volumetric methane concentrations ranging from 10 to 50 percent at a depth of 0.6 m (2 ft) below the soil surface (Esmaili, 1975).

A portable combustible gas meter is typically used to detect the presence of landfill gases. A commonly used combustible gas meter is an explosimeter that utilizes the Wheatstone bridge principle. One leg of the bridge consists of a catalytic unit that burns the combustible gases, which changes its resistance, thereby unbalancing the bridge and giving a reading on the galvanometer. This type of in situ monitoring is relatively simple and rapid. The explosimeter is much faster than using a gas chromatograph, but the explosimeter is used only for measuring methane concentrations and not for measuring concentrations of trace organics.

For situations where site conditions warrant the installation of a gas migration control system (for example, barriers, passive vents, and pumped extraction or injection systems), in situ field studies are not sufficient. In these types of situations, estimates of migration should be obtained only from mathematical models and appropriate data. The intent of the modeling is to simulate gas transport through the soil environment. It requires an understanding of how landfill gas is transported through soil materials and of the factors that influence this transport. The purpose of modeling gas migration will be to develop and understand:

1. The direction of the landfill gas migration
2. The speed or average velocity of migration
3. The concentration at a specific point and at a specific time
4. The flux (mass) of gas past a certain point and whether it may generate a potential problem (e.g., create an explosion hazard in an adjacent building)
5. The effectiveness of any proposed gas control program.

The remainder of this chapter will develop the principles that will allow such modeling assessments, including an understanding of the gas properties and of the soil environment and its influence on the migration pathways.

13.2   PROCESSES OF GENERATION AND MIGRATION

Many different phenomena determine the potential for migration of the landfill gas, including:

- Physical and chemical characteristics of the landfill gases and the refuse
- The processes that cause the gas to move within the landfill
- Those processes that affect the concentrations of the gas once it has left the landfill site.

The objective is to describe how these phenomena impact gas migration, to incorporate these phenomena into a model of the landfill site, and to develop the parameter values for use in the mathematical models.

Role of Moisture Content   An accurate representation of the moisture content distribution is important because it determines the pore space available for gas migration. As moisture content increases, gas permeability and diffusion coefficients decrease. The level of soil saturation also affects the partitioning of carbon dioxide between the gaseous and dissolved phases. In the unsaturated zone above the water table where landfill gas transport occurs, two immiscible fluids share the available pore space: the gas mixture and the residual moisture in the soil.

The hydraulic conductivity of a specified fluid then becomes a function of the fraction of the void space occupied by the fluid. By dividing the total porosity, n, into the porosity occupied by the gas mixture (ξ) and the porosity occupied by water (Θ), effective permeabilities become

$$K_g = f(\xi) \quad \text{and} \quad K_w = f(\Theta) \tag{13.1}$$

where:

n = $\Theta + \xi$ = total porosity,

K_g = effective hydraulic conductivity for the gas mixture,

K_w = effective hydraulic conductivity for the soil moisture.

Role of Absolute Viscosity   Absolute viscosity is the physical property that characterizes a fluid's resistance to flow. For gases at low density, viscosity increases as temperature increases. Values of viscosity are presented in Table 13.1 for methane, carbon dioxide, and atmospheric gas (after CRC, 1981).

Reid et al. (1977) described the following equation for adjusting the viscosity for temperature:

$$\mu = \frac{bT^{1.5}}{S + T} \tag{13.2}$$

where:

μ = absolute species viscosity (Pa·s),

T = temperature (K),

b, S = empirical constants.

Table 13.1   ABSOLUTE VISCOSITIES AS A FUNCTION OF TEMPERATURE

Gas	Temperature (·C)	Viscosity (Pa · sec) $\times 10^7$
Air	0	170.8
	18	182.7
	40	190.4
	54	195.8
Carbon dioxide	0	139.0
	15	145.7
	20	148.0
	30	153.0
Methane	0	102.0
	20	108.0

Source: CRC, 1981

This is known as the Sutherland equation. Specific values for b and S for methane, carbon dioxide, and atmospheric gas are given in Table 13.2.

Farmer et al. (1980) provided an equation for adjusting open-space diffusion coefficient data:

$$'D^{\circ}_{d_2} = 'D^{\circ}_{d_1} \left(\frac{T_2}{T_1} \right)^{\frac{1}{2}}$$
(13.3)

where the subscripts 1 and 2 denote the value of the unobstructed diffusion coefficient at temperatures T_1 and T_2 (K) respectively. Farmer et al. (1980a) also presented a method for interrelating the diffusion coefficients of different components in air, namely

$$'D^{\circ}_{d_A} = 'D^{\circ}_{d_B} \left(\frac{M_B}{M_A} \right)^{\frac{1}{2}}$$
(13.4)

where M_A and M_B are the molecular weights for compounds A and B.

Table 13.2 SUTHERLAND EQUATION EMPIRICAL CONSTANTS FOR METHANE, CARBON DIOXIDE, AND ATMOSPHERIC GAS

	Empirical Constants	
	b	*S*
Methane	9.1935	131.18
Carbon dioxide	14.841	208.61
Atmospheric gas	14.956	121.99

Source: After Reid et al., 1977

Currie (1970) presented an equation reflecting soil media, where, for dry sand

$$'D_d = D\overset{\circ}{_d} n^{\frac{3}{2}} \tag{13.5}$$

and for wet sand

$$'D_d = 'D\overset{\circ}{_d} n^{\frac{3}{2}} \tag{13.6}$$

where:

$D\overset{\circ}{_d}$ =   unobstructed diffusion coefficient (cm^2/s),

D_d =   effective diffusion coefficient (cm^2/s).

Millington and Quirk (1961) provided the alternative equation

$$'D_d = 'D\overset{\circ}{_d} \left(\frac{\xi}{n}\right)^4 n^{\frac{3}{2}} \tag{13.7}$$

Farmer et al. (1980b) and Sallam et al. (1984) found that the Millington-Quirk model provided good predictions of effective gaseous diffusion coefficients in partially -saturated porous media.

The preceding equations show that a decrease in the available pore space due to the addition of soil moisture has a different effect on the diffusion coefficient than increasing the volume of the solid phase by an equivalent amount. Table 13.3 lists unobstructed diffusion coefficient data from several sources for the gases in the vicinity of landfills.

Table 13.3 DIFFUSION COEFFICIENTS FOR BINARY MIXTURES

Binary Mixture	Temperature (·K)	$D_d^{\cdot}$ (cm^2/s)	Reference
$CO_2 - air$	273	0.139	CRC (1981)
	276.2	0.142	Reid et al., (1977)
	317.2	0.177	Reid et al., (1977)
$CO_2 - N_2$	298.0	0.167	Reid et al., (1977)
$CO_2 - O_2$	293.2	0.153	Reid et al., (1977)
$CO_2 - CH_4$	298.0	0.1705	Thorstenson & Pollock (1989)

EXAMPLE

Determine the CH_4 diffusion coefficient in a sandy soil at a gas-occupied porosity of 0.10 at 25°C, in soil with total porosity of 0.3.

Solution:

$'D_d^{\circ}$ for CO_2 in air at 317.2 K = 0.177 cm^2/s from Table 13.3

$'D_d^{\circ}$ for CH_4 in air at 317.2 K:

$'M_{A_{CH_4}}$ = 12 + 4 = 16

$'M_{B_{CO_2}}$ = 12 + 2 (16) = 14

$'D^{\circ}_{d_{CH_4}}$ = 0.177 $(44/16)^{1/2}$

= 0.294 cm^2/s

Temperature correction to 25°C or 298°K:

$$'D^{\circ}_{d_{T_2}} \;=\; 'D^{\circ}_{d_{T_1}} \left(\frac{T_2}{T_1}\right)^{\frac{1}{2}}$$

$=0.294 \,(298/317.2)^{1/2}$

$=0.277$ cm^2/s

Finally, the soil correction from Equation 13.7 is

$Dd_{CH_4} = 0.277 \,(0.1)^{10/3}/(0.3)^2$

$= 0.00143$ cm^2/s

Additional correction-adjustments for soil gas pressure and temperatures are presented in Currie, 1960; Campbell, 1985; and Rolston, 1986

13.2.1 Properties of Landfill Gas

Methane is a colorless, odorless gas that is less dense than air and combustible in concentrations in air in excess of approximately 5 percent by volume. Carbon dioxide, on the other hand, is denser than air and is not combustible. However, as the second major component in landfill gas, carbon dioxide represents a potentially serious problem because it produces acidic conditions by dissolving in groundwater (and forming H_2CO_3). This, in turn, can result in the release of alkaline earth metals from soil materials and production of a hardness halo downgradient from the landfill site (note that this hardness halo may also be caused by leachate).

At pressures of less than 3 kPa and the resultant pressure gradients, the landfill gas consisting of a mixture of CH_4 and CO_2 may be considered to be an incompressible gas.

Table 13.4   GAS PROPERTIES

	CH_4	CO_2	Landfill Gas	Air
Viscosity ($\times 10^{-5}$ Pa s)	1.03	1.39	1.21	1.71
Mass density, ρ (kg/m^3)	0.72	1.97	1.35	1.29
Molecular mass (M) (g)	16.0	44.0	30.0	28.9

Diffusion coefficient in air, D', (m^2/s) 0.157 $\times 10^{-4}$

Notes:

Landfill gas defined as 50% CH_4 and 50% CO_2; all properties reported at 1 atm and 0° C

Important gas properties for individual constituents and for landfill gas consisting of 50 percent methane and 50 percent CO_2 are listed in Table 13.4. It is apparent from the table that the mass density, ρ, of landfill gas is slightly greater than that of air.

From Table 13.4 it is also apparent that CO_2 is much denser than CH_4. However, since CH_4 and CO_2 are cogenerated within the landfill, the two gases will not separate (at least initially), although this does not preclude changes in relative concentrations of gases.

There may be CH_4 enrichment, since CO_2 is much more soluble in water than CH_4. Depending on the amount of water infiltrating within the soil profile and the extent to which it occupies the soil pores, CO_2 will enter the solution phase, leaving increased CH_4 concentrations in the gaseous phase, as noted in the previous paragraphs.

13.2.2 Influence of the Soil Stratigraphy and Migrational Conduits

To a large extent, the soil composition and structure determine the migration pathways that the gas will follow, as well as its velocity and spreading. The soil type as characterized by grain size, stratification, and variations of these in space is of initial importance. Any additional secondary physical features, such as crack or fissure formation, soil pore water content, and intrusion such as an excavation, are also of importance, since they may provide directional pathways that influence the migration. Boundary conditions such as the soil surface cover (e.g., snow, ice, pavement) and the ability of the gas to freely discharge, the elevation of the water table, and the existence of barriers or vents along the migrational paths also affect the direction and extent of landfill gas migration.

Specific soil properties of interest in migration include the intrinsic permeability k_S (m^2) of the soil, which can be used to estimate the conductivity of the gas, K_g:

$$K_g = \frac{\gamma}{\mu} k_s \qquad (13.8)$$

where:

γ = weight density of the gas (N/m^3)

    = ρg,

μ = absolute viscosity ($Pa \bullet s$),

g = gravitational constant (m/s^2).

The utility of Equation 13.8 arises from its ability to reflect different soil types (via the intrinsic permeability) (see Figure 6.1 for further details on magnitudes of this parameter) and different gases (via ρ and μ). The soil properties k_S and n (porosity) are independent of the fluid/gas and characterize the soil by reflecting the combined influence of grain size and shape and their distribution.

The properties of the soil are site-specific. Values of k_S can differ by as much as twelve orders of magnitude (10^{12}) between clay and gravel, as considered in Chapter 6. As a direct consequence, K_g also varies thereby requiring accurate assessment of soil conditions; otherwise, substantial errors in estimates of gas transport will occur.

Thus, it is apparent that the soil stratigraphy represents an important influence on the migrational pathways. The following general statements can be made:

- In homogeneous soils, transport is intergranular.
- In heterogeneous soils, most transport occurs through more permeable paths such as cracks and fissures, sand/gravel lenses, and/or inclusions

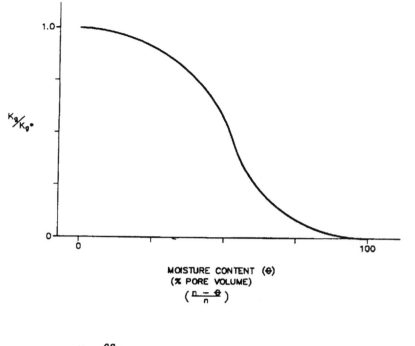

$$K_g = \frac{\rho g}{\mu} k_a$$

WHERE k_a = INTRINSIC PERMEABILITY OF SOIL
and Kg^* at zero percent moisture content

Fig. 13.3 Change in gas conductivity as a function of moisture content.

such as conduits and trenches.
- Gas conductivity changes dramatically with changes in moisture content. An indication of the magnitude of change is provided in Figure 13.3—as moisture increases the conductivity of the gas, Kg, decreases.

13.2.3 Pressure Buildup in Landfills

The processes of landfill gas generation produce the increased pressure that represents the driving force for advective flux. The pressure differential between the landfill and ambient conditions provides one of the two important migrational influences. Two questions must be answered.

1. To what extent does pressure build up?
2. How easily is the pressure dissipated?

Following from the discussion in Section 13.2.2, the essential factors are the interferences with the migrational pathways. For example, if the landfill has minimal cover, then the pressure buildup will be low, because any generated gas can migrate relatively easily through the surface cover. Examples of additional features influencing gas pressures within the landfill are summarized in Table 13.5.

Table 13.5 QUALITATIVE ASSESSMENTS OF FEATURES AFFECTING GAS PRESSURE BUILDUP WITHIN REFUSE

Landfill Feature	Tendency Toward Low Pressure (<3" H_2O)	Tendency Toward High Pressure (10-30" H_2O)
Cover	poor	good
Liner	unlined	lined
Depth	shallow	deep
Moisture	dry	wet
Gas volume produced	low	high

Note:

1" H_2O = .00246 atm

1" H_2O = 248.8 Pa = .249 kPa

McBean and Farquhar (1980) found that in sandy soil immediately adjacent to a landfill, hand-pumping extraction at one location influenced the pressure levels at another location 14 m away within 5 seconds. The recorded decrease in pressure after minimal pumping indicated a free flow of gas. Therefore, any pressure that might be generated at a specific location through gas production would be dissipated very rapidly in the sandy soil conditions.

It will be demonstrated that both advection and diffusion are important if pressure differences are less than 12 in. of water (3 kPa) for migration pathways. Under very high pressures, such as those developed using gas-extraction pumping, only advection will be important. If the pressures remain below 12 in. of water, then it is reasonable to assume that landfill gas behaves as an incompressible gas and to use Darcy's law to calculate gas flow.

13.3 MODELING GAS TRANSPORT IN SOIL

13.3.1 Elements of Modeling Approaches

The intent of modeling is to assess the extent of transport and the effectiveness of any control systems. Gas transport in soils is predictable with a variety of modeling processes. The accuracy of the predictions depends mainly on how well the model represents the real transport processes and how thoroughly the site conditions have been quantified. In all cases, the predictions must be considered to be approximate since even the most sophisticated of transport models are subject to the difficulty of parameter assignments in the context of soil permeability and dispersivity. This does not, however, negate the use of modeling as part of the total program to analyze landfill gas problems. In fact, some form of modeling is essential for predicting future conditions or for assessing the effectiveness of proposed venting or containment procedures.

Elements of Gas Transport in Soil   Transport or changes in concentrations result from the following four processes:

Process	Comments
Diffusion	in the direction of decreasing concentration in accordance with Fick's law
Dispersion	soil-specific
Advection	in the direction of decreasing pressure in accordance with Darcy's law
Reaction	gas- and soil-specific

Diffusion and Dispersion   Most modeling efforts in landfill gas migration have been characterized using Fick's law. *Diffusion* is the spreading of the concentrated contaminant plume due to concentration gradients. Bulk movement of the fluids is not necessary for this process to occur. Mechanical dispersion, on the other hand, is a direct result of the bulk movement or advection of fluid through porous media. The path the fluid takes as it passes through a porous medium is very tortuous due to the random orientation of solid particles or grains. This causes the "flow tubes" of the fluid to mix and disperse.

Advection   Advection of landfill gas is caused by a buildup in the gas pressure within the landfill as a result of the gas generation processes. Even when the pressure difference between the landfill and the adjacent regions is as small as 12 in. of water pressure, or 3 kPa (0.03 atm), advection is an important process in gas migration. For these low pressures it is possible to assume that the gas is incompressible, so Darcy's law for the flow of fluids is valid and may be used to estimate the gas flow.

Movement of the gas is slow, and a check of the Reynold's flow number shows that the gas flow is laminar, thus satisfying the requirement for Darcy's law to be applicable.

Reactions   Reactions of the gases are very gas-specific. When a gas has a high partitioning coefficient between the gas and solute form, substantial amounts of gas may go into solution. Carbon dioxide is one of the gases that dissolve in the pore water of the porous media through which the gas is passing. Methane is relatively insoluble, with a solute molar concentration less than 5 percent of the methane concentration in the gas phase. Thus, the combination of high loss of CO_2 and low loss of CH_4 results in an enrichment of CH_4 as the gas migrates.

Dissolution of a gas in a liquid obeys Henry's law, which relates the partial pressure of the gas to its equilibrium in the dissolved phase through the use of a proportionality constant.

Henry's law states:

$$P_g = Hx_g \tag{13.9}$$

where:

P_g =   partial pressure of gas (atm),

H =   Henry's law constant,

x_g =   equilibrium mole fraction of dissolved gas, defined as

$$\frac{\text{mol gas } (n_g)}{\text{mol gas } (n_g) \quad + \quad \text{mol water } (n_w)} = x_g \tag{13.10}$$

Representative values of Henry's law coefficients are listed in Table 13.6.

Table 13.6   HENRYS LAW CONSTANTS FOR METHANE AND CARBON DIOXIDE IN WATER

	H ($\times 10^4$ atm/mol fraction)	
$T°C$	CO_2	CH_4
0	0.0728	2.24
10	0.104	2.97
20	0.142	3.76
30	0.186	4.49
40	0.233	5.20
50	0.283	5.77

Source: Perry, 1963

Note:

1×10^4 atm/mol fraction = 1.013×10^6 kPa/molfraction

EXAMPLE

What are the saturation concentrations of carbon dioxide and methane in water in contact with landfill gas. Assume 1 atm and 20°C.

Solution   Assume that landfill gas is 50 percent carbon dioxide and 50 percent methane.

(a)For carbon dioxide, $p_g = 0.50$. From Table 13.6, $H = 0.142 \times 10^4$ atm/mol fraction.

From Equation 13.9,

$$'x_g = \frac{'P_g}{'H} = \frac{0.50}{0.142 \times 10^4} = 3.52 \times 10^{-4} \qquad (13.11)$$

One liter of water contains 1000 g/18 g/mol = 55.6 mol (where H = 1 and O = 16, so H_2O = 1 + 1 + 16 = 18 g/mol).

From Equation 13.10,

$$\frac{'n_g}{'n_g + 'n_w} = 3.52 \times 10^{-4}$$

but since $n_w \gg n_g$

$$n_g = 1.95 \times 10^{-2} \text{ mol/L}$$

The gram molecular weight of carbon dioxide (atomic weights of C = 12 and O = 16) is 12 + 2 × 16 = 44 g/mol. The saturation concentration for carbon dioxide is

$$
\begin{aligned}
C_s &= 1.95 \times 10^{-2} \text{ mol/L} \times 44 \text{ g/mol} \times 10^3 \text{ mg/g} \\
&= 858 \text{ mg/L}
\end{aligned}
$$

(b) For methane, p_g = 0.50, and from Table 13.6, $H = 3.76 \times 10^4$ atm/mol fraction. From Equation 13.10,

$$x_g = 0.50/(3.76 \times 10^4) = 1.33 \times 10^{-5}$$

From Equation 13.10:

$$
\begin{aligned}
n_g &= (n_g + 55.6)\, 1.33 \times 10^{-5} \\
&= 7.3 \times 10^{-6} \text{ mol/L}
\end{aligned}
$$

The gram molecular weight of methane (atomic weights of C = 12 and H = 1) is 12 + 4 × 1 = 16 g/mol. The saturation concentration for methane is

$$
\begin{aligned}
C_s &\cong 73 \times 10^{-5} \text{ (mol/L)} \times (16 \text{ g/mol}) \times (103 \text{ mg/g}) \\
&\cong 11.8 \text{ mg/L}
\end{aligned}
$$

The ratio of saturation concentration for carbon dioxide to methane is 858/11.8 = 73.

13.3.2 Mathematics of Landfill Gas Migration

The following is a brief examination of one of the alternative modeling approaches. The objective of the modeling approach is to study (1) the extent to which the gas will be transported through the soil and (2) how gas transport will be affected by the implementation of control systems such as barriers, passive vents, and pumped extraction or injection systems.

It is not always practical or even feasible to carry out sufficient field experiments to make these judgments. In such cases, the use of mathematical models to simulate gas transport is required.

To assess the extent of transport and the effectiveness of gas control systems, the processes described in Section 13.3.1 are characterized by a mathematical equation. The physical processes that result in chemical transport in groundwater are basically the same as those that cause gas migration through the unsaturated zone. Various methods for modeling gas migration in porous media exist.

The complexity of the model for a particular application depends on the configuration and constituents being modeled. The complexity of the model is particularly sensitive to the number of dimensions being modeled, as indicated in the following:

- One-dimensional modeling, as depicted in Figure 13.4(a), is appropriate to situations such as soil layers, excavations, and sewers.
- Two-dimensional modeling, as depicted in Figure 13.4(b), is appropriate to situations of gas migration through the soil close to the landfill.
- Three-dimensional modeling is obviously considerably more cumbersome but must be employed when assumptions necessary for one- and two-dimensional modeling are inappropriate.

Regardless of the number of dimensions needed in the modeling, the concern is with the migration of a multicomponent gas mixture through a variably saturated heterogeneous porous medium. Three categories of equations are necessary to accomplish the modeling:

1. *Constitutive equations* or equations of state, which describe the gas mixture properties as a function of the relative concentrations of the component gases
2. *Advection equations*, which describe the movement of the gas mixture in response to spatial variabilities in mechanical energy
3. Equations that describe the mass conservation of the diffusing species.

The universal gas law states that

$$p_v V = \frac{m}{M} RT \tag{13.12}$$

where:

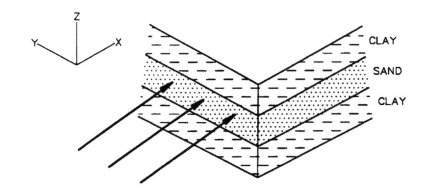

Fig. 13.4 (a): One-dimensional modeling.

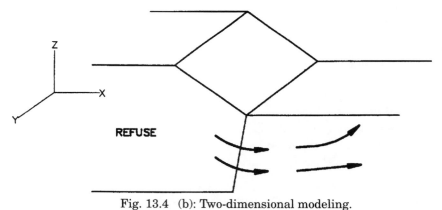

Fig. 13.4 (b): Two-dimensional modeling.

p_v = equilibrium vapor pressure (atm),

V = volume (L),

m = mass of substance (g),

M = gram molecular mass of substance (g/mol),

R = universal gas constant, 0.082 L·atm/(mol·K),

T = temperature (K).

The vapor density (mass/unit volume), c_g, is

$$c_g = \frac{m}{V} \tag{13.13}$$

where: c_g = the concentration of the component in the gas phase.

Combining Equations 13.11 and 13.12 results in the relationship among the gas concentration, the pressure, and temperature:

$$c_g = \frac{M p_v}{RT}$$

(13.14)

Equation 13.13 indicates the sensitivity of the gas concentration calculations to changes in pressures and temperatures, for example:

- For a change in pressure of 0.03 atm (or 3 kPa) at 1 atm pressure, the change in concentration is approximately 3 percent.
- For a change in temperature of 10°C at an ambient temperature of 20°C (or 293 K), the change in concentration is approximately 3.5 percent.

The implication of the examples is that the variations in mass concentration with changes in pressures and temperatures as experienced in field conditions are relatively small. These findings support the concept that relatively simple modeling approaches for landfill gas migration will provide reasonable assessment capabilities.

As determined previously, at pressure differentials between locations of less than 3 kPa, the landfill gas consisting of a mixture of CH_4 and CO_2 normally can be considered an incompressible gas. Also, the velocities produced in soil as a result of such gradients generally result in Reynolds numbers (Re) less than unity, which allows the gas flow to be treated as laminar.

Advection Equation   Darcy's law written in one dimension is

$$q = -K_g \frac{dh}{dx}$$

(13.15)

where:

q = specific discharge or macroscopic velocity (m/s) gas,

K_g = conductivity (m/s) (see Equation (13.8)),

x = length in the direction of flow (m),

h = piezometer head

m = $\frac{'P}{\delta} + 'z$

P = pressure (Pa),

z = elevation above a datum (m)

δ = unit of weight of water.

Darcy's law can be used to model gas movement through soil in one, two, or three dimensions provided that the flow is laminar and that diffusion and property variations in x, y, or z can be neglected. As such, it is particularly useful for approximate estimates of gas transport when property variations in the direction of the flow are small. It is the simplest of models but useful in certain cases, particularly when gas pumping is involved.

Darcy's law has often been used to describe laminar flow of fluids through process media, as described in Chapter 6. Laminar, as opposed to turbulent, flow means that inertial forces are negligible in comparison with viscous forces.

To determine whether Darcy's law is valid for application in a specific situation, the Reynolds number is calculated (where the Reynolds number is the ratio of inertial forces to viscous forces). The Reynolds number, Re, is defined as

$$Re = \rho \frac{\upsilon\theta}{\mu} \tag{13.16}$$

where:

ρ =   density of the fluid (kg/m^3 or slug/ft^3)

μ =   viscosity of the fluid (N·s/m^2 or lb·s/ft^2),

υ =   macroscopic velocity of flow (m/s or ft/s),

θ =   a characteristic dimension of the system (m or ft).

For flow of gases through porous media, υ is the bulk velocity (as opposed to interstitial velocities). The characteristic dimension, θ, is generally taken as the mean grain diameter of the porous medium. Laminar flow conditions are typically assumed for Re less than 1 to 10.

EXAMPLE

At 0°C and 1 atm,

$\rho_{CH_4} = 0.714 \times 10^{-4} \text{kg/m}^3$

$\mu_{CH_4} = 1.04 \times 10^{-5} \frac{\text{N sec}}{\text{m}^2}$

$\rho_{CO_2} = 1.95 \times 10^{-4} \text{kg/m}^3$

$\mu_{CO_2} = 1.39 \times 10^{-5} \frac{\text{N sec}}{\text{m}^2}$

If $\theta = 0.20$ cm, then for landfill gas that is 50 percent CH_4 and 50 percent CO_2, is $\rho_{mix} = 1.33 \times 10^{-4}$ and $\mu_{mix} = 1.22 \times 10^{-5}$. Assuming a macroscopic velocity of 1 cm/s gives

$$Re = \frac{\left(1.33 \times 10^{-4} \frac{\text{kg}}{\text{m}^3}\right)\left(1 \frac{\text{cm}}{\text{s}}\right)(0.20\text{cm}) \times 10^{-4} \frac{\text{m}^2}{\text{cm}^2}}{1.22 \times 10^{-5} \frac{\text{Ns}}{\text{m}^2}\left[\frac{\text{kg m}}{\text{N s}^2}\right]}$$

$= 0.00022$

The conclusion is that the flow is laminar.

As a general rule, for flow rates typical of landfill gas migration and recovery, Darcy's law will apply if the characteristic grain sizes of the refuse or soils are smaller than 0.2 cm, the grain size that generally distinguishes gravels from coarse sands.

Models with Advection and Diffusion   It was stated previously that gas migration at most landfill sites includes both advection and diffusion.   Thus, a satisfactory model in these cases must account for both processes.   The general modeling approach combines Darcy's law with a mass conservation equation. The typical form of this equation written in one dimension is

$$\frac{\partial C}{\partial t} = \frac{\partial^2 (DC)}{\partial x^2} - v\frac{\partial C}{\partial x} \pm S \tag{13.17}$$

where:

C=gas concentration (mole fraction/m^3),

D=dispersion coefficient (m^2/s),

S=source, sink, or decay term,

and other variables are as defined previously.

These equations account for diffusion (the first term on the right side of Equation 13.16) and can also be used to include property variations both during transport and with respect to time.   However, as these variations become more complex, it becomes more difficult to solve the equations analytically, and numerical solutions must generally be used, especially in two and three dimensions.

13.4   GAS MIGRATION MODELING

Understanding the transient behavior of gas migration in soil is nontrivial, partly because of the anisotropy of the medium and the lack of homogeneity in both the medium and the migrating fluid. Gas migration modeling is useful in predicting the gas migration at existing sites and in evaluating gas migration control alternatives.

A number of models have been proposed to represent the movement of gases within the unsaturated zone. The key considerations in the development of mathematical models to describe the physical phenomena are the selection of the model dimensionality, the situation to be modeled, and the numerical solution technique.

Mathematical models for the simulation of gas transport in porous media have been developed by Moore (1979), Moore et al. (1979), Findikakis and Leckie (1979), Mohsen (1975), and Mohsen et al. (1978, 1980). All the models developed by Moore and his co-workers are based on the representation of a porous medium by an aggregation of parallel tubes, each of constant radius, that simulate the pore-size distribution of the medium. The transport equations are therefore founded on the theory of flow through capillary tubes, with a tortuosity factor included to account for the circuitous flow paths of the gas molecules. The majority of Moore's models employ a finite difference solution. Th models have been formulated for both an axisymmetric onfiguration and a two-dimensional Cartesian-coordinate configuration.

Mohsen's model differs significantly from Moore's work by considering gas flow as intergranular, based on the interpenetrating continuum approach of Bear (1972). The model consists of the Darcy equation and continuity equations for the gas mixture and the migrating species. Bear's work was, however, intended primarily for groundwater flow. Thus, Mohsen's equations make use of mass-averaged quantities, which are not so well suited for describing gas flow as are molar-averaged quantities. The equations consist of mass continuity and Darcy's law with a finite element solution.

The equations can assume many different forms. Metcalfe and Farquhar (1987) used a mixture of traditional groundwater flow–contaminant transport equations with gaseous flows in molar quantities to represent gas migration through the unsaturated zone. The model accounts for gas migration due to gas pressure, concentration, and velocity gradients.

13.5   AVAILABLE CONTROL METHODOLOGIES

Since its implementation date of November 6, 1980, the U.S. Resource Conservation Recovery Act has specified that the combustible gas concentration shall not exceed 5 percent at or beyond the property boundary. A number of methods have been proposed to control or prevent the migration of gas from sanitary landfills. The control technologies must operate for a lengthy time, typically on the order of decades. Pumping is feasible but expensive, since it must operate for a long time period. Gas control systems use natural barriers when feasible and constructed barriers such as trenches, membranes, wells, and vents. Natural barriers to gas migration include moist fine-grained soils and saturated coarse-grained soils. Construction barriers can be categorized into either active or passive protection systems, which are differentiated as follows:

- Passive systems rely either on the use of low-permeability materials, such as barriers and vents, or on large differences in conductivity between refuse and adjacent media. These are designed to produce selective venting of the

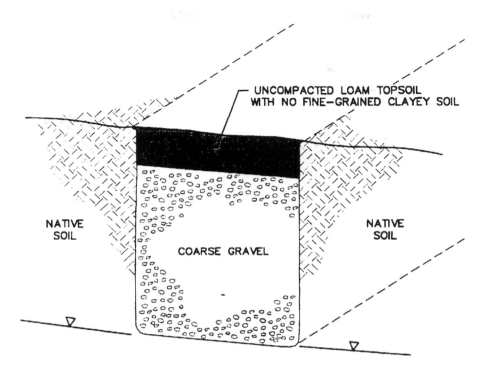

Fig. 13.5   Gravel and dirt trap trench.

landfill gas to the atmosphere.   The systems may consist of trenches, ditches, vent pipes, membranes, or the like.   In passive gas control systems, the pressure of the gas generated within the landfill serves as the driving force for the movement of the gas.

* Active systems employ energy expenditure to induce a vacuum to control the flow of gas.   For example, a centrifugal blower may be used to draw the gas from the area to be protected.

Selection of the system is site-specific, based on economics, degree of protection required, and system reliability.   Generally, active systems are preferred whenever any of the following conditions apply: the refuse age is less than twenty years, the depth of refuse is greater than 10 m, and development to be protected is less than one-half kilometer away from the landfill or property boundary. Among the many control methodologies are the following:

1. A simple trench filled with granular backfill, as depicted in Figure 13.5, acts as a passive system.   The depth of the trench must be adequate to intercept migrating gas.   The trench extends from the saturated zone to the

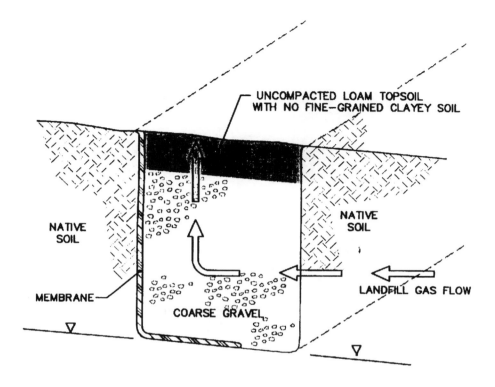

Fig. 13.6 Gravel and dirt trap trench with membrane on back wall.

surface. This type of system is effective only when the quantities of gas are
small.

The construction of such trenches is constrained to depths of approximately
7 m because of limitations on the depth to which backhoes can reach, cav-
ing conditions, the cost of shoring, and the cost of earthwork. The considerable differences between the conductivity of the gravel and that of adjoining
materials allows the gas to vent from the trench.

2. A trench backfilled with gravel and a membrane installed along the back
   wall of the trench, as depicted in Figure 13.6, is another passive system,
   but it represents a more effective interceptor. The membrane is much more
   resistant to gas flux and thus is a more effective barrier than the trench
   alone.

   Some of the first systems utilizing the membrane approach involved
   emplacement of a bentonite slurry wall, a curtain of concrete grout, or a

thin plastic membrane. Each of these approaches has inherent limitations, as outlined:

- Concrete is permeable to gas under pressure and thus induces a change in flux rate without actually stopping the gas.
- When grout is injected at periodic intervals along a perimeter, there is no assurance that every linear foot of perimeter is adequately protected.
- A bentonite or clay slurry is not effective because desiccation cracks form if it is allowed to dry out. Hence, slurries are more commonly used in groundwater interception projects. High or low pH liquids may adversely affect the clay as well.
- The longevity of the membrane under different environmental conditions may be in question.

3. A gravel trench with vertical vent pipes allows easy exhaust to the surface, as shown in Figure 13.7. The typical diameter, spacing, and stickup (the length above grade of the vent) are as noted in the figure. Note that the stickup is required to prevent contact with any passerby and to minimize opportunities for vandalism. The stickup promotes mixing with the ambient air.

4. A gravel trench with horizontal collector and vertical vent pipes, as shown in Figure 13.8, provides an easier mechanism or conduit for exhaust to the atmosphere. Also, if at a future point it is necessary to go to an active control system, this system is easily extended.

   Additional design features sometimes involve use of wind vanes at the surface to help draw off or draw out the landfill gas. Alternatively, the landfill gas can be drawn out by pumping. A mechanical blower system can be connected to the vent pipes to exert a vacuum on the pipe/trench system to further increase the amount of gas being withdrawn. Use of the pumping system represents a transition from the passive type of system to the active type of system.

5. Because passive systems are limited by depth, extraction wells are placed in a line at discrete intervals and connected by a common header pipe. The header pipe is connected to an electrically driven centrifugal blower which induces a vacuum. The reduced pressure in this type of system develops a zone of influence and, in essence, causes a flow of gas to the well. Pumping of the gas extraction system significantly increases the area influenced. The spatial extent influenced can be adjusted or controlled by varying the vacuum pressure or by installing the extraction wells at different depths within the unsaturated zone.

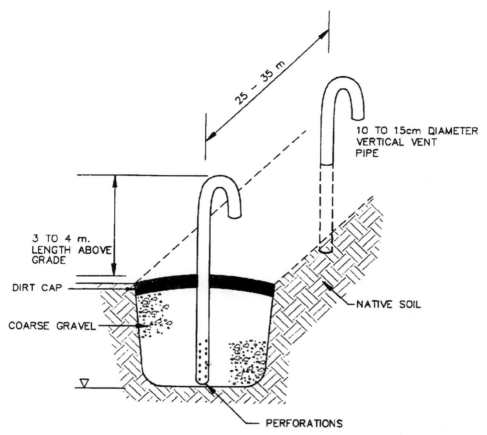

Fig. 13.7 Gravel and dirt cap trench with vertical vent pipes.

The main objective of perimeter ventilation systems is to intercept the
gases flowing outward from the landfill and to carry these gases away from
the soil pore space. Forced ventilation systems to control gas flow from sanitary landfills can be designed in a variety of configurations. In general, a
forced ventilation system consists of a series of perforated pipes that are
encased in gravel and are laid either on top of or along the side of the landfill or installed within a dry well drilled adjacent to the landfill for gas ventilation purposes. An example of a gas extraction well is shown in
Figure 13.9. The perforated pipes are connected to a system of collector
pipes, which are in turn connected to the suction side of a blower type of
positive-action suction pump.
Some of the design parameters for a forced ventilation system are the diameter of perforated pipes, depth of the gravel casing, diameter of the ventilation well, depth of ventilation well, and spacing of pipes or ventilation
wells.

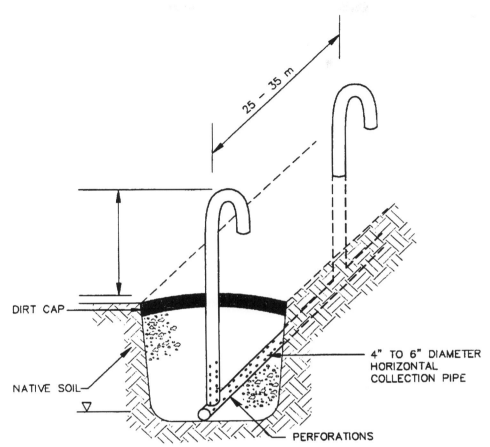

Fig. 13.8 Gravel and dirt cap trench with horizontal and
vertical vent pipes.

The area of influence is a function of a number of variables, including the
porosity of refuse, the length of well screen, depth of well, and applied well
vacuum. An advantage of the active extraction system is that the negative
pressure gradients help to suppress the surface emanation of odors. In
addition, gas extraction wells have the positive attributes that they significantly increase the width of the hypothetical "trench" from 1 m to perhaps
10 to 50 m depending on the applied well-head vacuum and the depth and
length of well screen. However, the collection of these gases creates a point
source discharge of odors and air pollution. This requires burning of the
gases in a suitably designed flare.

As the landfill gas is withdrawn and brought to the surface for transport to
some central collection point, the gas cools in the pipeline and forms a condensate. The condensate droplets coalesce and create a stream of liquid in

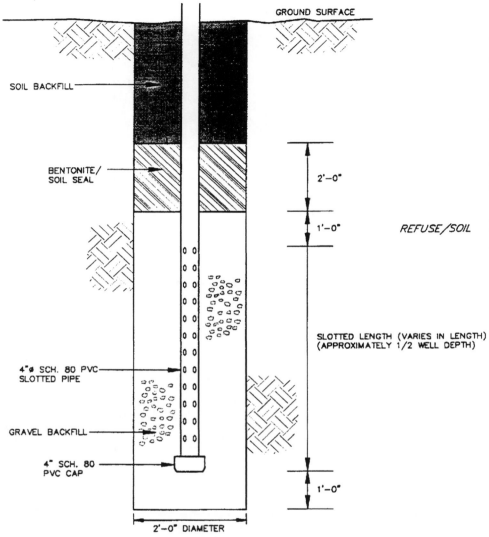

Fig. 13.9 Details of a landfill gas extraction well.

the collection pipelines. The larger the temperature difference between the landfill and the atmosphere, the greater the amount of condensable liquids. Landfill gas condensate typically appears in two places: in the gas collection pipe system and, if the gas is processed, in a treatment facility. The condensate must be removed from the gas collection system to prevent freezing or saturation. The collected moisture fills the pore space of the venting sys-

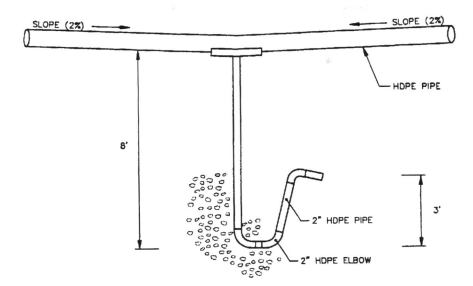

Fig. 13.10   Condensate trap detail.

tem and prevents the free passage of the gases. Historically, landfill gas
condensate collected within the collection pipe system has been discharged
back into the landfill at periodic disposal points. Condensate removed with
the landfill gas extraction has been either pumped into sumps or wells, or
dispersed over the landfill. In some locales now, however, the landfill gas
condensate cannot be returned to the landfill but must be treated and discharged to the sewer system or a surface discharge point. A schematic of
condensate collection is provided in Figure 13.10.

6. Air injection systems provide an alternative to gas extraction wells.
Although in many instances gas extraction wells are superior to passive
trench or barrier systems, it is not always possible to use them (Lofy, 1992).
If the water is deeper than the screen interval, the system will not create a
vacuum. As an alternative, Lofy (1992) reported on the use of the positive
pressure of injected air of 5 to 15 cm (2 to 6 in.) of water column to counter-
balance the diffusive flow component at the Operating Industries Landfill
in Monterey Park, California. Two Gardener-Denver compressors capable
of pumping up to 1200 ft^3/min at a maximum of 40 psig were installed. Gas
concentrations were stabilized within 3 to 5 days. The average amount of
air required per lineal foot of protected perimeter appears to be approximately 0.5 ft^3/min per lineal foot for the system of 21 wells spaced along a

2000 ft perimeter. The use of air/pressure curtains has involved no condensate collection and/or need for a flare.

Concerns regarding landfill air injection systems include the following:

1. The introduction of air into a formation may provide sufficient oxygen to cause spontaneous combustion or, if there is already a smoldering fire within the landfill, provide sufficient oxygen to sustain a true fire in the landfill.
2. Introduction of air may destroy the anaerobic methane formers, which is only of concern if there is interest in recovering the landfill gas as a resource. If aerobic conditions are established, the organic content of the refuse will stabilize more quickly.
3. Severe seasonal changes in climatic conditions may cause erratic performance, by changing the surface conditions.
4. The introduction of air may push the landfill gas to an undesirable location.

13.5.1 Statistical Modeling

An alternative type of modeling is based on the analyses of data. Because landfill gas generation and movement are necessarily stochastic phenomena, such analyses of data must reflect their statistical nature. Examples of such models include McBean et al, (1984), and Crutcher et al. (1981, 1982), but their generality to other sites must be carefully evaluated, since the models are statistical fits to monitored data.

13.6   TRACE LEACHATE GASES

Landfill gas is typically thought of as being about 50% methane and about 50% CO_2 gas, whereas, in reality, it consists of many other trace gases and liquids.

Associated with the methane, carbon dioxide, and other trace gases are aqueous vapors which derive from:

- Precipitation that has percolated into the landfill
- Various liquids and solutions that have been buried in the landfill, such as chlorinated or petroleum hydrocarbons or pesticides
- Water that is a product of decomposition.

Landfill gas can hold between 400 and 1600 gallons of condensable liquids per million cubic feet of gas. These condensable liquids consist primarily of water and various hydrocarbons. For example, at the Brock West Landfill in Toronto, they have monitored such compounds as dichloroethane, dichloroethene, trichloroethene, toluene, acetone, xylene, vinyl chloride, and benzene.

Fig. 13.11 Landfill gas
flare at Yolo County Landfill, California.

The most frequently encountered trace gas within landfill gas is benzene. In general, toluene, xylene and vinyl chloride production appear to be continuous for a long period of time from landfills and are relatively independent of the age of the waste. Young and Parker (1983) found that the domestic wastes they investigated did not represent any significant hazard due to the trace gases. Odors are worst during the first year after deposition, and organosulfurs and esters play particularly important roles. In response to concern with odors and trace gases, collection and ultimate flaring of landfill gas is utilized, such as that depicted in Figure 13.11.

Modern flaring facilities are designed to meet rigorous operating specifications to ensure the effective destruction of VOCs. For example, a minimum requirement might be a minimum combustion temperature of 1500°F and a residence time of 0.3 to 0.5 s.

References

Bear, J. 1972 *Dynamics of Fluids in Porous Media*. American Elsevier, New York.

Campbell, G. S. 1985. *Soil Physics with Basic Transport Models for Soil Plant Systems*. Elsevier, New York.

CRC Press, Inc. 1981. *CRC Handbook of Chemistry and Physics*. ed. R. C. Weast. 62nd ed., Boca

Raton, Florida.

Crutcher, A., McBean, E., and Rovers, F. 1981. The Impact of Gas Extraction on Landfill-Generated Methane Gas Levels. *Journal of Water, Air and Soil Pollution* 16:55–66.

Crutcher, A., McBean, E., and Rovers, F. 1982. Temperature as an Indicator of Landfill Behaviour. *Journal of Water, Air and Soil Pollution* 17:213–223.

Currie, J. 1960. Gaseous Diffusion in Porous Media, Part I. A Non-Steady State Method. *British Journal of Applied Physics* 11:314–317.

Currie, J. 1970. "Movement of Gases in Soil Respiration." In Sorption and Transport Processes in Soil. *Monogr. Soc. Chem. Ind.* no. 37:152–171.

Esmaili, H. 1975. Control of Gas Flow from Sanitary Landfill. *ASCE Journal of the Environmental Engineering Division* 101, no. EL4 (August):555–566.

Farmer, W., Yang, M., and Letey, J. 1980a. *Land Disposal of Hexachlorobenzene Wastes: Controlling Vapor Movement in Soil*. U.S. Environmental Protection Agency, Cincinnati, Ohio.

Farmer, W. J., Yang, M. S., Letey, J. and Spencer, W. F. 1980b. Hexachlorobenzene: Its Vapor Pressure and Vapour Diffusion in Soil. *Soil Science Society of America Journal* 44:676–680.

Findikakis, A. N. and Leckie, J. O. 1979. Numerical Simulation of Gas Flow in Sanitary Landfills. *ASCE Journal of the Environmental Engineering Division* 105 (October):927–945.

Lofy, R. 1989. *The Study of Zones of Vacuum Influence Surrounding Landfill Gas Extraction Wells*. Report of Lockman & Associates, Monterey Park, Calif.

Lofy, R. 1992. *Evolutionary Development of Landfill Gas Migration Control Systems*. Report of Lockman & Associates, Monterey Park, Calif.

Los Angeles, County of, 1972. *Development of Construction and Use Criteria for Sanitary Landfills*. Final Report to Department of Health, Education, and Welfare, Public Health Service, Solid Wastes Program.

Macfarland, I. 1970. Gas Explosion Hazards in Sanitary Landfill. *Public Works*. May 1976.

McBean, E., Crutcher, A., and Rovers, F. 1984. Influence Assessment of Landfill Gas Pumping. *Water, Air and Soil Pollution* 22:227–239.

McBean, E., and Farquhar, G. J. 1980. An Examination of Temporal/Spatial Variations in Landfill-Generated Methane Gas. *Water, Air and Soil Pollution* 13:157–172.

Metcalfe, D. E., and Farquhar, G. J. 1987. Modeling Gas Migration Through Unsaturated Soils from Waste Disposal Sites. *Water, Air and Soil Pollution* 32, no. 1/2 (January):247–259.

Millington, R. J. 1959. Gas Diffusion in Porous Media. *Science* 130:100.

Millington, R., and Quirk, J. 1961. Permeability of Porous Solids. *Transactions of the Faraday Society* 57:1200–1207.

Mohsen, M. F. N. 1975. "Gas Migration from Sanitary Landfills and Associated Problems." Ph.D. diss., Department of Civil Engineering, University of Waterloo, Waterloo, Ontario.

Mohsen, M. F. N., Farquhar, G. J., and Kouwen, N. 1978. Modeling Methane Migration in Soil. *Applied Mathematical Modelling* 2:294–301.

Mohsen, M. F. N., Farquhar, G. J., and Kouwen, N. 1980. Gas Migration and Vent Design at Landfill Sites. *Water, Air and Soil Pollution* 13:79–97.

Moore, C. A. 1979. Landfill Gas Generation, Migration, and Controls. *CRC Critical Reviews in Environmental Control*. 9 (2):157–183.

Moore, C. A., Rai, I. S., and Alzaydi, A. A. 1979. Methane Migration around Sanitary Landfills. *ASCE Journal of the Geotechnical Engineering Division* February:131–144.

Perry, J. H. 1963. *Chemical Engineer's Handbook*, 4th ed. McGraw-Hill, New York.

Reid, R. C., Prausnitz, J., and Sherwood, T. 1977. *The Properties of Gases and Liquids*, 3d ed. McGraw-Hill, New York.

Rolston, D. E. 1986. *Gas Diffusivity in Methods of Soil Analysis Part 1, Physical and Mineralogical Methods*. Agronomy Monograph no. 9, 2d ed. American Society of Agronomy, Soil Science Society of America, Madison, Wisconsin.

Sallam, A., Jury, W. A., and Letey, J. 1984. Measurement of Gas Diffusion Coefficient under Relatively Low Air-Filled Porosity. *Soil Science Society of America Journal* 48:3–6.

Thorstenson, D. C., and Pollock, D. W. 1989. Gas Transport in Unsaturated Zones: Multicomponent Systems and the Adequacy of Fick's Law. *Water Resources Research* 25:447–507.

Young, P., and Parker, A. 1983. The Identification and Possible Environmental Impact of Trace Gases and Vapors in Landfill Gas. *Waste Management & Research* 1:213–226.

Zabetakis, M. G. 1962. *Biological Formation of Flammable Atmospheres.* U.S. Dept. of the Interior, Bureau of Mines, Report of Investigations, no. 6127.

13.7 PROBLEMS

13.1 Assume that you are a consultant preparing a legal case involving methane excursion that has required abandonment of a series of buildings due to explosive methane concentration levels in the basement. The buildings are adjacent to a landfill at a distance of 25 m. Complicating the issue is the suspicion that buried organic material (trees, etc.) below and between the landfill and the buildings may be the source of the methane problems.

The lawyer has asked you to provide a conceptual flowchart describing the contents of a study you would undertake to identify where the gas in coming from. Your study program is not intended to be an expensive field program (approximately $20,000).

13.2 A new landfill is to be placed in a valley with relatively steep side slopes. The hillside is a coarse sand. At the top of the hillside there is existing residential housing, 35 m away. Your client (who owns the landfill property) does not want to spend much money. What would you recommend as a gas control system, and why?

13.3 The following questions deal with pump tests at a landfill site in Minnesota to determine gas yield for recovery:

    (a)   Sketch the general shape of iso-pressure contours produced by a gas extraction well under the following conditions:

       - summer and winter with the zone of saturation at the base of the landfill;

       - summer and winter with the zone of saturation well below the base of the landfill.

    Which is likely to yield the higher CH_4 concentration?

    (b)   Why is it difficult to determine the extent of the zone of influence of the extraction well, and what problems does this cause?

    (c)   At the limit of the zone of influence, pressures at a point do not fluctuate significantly and often alternate between positive and negative values. Explain how this might happen.

13.4 Discuss the production of gas in MSW landfills, considering the following:

    - phases of gas production and composition,

    - factors affecting the production of CH_4,

    - rates of CH_4 production expected in Michigan.

After the addition of a metallic sludge to a section of a landfill, the CH_4 production dropped off, and the pH of the leachate from this section dropped to below 5. Explain what might have happened, using your understanding of the biochemical processes involved.

13.5 A landfill is being planned to occupy a roughly rectangular space, 300 m x 425 m, with an average finished refuse depth of 20 m. It is estimated that the average deposition rate will be 120,000 t/yr and that the compaction equipment will achieve approximately 35 lb/ft^3 (conversion needed) throughout the site. A subdivision exists along one of the longer sides and is to be protected by an active gas extraction system. The groundwater table is at an average of 15 m. Prepare a preliminary design for the extraction system by estimating the numbers and locations of extraction wells and the pump capacity required.

CHAPTER **14**

LANDFILL GAS COLLECTION AND RECOVERY

14.1 INTRODUCTION

Landfill gas (LFG) is a collective term referring to the gases generated within the landfill. LFG, as generated, is a saturated gas consisting of methane (CH_4) and carbon dioxide (CO_2) with other trace contaminants. The generated gases may migrate into adjacent soils and create explosion hazards, as described in Chapter 13. The concern with the potential for explosion, in combination with the increasing costs of conventional sources of energy, has resulted in considerable interest in the possibility of recovery of the landfill gas. The intent of the recovery of the LFG may therefore be (1) to prevent migration onto adjacent properties and (2) to use it as an energy resource. The primary focus of this chapter is, however, recovery of the LFG as an energy source.

Exploiting landfill gas as an energy resource has long been promoted as "good housekeeping", along with other good waste management practices. Thus, extracting the gas and putting it to beneficial use as a fuel prevents it from entering the environment in an uncontrolled manner.

The typical composition of the landfill gas is approximately 50 percent methane and 50 percent carbon dioxide, thereby making it a potential fuel supply. Landfill gas recovery is a relatively new technology that has created significant interest over the last twenty years. Because of the nature of the trace gases and the varying composition of the landfill gas over time, any recovery facility must involve an adaptable design to accommodate concerns with collection, processing, and utilization of the fuel source. Thus, there are unique problems associated with the subsequent use of LFG that must be considered in the assessment and design. Although not a disposal option or a method to prolong landfill life, methane recovery provides public and private entities an opportunity to convert a previously untapped resource and landfill hazard into both a productive fuel and a revenue source.

Initially, the direct firing of low to medium energy content gas in boilers located within a few miles of the landfill, or the production of pipeline quality gas, were the only practical known uses. However, as a result of enactment of Public Utility Regulatory Policies Act of 1978 (PURPA), the generation of electricity using internal combustion engines became a viable use for landfill gas in the United States. Utilities are mandated by PURPA to purchase all cogenerated electricity and pay the fair market value for that electricity based on cost avoided by the utilities. Prior to the enactment of PURPA, there were enormous institutional and legal obstacles; PURPA made it possible for private individuals and firms to require utilities to accept generated electrical power at an economically acceptable price. Reliability of supply is the major reason for resistance on the part of the utilities.

Besides the implementation of the Resource Conservation and Recovery act of 1976 (RCRA), and the energy scarcity and escalating price of energy, no other factor had the same impact as the passage of PURPA in providing a sound legal and economic basis for the development of LFG-generated electrical power.

Nevertheless, the feasibility of recovering landfill gas is very site-specific, depending on a number of considerations, including the quantity and quality of gas recoverable, the availability of a market for the recovered gas within an economically viable transportation distance, and the unit price obtainable for the energy product. A general guideline for a recovery system to be viable is that it must involve a landfill with a minimum in-place waste quantity of 500,000 to 1,000,000 t and a minimum depth of 15 m. The likelihood of meeting these conditions in the future is better than the existing situation because there is a trend toward the regionalization of disposal sites, which will result in landfill development in large urbanized areas, where substantial quantities of refuse are deposited at great depths. These types of trends, in combination with escalating traditional energy costs, are making the recovery of gas as an alternative energy source all the more viable. On the negative side, extensive recycling of biodegradable components of the waste stream, as considered in Chapter 2, will act to decrease the quantities of landfill gas produced and, consequently, the economic viability of gas recovery. In this chapter we review aspects pertinent to the recovery of the landfill gas as an energy source.

14.2 FINANCIAL VIABILITY OF THE RECOVERY SYSTEM

Proper landfill management requires the monitoring and control of methane to minimize fire and explosion hazards on the landfill property and to adjacent areas. However, the type of field instrumentation for recovery of the landfill gas

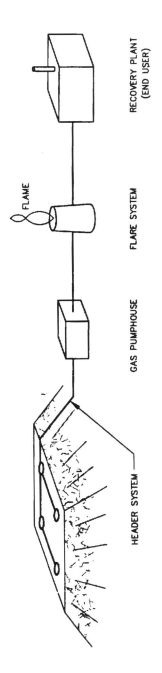

FLAME

RECOVERY PLANT
(END USER)

FLARE SYSTEM

GAS PUMPHOUSE

HEADER SYSTEM

NOTE: INDIVIDUAL WELLS PLACED WITHIN THE REFUSE

Fig. 14.1 Schematic of gas recovery system.

for an energy resource is quite different from that necessary for migration control. The components of the gas recovery system, as depicted in schematic form in Figure 14.1, include:

1. One or more wells placed within the refuse
2. A header system to connect the wells to the gas pumphouse system creating the suction
3. A flare system providing the opportunity to combust the landfill gas in the event that the gas is not needed
4. An end user of the gas. This end user might assume many different forms, but examples include an engine used to generate electricity, or a facility that processes the LFG to separate the CO_2 from the CH_4, to upgrade the CH_4 concentration. The upgraded LFG could be placed in distribution lines as a "natural gas."

Each of the individual components of the overall recovery system serves a specific role and reliability and performance criteria that must be considered in the project evaluation. However, the viability of the system is directly dependent on the value of the resource (the landfill gas) recovered in direct relationship to the use to which the gas can be put. For example, if the landfill gas is to be utilized to generate electricity, it is possible that more electricity will be produced than the landfill facility itself needs, thereby requiring energy to be marketed to outside users, usually to a power utility.

Because a power utility has a number of alternative sources of energy generation (e.g., thermal, hydroelectric, nuclear facilities), the price that is paid for the electric energy generated by the use of the landfill gas is subject to negotiation. To some extent, the energy price obtained has been a major hindrance to the development of landfill gas as a resource because many utilities consider the energy available from landfill gas not worthy of the trouble; the technology is novel. In many situations, access to the market has been obtained only following significant political pressure.

As an example, in determining the price that will be paid for electrical energy produced from landfill gas–fired facilities, Ontario Hydro in the Province of Ontario, Canada, specifies a distinct cutoff limit at 5 MW. Projects that do not exceed this production limit receive many advantages that improve project viability. Table 14.1 summarizes the major distinctions between the small generator and the large (>5 MW) generator. The table shows that Ontario Hydro offers standard purchase rates of 3.76 ¢/kWh to small generation facilities (<5 MW).

Ontario Hydro's purchase rate is based entirely on its estimated savings on the principles of avoided cost. The price paid for the electricity also reflects that, even today, the reliability of the landfill gas facility to generate power over the long term (avoiding system failure) is still a relatively new concept.

Further arguments regarding the price paid for the electricity are related to the difficulties of transmitting the power off site. A larger facility requires dedi-

Table 14.1 ONTARIO HYDRO PURCHASE RATES

Rate	Conditions
3.76¢ / kWh*	Standard rate (1988)
	>65% capacity factor
2.54¢ / kWh*	Standard rate (1988)
	<65% capacity factor
Negotiated rate for generators in excess of 5 MW	Upper limit is Ontario Hydro's avoided cost
Fixed 10-year rate 4.9¢ / kWh	<5 MW
	Provides greater benefits in the early years
Fixed rate for generators in excess of 5 MW	Negotiable

*A new rate is determined on a yearly basis.

cated feeders to be installed to export the power from the site and may represent a significant expenditure.

14.3   THE GAS RECOVERY SYSTEM

Each component of Figure 14.1 must be designed in accord with the potential behavior and problems that may develop; the landfill gas is a troublesome product of biological action that can cause many problems. Any utilization of the landfill gas must demonstrate that there are no unacceptable air or water emissions.

14.3.1 Gas Extraction Wells

Both horizontal and vertical well collector systems exist, such as those depicted in Figures 14.2(a) and (b). They are quite different in their physical layout, yet in concept remain the same. Figure 14.3 is an example of a vertical gas extraction well. Suction pressure provided by the gas pumping draws the landfill gas into the vertical/horizontal piping through the gravel backfill and the slotted pipe.

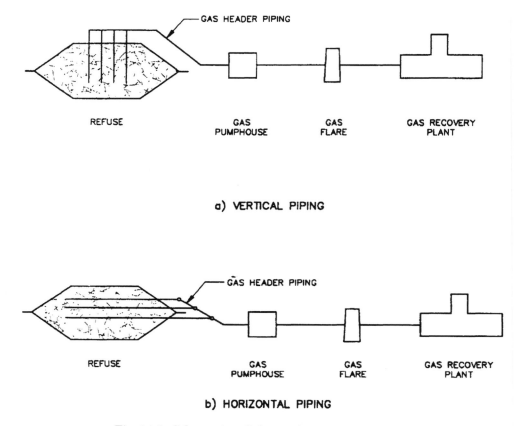

a) VERTICAL PIPING

b) HORIZONTAL PIPING

Fig. 14.2 Schematics of alternative gas recovery systems.

when the collector pipes are installed following the refuse placement. Alternatively, the horizontal piping systems are more frequently used when the piping is installed as the refuse is placed.

For both vertical and horizontal piping systems the suction pressure and the permeability of the refuse establish the zone of influence. If the suction pressure as a result of pumping is large (in which considerable energy is expended on creating the suction), a larger sphere of influence of the pumping is created and fewer pipes are necessary. If the spacing of the pipes is too close, there will be interference in terms of gas collection between the individual pipes. If the imposed pumping rate is too great, atmospheric leakage into the landfill will dilute the energy content per volume of gas extracted. Designers of gas extraction wells must therefore consider such items as the spacing, location, and depth.

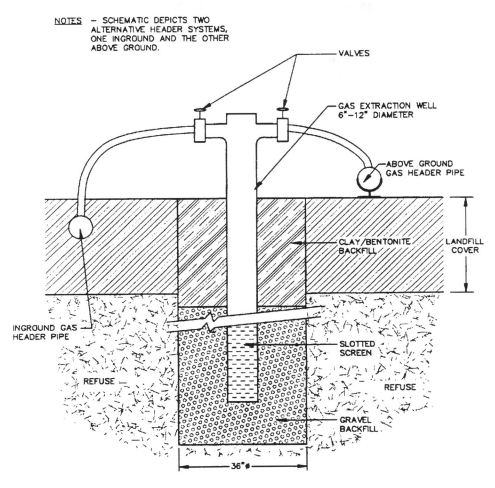

NOTES – SCHEMATIC DEPICTS TWO
ALTERNATIVE HEADER SYSTEMS,
ONE INGROUND AND THE OTHER
ABOVE GROUND.

VALVES

GAS EXTRACTION WELL
6"–12" DIAMETER

ABOVE GROUND
GAS HEADER PIPE

CLAY/BENTONITE
BACKFILL

LANDFILL
COVER

INGROUND GAS
HEADER PIPE

SLOTTED
SCREEN

REFUSE

REFUSE

GRAVEL
BACKFILL

36" ∅

Fig. 14.3   Vertical landfill gas extraction well.

Table 14.2 gives the zones of influence for a series of installations. It must be acknowledged that the "apparent" zone of vacuum influence around a well is continually changing in response to variables including barometric pressure, availability of gas, ratio of gas draw to gas generation rate, preferential pathways for gas movement, and percolation rate into the landfill (Lofy, 1989). McBean et al. (1984) found isotropic effects of pumping.

The resulting configuration of pumping wells may assume many different forms, one of which is depicted in Figure 14.4. Considerations of location include site topography, age of refuse, and system expansion over time.

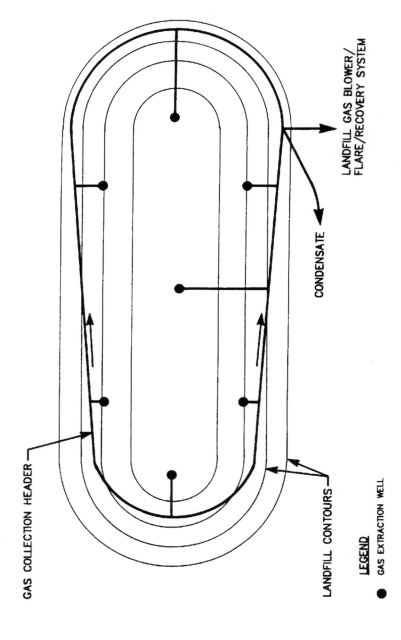

Fig. 14.4 Plan view of possible landfill gas system layout.

Table 14.2 PERFORMANCE OF SOME LANDFILL GAS EXTRACTION WELLS

Gas Flow (scfm)	Pressure in Well (in. of H_2O)	Well Depth (ft)	Radius of Influence (ft)	Medium	Location
30	-7.5	40 (4"/8")[*]	200	refuse	Winnebago, Wis.
36	-6.5	45 (6"/36")	150	refuse	Kitchener, Ont.
41	-7.0	27 (12"/24"	100	sand	Kitchener, Ont.
45	—	27 (4"/18")	200	refuse	Winnebago, Wis.
235	-39	— (6"/—)	500	refuse	Seattle, Wash.
240	-40	40 (6"/—)	—	refuse	Seattle, Wash.
320	-14	110 (—)	500	refuse	Palos Verdes, Ca.

* Well pipe diameter/borehole diameter

14.3.2 Features Affecting the Viability of the Gas Extraction System

The rate at which the landfill gas is generated was discussed in Chapters 4 and 5, wherein the numerous features affecting the rate of gas generation were described. Knowing the rate of gas generation is essential to determining the quantity of gas that can be extracted from the site.

Pumping an individual well at a greater vacuum will give it a wider zone of influence, which is acceptable, but obviously there are points of diminishing marginal returns. Larger suction pressures influence a larger region but involve more energy expended in the pumping. Pumping at greater vacuum also increases the potential for drawing in atmospheric air if the pumping rate is set too high. Crutcher et al. (1982) reported a significant air intrusion into a landfill that resulted in elevated temperatures and an underground fire. The changes in temperature in response to the fire are graphed in Figure 14.5. Air intrusion would lead to elevated levels of oxygen and nitrogen, lowering the landfill gas concentrations and detracting from its usefulness as an energy resource. In addition, the presence of oxygen will increase the risk of explosion and spontaneous combustion and decrease the activity of the methanogens that produce the gas.

The performance of the gas extraction system is affected by:

- Daily cover, which inhibits free movement of landfill gas
- Sludge or liquid wastes, which affect the ease with which landfill gas will move
- Elevated or perched liquids, which affect the ease with which the gas will move

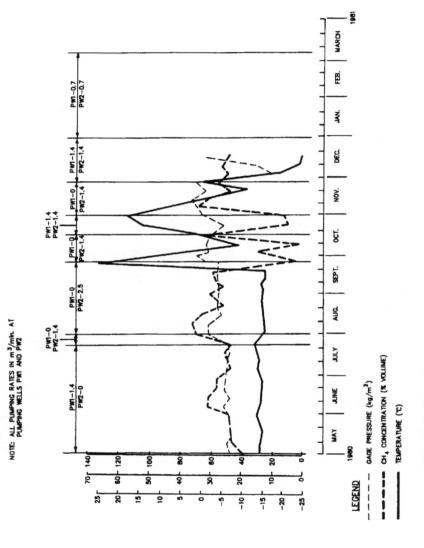

Fig. 14.5 Plot of temperature, methane concentration and pressure with time.

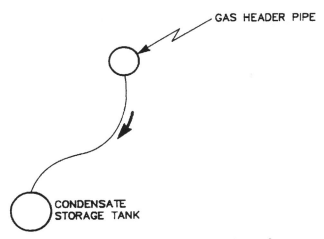

Fig. 14.6 Schematic of condensate drainage from gas
header pipe to storage tank.

- Shallow depth, which makes it difficult to extract the gas, because atmospheric air will be drawn in during the pumping. In addition, a deep landfill is preferable for methane gas recovery because any oxygen in the
  infiltrating air will be consumed in the upper portions of the landfill, leaving the larger depths to continue in anaerobic decomposition
- Permeability of final cover—the more impermeable the final cover, the
  greater the resistance against atmospheric air being drawn in.

14.4 POTENTIAL PROBLEMS IN GAS RECOVERY SYSTEMS

The design of gas recovery systems must consider the following factors:

1. The extraction must be completed in a difficult environment. For example,
   planning must reflect differential settlement as a result of the compression
   of the wastes over time. Pipe failure through sag development and failure
   at the condensate outlets are particular concerns.
2. Condensate blockage in header pipes is a major concern. The gas extracted
   from the landfill is saturated, which causes moisture to collect within the
   pipes. Therefore, the condensate within the pipes must be dealt with, otherwise it will "blind" the pumping suction pressure. Gas collection headers
   are usually installed with a minimum slope of 3 percent to allow for differential settlement. Because headers are constructed in sections that slope
   up and down throughout the extent of the landfill, condensate traps are

installed at the low spots in the line. In locations where the practice of
returning the condensate to the landfill is not allowed, the condensate traps
are connected to holding tanks, as depicted in Figure 14.6. Condensate
from the holding tanks is pumped out periodically and either transported to
an authorized disposal facility or treated on site prior to disposal or discharged to a local sewer.

EXAMPLE

Estimation of condensate water quantities.

Landfill gas is usually saturated with water vapor. The quantity of water vapor
leaving the landfill with the gas may therefore be estimated using the perfect gas
law, as given in Equation (13.11).

The mass of water contained per liter of landfill gas at 32°C is then

P_v =   490 kg/m^2 or .048 atm

V =   1.0 m^3

T =   273 + 32 = 305° K

R =   universal gas constant, 0.082 L·atm/(mol°K)

and, using Equation (13.11),

$$n = \frac{(.048 \text{atm})\ (1'\text{L})}{\left(0.082\ \dfrac{\text{L atm}}{\text{mol K}}\right)(305°\ \text{K})}$$
or

= .0019 kg-moles

or (0.0019 kg-moles) (18 gm H_2O/gm-mole) = 0.0345 kg H_2O per L of landfill gas.

---

3. The extraction system must balance where the gas is being pumped from,
   in accordance with the spatial variability in the gas generation rates.
4. Any existence of substantial water in gas wells makes extraction very difficult if not impossible. In significantly deep water, individual gas bubbles
   will rise ve. tically rather than respond to suction pressure.
5. Air can intrude because of suction pressures through the landfill surface.
   This situation can be identified by monitoring for elevated oxygen and
   nitrogen concentrations in the landfill gas.
   EXAMPLE

Estimate the quantity of condensate arising from gas pumping. Assume that the
landfill gas is saturated. The gas leaving the landfill is at 100°F and then cools to
40°F in the piping.

Solution:

Water content at 100°F   =   340 gal/million ft^3 of gas

Water content at 40°F   =   50 gal/million ft^3 of gas

Condensate production=340 - 50 = 290 gal/million ft^3 of gas

---

6. Breaks in collection lines allow air intrusion, causing deterioration in gas quality. Precautionary responses to minimize problems include:

- Using steep pipe grades (2% or better) where possible
- Utilizing many condensate traps (e.g., 1 per 300 m)
- Screening openings in the collection system to filter out particulates and mud.

14.5   OPTIONS FOR LANDFILL GAS UTILIZATION

The opportunities for utilization of the landfill gas directly depend on the degree to which the gas is cleaned, which is partly a function of the economics of the application. The alternatives for treating the gas range from simple moisture removal—to reduce the potential for problems associated with the formation of condensate—to the complex cleanup required to meet the protocol needed for inserting the resulting gas into a natural gas pipeline. An extensive range of possibilities exists between these two ends of the spectrum of treatment. The extent of processing of landfill gas is determined by market conditions. Raw landfill gas has a heating value of 400–600 Btu/ft^3, in comparison with approximately 1000 Btu/SCF for natural gas. The nature and extent of processing systems include condensate and particulate removal, dehydration of the gas, carbon dioxide removal, hydrogen sulfide removal, and nitrogen removal. Potential problems with the mechanical and processing system components include corrosion, liquid carryover into the equipment, and fluctuations in landfill gas quantities. The nature of the trace gases and the variable composition of landfill gas require an adaptable design for the processing and utilization.

The various options for use of the landfill gas can be grouped as follows:

1. Incineration — combustion of the landfill gas as extracted;
2. Low BTU gas — process removes only free moisture; the gas is essentially unrefined or raw. (Fuel gas of approximately 450 Btu/ft^3 heating value is obtained);
3. Medium Btu gas — process consists of compression and removal of moisture and heavy-end hydrocarbons, leaving a medium Btu fuel gas, carbon dioxide, and some remaining trace contaminants;
4. High Btu gas — process requires removal of all moisture, trace gases, and carbon dioxide, leaving a pipeline grade fuel (± 1000 Btu/ft^3);
5. High Btu gas/CO_2 recovery — process is similar to that used to produce high Btu gas but also includes CO_2 recovery; and
6. Chemical products — cleanup processes involve conversion of LFG into chemical fractions such as methanol.

Choosing which options to pursue requires reliable projections relating to product costs, risks, and returns.

14.5.1 Incineration

Incineration of the landfill gas is a control measure intended to limit the gas migration from a landfill site and to mitigate landfill gas odors that are a result of emissions to the atmosphere. An active gas collection and control system is used, but there is no recovery of the energy resource provided by the LFG.

14.5.2 Low Btu Gas

The only viable direct use of the raw landfill gas is as a heating fuel, for example, a steam boiler to generate electricity, or to satisfy for the heating requirements of a close-by area (on- or off-site, for example, a cement kiln).

Landfill gas has proved acceptable for use in steam power plants, and this may prove to be a viable choice for the utilization of the gas. The steam power plant has the capacity to meet all existing emission standards and is capable of handling variability in both the quantity and nature of the gas supply.

However, the generating station has limited flexibility to alter the plant size based on the variability in the actual landfill gas production. Commitment to a project of this nature requires a considerable capital investment at the outset.

Raw landfill gas is hard on internal combustion engines. For example, a complete overhaul will probably be needed every 35,000 to 40,000 hours. The oil typically will need to be changed frequently as well (e.g., every 700 to 800 operating hours).

Landfill gas typically contains trace levels of volatile organic compounds (VOCs) and chlorofluorocarbons (CFCs). The VOCs in most cases contain some chlorine.

When the VOCs and CFCs are burned within an engine, the chlorine and fluorine are released and form hydrochloric and hydrofluoric acids, which are corrosive (Stecker and Marsden, 1990). Corrosive effects may also occur because of the water vapor; the water vapor in the gas combines with organic compounds to form organic acids such as carbonic acid.

14.5.3 Medium Btu Gas

Table 14.3 summarizes some of the uses for landfill gas as medium and high BTU gas.

One option is to sell LFG to a local industry for use as a fuel. A fuel of this type has approximately half the average heat content of pipeline natural gas. In assessing the proposed utilization of landfill gas as a backup or replacement fuel for natural gas, it is essential to determine whether local industries constitute a suitable market. There must be a match between supply and needs of the industries.

Table 14.3 OPTIONS FOR USE OF LANDFILL GAS

Medium-energy gas use (no CO$_2$ removed)	sale to industry/utility
	steam production
	power generation
	internal combustion engine
	gas turbine
High-energy gas use (CO$_2$ removed)	sale to natural gas pipeline
	possible use in vehicles
	CO$_2$ recovery

There are presently several well-developed technologies that can upgrade the LFG to medium Btu gas. These technologies utilize compression, refrigeration, and chemical processes to remove the moisture and some of the trace contaminants (heavy-end hydrocarbons) from the landfill gas. These processes essentially yield a relatively clean, dry gas that is a mixture of primarily methane and CO$_2$ with some light-end trace contaminants.

Partially cleaned up gas is versatile and can be used for: (1) a heating fuel—the processed gas can be used directly by local industrial plants for their various heating requirements—and (2) for electrical energy generation—the medium Btu gas has been successfully used to operate both gas turbines and reciprocating engines for electrical energy generation.

The electrical energy generation uses various existing technologies. The equipment required is commercially manufactured and can be selected to efficiently enlarge the plant size as the gas resource develops.

Currently, four technologies or combinations of technologies are available for the conversion of landfill gas for power generation, namely, reciprocating engines, gas turbines, steam turbines, and a combination of gas and steam turbines.

Reciprocating Engines It is noteworthy that experience at sites has shown that contaminants in the landfill gas can produce rapid engine wear, sufficient to cause engine failure for operating times as short as 10 to 30 days (Stecker and Marsden, 1990). Engine corrosion can be reduced by dehydrating the gas with a refrigerated dryer.

Gas Turbines Both gas turbine and reciprocating engine systems have inherent advantages and disadvantages. Table 14.4 lists the various technical and cost considerations. A gas turbine system is inherently more resistant to corrosive effects from landfill gas because of the high-temperature alloys used, has lower air emissions, has a dual oil system to prevent shutdown, and is potentially simpler to operate (i.e., has greater flexibility in utilizing LFG despite changes in LFG volumes and characteristics). On the negative side, gas turbines have a higher capital cost and lower energy efficiency (typically 15 to 25%). However, the operating features of gas turbines have been found to more than offset the lower capital cost and higher energy conversion efficiency of reciprocating engines.

Table 14.4 GENERAL COMPARISON OF GAS TURBINE AND RECIPROCATING ENGINE SYSTEMS

	More Advantageous System	
Consideration	Gas Turbine	Reciprocating Engine
Size options available		X
Capital cost		X
O&M cost	X	
Energy efficiency and revenue		X
Overall cost		X
Resistance to corrosion	X	
Air emissions	X	
Need for specialized maintenance		X
Need for operations attention	X	

Source: Modified from Stecker and Marsden, 1990.

Gas turbines also yield a higher quality of waste heat for recovery (e.g., waste temperature of 800°F versus less than 250°F), which means that only gas turbine technology can economically produce the waste heat required for combined cycle power generation systems.

Steam Turbines   Selection of either the steam turbine or the combined-cycle technology (to be discussed next) usually implies a large-scale, long-term power generation project. A substantial capital outlay is associated with each of these options, which typically have a generation capacity in excess of 10 MW.

The steam turbine option uses the LFG to fire a boiler that produces the steam necessary to operate the turbine. This technology can accommodate and utilize a significant range in gas quantity and quality and should yield an overall efficiency of LFG fuel conversion to power generation of approximately 25 percent. The system needs cooling towers and continuous supervision by high-pressure-boiler operators.

Combined Cycle   Combined-cycle technology uses both a steam turbine and gas turbine(s). A significant improvement in energy conversion efficiency to over 35 percent can be realized by recovering and utilizing large volumes of high-quality waste heat from the gas turbines in a waste heat boiler. This waste heat can be used to produce steam for the steam turbine, which reduces the volume of LFG required to fire the boiler. Selection of any of these technologies implies rejection of all other options for utilization of LFG, including high Btu and CO_2 processing.

Summary   Reciprocating engines and gas turbine systems typically involve a series of generating units with net outputs ranging from approximately 0.7 to 3.0 MW per unit and with LFG requirements ranging from less than 330 scfm to 1650 scfm per unit. There has been considerable operating experience with these types of power generation systems in the United States.

Steam turbine and gas/steam turbine combination systems require a significant steam boiler and steam turbine component and thus are generally feasible only for large-scale projects. Each of the four power generation technologies requires a significant capital investment (e.g., $1.4 million/net MW in 1989 dollars) (CRA, 1989).

Table 14.5   TYPICAL EFFICIENCIES OF POWER GENERATION

Gas turbines	23%
Reciprocating engines	28%
Combined gas and steam cycle	35–40%

Table 14.5 lists typical efficiencies of power generation. Reciprocating engines are generally smaller than gas turbines. Thus, reciprocating engines are ideally suited for small LFG recovery levels as well as for small-scale increases or decreases in power capacity. Experience in the U.S. confirms that landfill site operations tend to consider gas turbines more practical and economical for use on larger LFG sites with stable gas recovery in excess of 1500 scfm.

14.5.4 High Btu and CO_2 Recovery

Production of pipeline grade natural gas (approximately 1000 Btu/ft^3) and CO_2 for commercial sale involves separation, compression, moisture removal, and cleanup of the LFG. The end products must meet very strict standards over the long term. A number of production alternatives exist, namely,

- Physical processes in which CO_2 is removed by dissolution in water (e.g., Selexol by Allied Chemical and BINAX by Central Plants Inc.),
- Chemical removal by bonding between solvent and CO_2 (MDEA process by BASF),
- Adsorption of a thin layer of molecules to the surface of solids such as activated carbon; and
- Membrane removal, which allows specific gas species to permeate (pass through) faster than others (CO_2 is faster than CH_4).

The preprocessing stage for high Btu gas refinement (development of the medium Btu process) is compatible with the possible development of a future high Btu gas recovery plant. Pre-treatment is an essential step in the processing of the landfill gas for pipeline quality upgrading.

The refinement of carbon dioxide as a part of this process may improve the outlook for this option in the future, but major tasks still remain in analyzing the feasibility. No known landfill gas plant for full CO_2 recovery has been identified.

CO_2 has a wide range of applications, however, a suitable market for large volumes of CO_2 is needed. The non-food market includes the manufacture of urea, pharmaceuticals, dyes, pigments, and other chemicals as well as pressurizing gas to promote recovery of oil. Food applications include its use in carbonating beverages and as a refrigerant.

14.5.5 Manufacture of Chemical Products

The principal components of LFG—CH_4 and CO_2—make it attractive as a basic raw material for the manufacture of certain chemical products such as methanol. However, the synthesis of chemical products from LFG does not appear to have been seriously considered yet in the LFG utilization industry.

14.6   CASE STUDIES

Landfill gas will be successfully exploited only if it is competitively priced against other competing sources of energy. Further, the industry is producer-led, so gas will be exploited only if the producer can make a profit.

Historically, the quarrying industry, uniquely placed as producer and potential user, has nurtured landfill gas exploitation for energy. This industry has seized the opportunity of using locally produced low-cost gas as a replacement fuel in kilns—a highly attractive option and one likely to remain so for the foreseeable future. Other local, highly energy intensive industries have also been able to obtain gas from producers via pipelines directly from the landfill to their facility.

Increasingly, operators have turned to power generation and export of electricity to third parties or the national grid as a means of exploiting the gas, particularly from the more remote and rural sites.

A number of field applications for energy recovery have been described in the technical literature. Dinsmore (1987) described an application of high BTU landfill gas recovery. Table 14.6 lists examples of some of the earlier field-scale applications.

In Husum, Germany, electricity is generated directly from combustion of the landfill gas. In addition, the energy recovered from the engine coolant radiator system is utilized to heat greenhouses.

Table 14.6 EXAMPLES OF LANDFILL BIOGAS PROJECTS

Project/Project Manager	Commercial Use	Date of First Operations
C.I.D. Chicago, Ill/Getty Synthetic Fuels, Inc.	The National Gas Pipleine Co. of America purchases the gas for blending with pipeline gas supplies.	December 1980
Fresh Kills, Staten Island, New York/Getty Synthetic Fuels, Inc., Methane Development Corp.	Brooklyn Union Gas Co. uses the gas to blend with pipeline gas supplies.	July 1982
Palos Verdes Ca/ Getty Sythetic Fuels, Inc.	Southern California Gas Co., purchases the gas to blend with pipeline gas supplies.	June 1975
Monterey Park, Ca/Getty Synthetic Fuels, Inc.	Southern California Gas Co., purchases the gas to blend with pipeline gas supplies.	August 1979
Mountain View, Ca./ Pacific Gas & Electric Co.	PG&E uses the gas to blend with pipeline gas supplies.	August 1978
Cinnaminson N.J./ Public Service Electric Gas Co.	Public Service Electric & Gas Co. sells the gas for use in manufacturing.	August 1979

source: Extracted from Wingewoth and Bohn, 1983

Berenyi and Gould (1986) surveyed methane recovery facilities within the United States. They found that over half of them (54.7 percent) are located in the West, 29.5 percent in the Northeast, 9.3 percent in the South and 14.7 percent in the North-central region. The low and medium BTU recovery systems either utilize the raw gas as fuel to drive turbines that generate steam for space heating or partially refine the landfill gas by removing particulate matter and/or condensate water. The high Btu recovery systems typically remove most of the carbon dioxide and other contaminants. Over half the facilities (54.7 percent) use the gas produced to generate electricity on site. The average facility produces 1.9 MW of electric power. It is noteworthy that methane recovery is viewed as a commercial venture, with private firms competing with each other to develop and operate the most profitable processes.

Crutcher et al. (1981) demonstrated that it is both feasible and economical to utilize landfill gas in an unprocessed state to heat a greenhouse. Gas recovery at Outagamie County, Wisconsin, was described by Stecker and Marsden (1990). This recovery system involves sale of electric power to the local utility and on-site use of generated electric power. The facility is 70 acres in area with a

50 ft depth and 1000 t/d loading rate. The solid wastes disposed of included 250 to 30C tons of paper mill sludge per day. The concomitant high water input resulted in high rates of decomposition and gas generation and development of gas pressures of over 400 in. of water column. The high rate of decomposition was causing odors of concern to neighbors. In February 1984, a landfill gas recovery system supplying methane gas for heating its on-site service building was implemented, recovering 800 to 1000 ft^3/min. Energy is now being produced for sale using reciprocating engines.

Richards (1988) indicated the likely dramatic rise in energy savings in the United Kingdom over the period 1986 to 1989 was due to landfill gas use in a variety of ways, including in kilns, in boilers, and for power generation.

An important variable in determining the economic viability for methane recovery is the design life of the gas collection system due to availability of the gas. The design life for most utilization projects is 10 to 20 years. In regions of drier climate and uniform temperatures, the design life may reach 25 to 40 years.

References

Berenyi, E. B., and Gould, R. N. 1986. Methane Recovery from the Municipal Landfills in the USA. *Waste Management and Research* 4:189–196.

Brown and Caldwell. 1980. *Landfill Gas Recovery, Processing and Utilization in California.* Report to California State Solid Waste Management Board, Sacramento.

Buivid, M. G., Wise, D. L., Blanchet, M. J. Remedios, E. C., Jenkins, B. M., Boyd, W. F., and Pacey, J. G., 1981. Fuel Gas Enchancement Controlled Landfilling of Municipal Solid Waste. *Resource Recovery and Conservation* 6:3–20.

CRA, Ltd. 1989. *Summary Evaluation of Landfill Gas Utilization, Keele Valley Landfill Site.* Report to the Municipality of Metropolitan Toronto, Waterloo Ontario, May.

Crutcher, A., Rovers, F., and McBean, E. 1981. Methane Gas Utilization from Landfill Site. *ASCE Journal of the Energy Division* 107, no. EY1 (May):95–102.

Crutcher, A., Rovers, F., and McBean, E. 1982. Temperature as an Indication of Landfill Behavior. *Water, Air and Soil Pollution* 17:213–223.

Dinsmore, H. L. 1987. High Btu Landfill Gas Recovery Utilizing Pressure Swing MDEA Process. *Waste Management & Research* 4:133–139.

Lofy, R. 1989. *The Study of Zones of Vacuum Influence Surrounding Landfill Gas Extraction Wells.* Report of Lockman and Associates, Monterey Park, Calif.

Lofy, R., 1990. Methane Gas Recovery/Treatment for Use as a Fuel. Notes from Sanitary Landfill Gas and Leachate Management. University of Wisconsin at Madison.

McBean, E. A., Crutcher, A., and Rovers, F. 1984. Influence Assessment of Landfill Gas Pumping. *Water, Air and Soil Pollution* 22:227–239.

Richards, K. M. 1988. *The UK Landfill Gas and Municipal Solid Waste Digestion Industry.* Energy Technology Support Unit, Harwell Laboratory Didcot, Oxon, England.

Snyder, N. W. 1981. *Biogas Treatment to High-BTU Gas: Technical and Financial Analysis.* Report of the Ralph M. Parsons Company, Pasadena, Calif.

Stecker, P. O., and Marsden, M. 1990. "Comprehensive Landfill Gas Control and Recovery at Outagamie County, Wisconsin, Landfill." Paper presented at GRCDA 13th Annual International Landfill Gas Symposium, Lincolnshire, Illinois, March 27–29.

Wingenroth, J. L. and Bohn, A. A. 1983. Status of Landfill Biogas Projects *Gas Energy Review* 11, no. 10 (October):18–23.

14.7 PROBLEMS

14.1 The enhancement of gas generation within landfills has been the subject of some recent experimental work. Identify five different design and/or operational modifications that have been used to enhance gas generation. Explain how these could influence the yield of methane gas.

12.2 An apartment complex involving 1500 residents is considering using a large lined pit to put their refuse in and heating the water for their swimming pool with the resulting gas that will be generated. What are your estimates of the gas that (a) will be generated and (b) can be effectively used if they contain within each cell the refuse for 3 months?

14.3 A city with population of 50,000, located in Idaho, is considering use of methane gas produced by its landfill as an energy source to heat the aquarium water for its adjacent year-round recreational complex. The landfill first received refuse in 1972. Only municipal refuse was placed in the site. The adjacent soil is all a fine sand. Design (conceptually) a first-step field gas-extraction system that could be used to determine if the concept is viable.

14.4 An 82.5 ha landfill, 30 m deep on average, was equipped for gas recovery and electrical production. The total cost of the project was $7.5 million at 14.5 percent interest compounded annually. At $0.06/kWh, estimate how long it will take for the project to show a profit. Use average figures and assume an annual operation and maintenance cost of $0.015/kWh.

14.5 Why is moisture a problem in a landfill gas collection system? What can be done about it, and how do the methods work? What methodology to deal with these concerns would you suggest for a community such as Minneapolis, Minnesota, which has a relatively lengthy and severe winter?

14.6 What factors should be considered in evaluating the technologies for the conversion of landfill gas for power generation?

14.7 List three reasons why an LFG extraction system should minimize the drawing in of atmospheric air.

14.8 What design features will affect the performance of an LFG extraction system?

14.9 Consider a community of 200,000 people using a single landfill for the disposal of its municipal solid waste. Estimate the percentage of the community's electrical energy needs that could be met from power production using the landfill gas. Provide optimistic, average, and pessimistic projections.
The following elements will be needed in the problem:
   (a)   production rate of municipal solid waste in the community
   (b)   compaction ratio in landfill and soil municipal solid waste
   (c)   landfilling sequence and geometry, assuming landfill development of two 15 m lifts,
   (d)   time required to initiate methane production
   (e)   amounts, rates, and duration of CH_4 production
   (f)   typical per capita electrical energy consumption

(g)   conversion efficiencies.

List assumptions made and parameters used.

14.10 A landfill is being planned to occupy a rectangular space of 300 m x 450 m with an average finished refuse depth of 20 m. If it is estimated that the average deposition rate will be 120,000 t/yr and that the compaction equipment will achieve 320 kg/m^3 throughout the site. It is proposed that landfill gas will be extracted from the landfill for energy recovery as a low-grade fuel. The site will be filled in cells taken directly to final elevation. Gas will be extracted through vertical extraction wells as soon as a cell is completed. Make preliminary estimates of the amount of gas and energy that will be produced, the period of production, and the number of gas extraction wells required. Identify and comment upon the major weakness of your preliminary design (1 ft^3 CH$_4$ = 1000 Btu; 1 scfm = 4.72 x 10^{-4} m^3/s).

14.11 Driplegs and water knockout drums are required in landfill gas recovery piping systems. Why are these needed? Describe how they function.

CONSTRUCTION AND
OPERATION PRINCIPLES
IN LANDFILLING

15.1 CELL CONSTRUCTION AND OPERATIONS

15.1.1 Cell Construction

The building block common to all landfills is the cell. All solid wastes received are spread and compacted in cells or layers within a confined area. At the end of each working day, or more frequently if needed, the area is covered completely with a thin, continuous layer of daily cover material that is then compacted. The compacted wastes and daily cover material constitute a cell. A series of adjoining cells, all of the same height, make up a *lift*. The completed landfill consists of one or more lifts, as depicted in Figure 15.1. Lift heights are normally maintained in the 3 to 5 m range.

The dimensions of a cell are determined by the volume of compacted wastes, which, in turn, depend on the density of in-place solid wastes. The desirable field density of wastes within a cell is 600 kg/m^3 (1000 lb/yd^3), but the density obtained in the landfill is in part a function of the degree of compaction effort extended and the nature of the refuse material. For example, a density of 600 kg/m^3 may be difficult to achieve if the refuse consists of bushes, trees, plastic turnings, synthetic fibers, construction wastes, or rubber powder. (These materials resist compaction, so they should be spread in layers up to 0.7 m (2 ft) thick and covered with 15 cm of soil over which mixed solid wastes should be spread and compacted.)

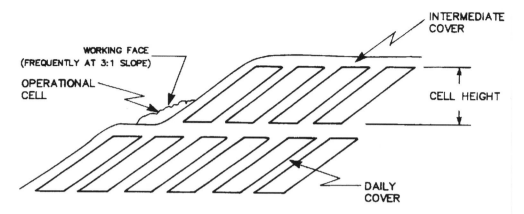

Fig. 15.1 Configuration of daily cell and lifts.

Fig. 15.2 Photograph of working face at Yolo County Landfill,
California.

Orderly operations are achieved by maintaining a narrow *working face* (that portion of the active cell onto which additional solids wastes are deposited, spread, and compacted), such as depicted in Figure 15.2. The working face should be sufficiently wide to minimize the waiting times for trucks waiting to unload, but not so wide that it becomes difficult to manage the ongoing operations.

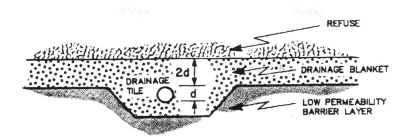

Fig. 15.3   Drainage blanket protection for a leachate collection system.

Cells are approximately 3 m (8 ft) high because this height will not cause severe settlement and slope stability problems. If the land and cover materials are readily available, a 3 m height is typically used, but heights up to 10 m (30 ft) have been reported. The objective, in part, is to minimize cover material volume due to its cost and availability. Any unnecessary volume of cover material decreases the volume of refuse that can be placed in the site.

Cover material volume requirements depend on the surface area of wastes to be covered and the thickness of the soil needed to perform the particular functions. Cell configuration can greatly affect the volume of cover material needed by minimizing the ratio of surface area to volume.

In general, a cell is rectangular in surface area, its sides sloped as steeply as practical operation will permit. The best compaction occurs at about 10 percent slope, but slopes up to 30 percent may be employed. Relatively steep slopes assist in minimizing the ratio of surface area to volume. Hence, the volume of cover material is minimized. Steep slopes also aid in shredding and obtaining good compaction of the solid waste. Solid wastes are spread in layers not greater than 0.7 m (2 ft) thick on the incline over the cell height and worked from the bottom of the slope to the top. Care must be taken during the emplacement of wastes to avoid damaging or disrupting the drainage system or liner. For example, sharp items at the bottom of the cell may be pushed into the liner, and/or silt and clay fines may be washed over from the refuse and daily cover into the drainage layer and create a low-permeability seal.

Protection may be afforded by ensuring that the first 2 m of waste are left lightly compacted above the drainage blanket. Not only will this protect the blanket from equipment damage but will enhance drainage in the lower horizons of waste and provide extra filtering for suspended material being transported by leachate. Another degree of protection is provided by placing the leachate collection system piping and its surrounding stone filter within a trench in the base of the landfill. An additional element of protection may be obtained by ensuring that the thickness of the filter medium above the pipe is at least twice the diameter of the pipe, as illustrated in Figure 15.3.

The two basic methods of operation for landfilling, or sequencing, of the daily cells are the *trench method* and the *area method*. Figures 15.4(a) and (b)

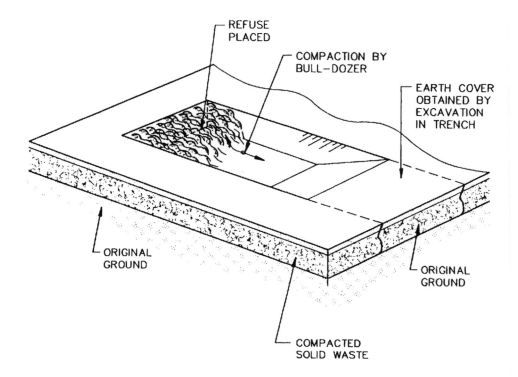

Fig. 15.4 (a) Trench method of sanitary landfilling.

are schematics of the two procedures. In the trench method, waste is spread and compacted in an excavated trench. Cover material is taken from the excavation and spread and compacted over the waste to form the basic cell structure. Excavated material not needed for daily cover may be stockpiled and later used as a cover for an area of fill operation on top of the completed trench fill. In the area method, a bulldozer spreads and compacts the waste, and a scraper is used to haul the cover material at the end of the day's operations.

There is no "best method" that applies to all sites. The method selected for a particular site depends on the specific physical conditions involved and the amount and types of solid wastes being handled. However, the trench method has become difficult to use, since liners and leachate collection systems must be placed in advance. As a result, the area method is now more commonly used.

15.1.2 Cover Materials and Frequency of Application

Cover material is classified as daily, intermediate, or final cover in accord with the frequency with which the material is applied.

Daily Cover As the name implies, daily cover is placed at the end of the working day (or more frequently, as the situation warrants). The primary func-

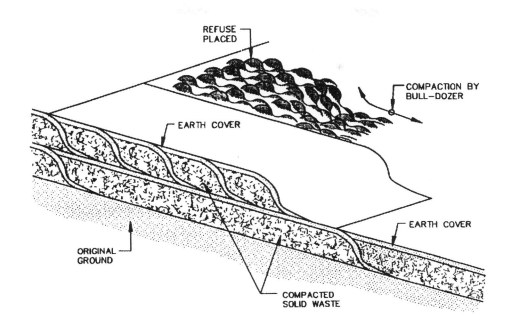

Fig. 15.4   (b) Area method of sanitary landfilling.

tions of daily cover are listed in Table 15.1. It is generally accepted that the application of 15 cm (6 in.) of compacted cover on a daily basis will achieve these desired functions.

In general, should this cover material be exposed for more than one week but less than 1 year, daily cover is sufficient. Coarse-grained soils can be compacted to 60 to 80 kg/m^3 (100 to 135 lb/ft^3) to minimize erosion.

An important trend is the increasing acceptance of alternative daily cover materials in lieu of compacted soil. Examples include mulches of leaves and wood, foam, geotextiles, and plastic sheets. In the case of geotextiles and plastic sheets, covers are removed prior to placement of the subsequent days' solid waste. The removal of the geotextiles and plastic sheets saves capacity that would be lost if 15 cm (6 in.) of soil were to be used. Using mulches of leaves and wood instead of discarding them as part of the refuse material saves on the costs of acquiring cover material and optimizes use of disposal space. Figure 15.5 shows mulched leaves prior to their use at the Keele Valley Landfill Site in Vaughan. The mulch is created, turned over approximately once per month using a backhoe, and then used in the following year. In other situations, selected solid wastes are shredded, and the shredded material is placed in windrows for composting. Leachate from the landfill is sprayed onto the shredded waste to increase the moisture content for optimum composting. The composted material is later used as daily cover material, but there is the potential for odor problems associated with this procedure.

Table 15.1 PURPOSES OF DAILY COVER MATERIAL

Moisture control

Capping of refuse as a litter control measure

Helps reduce odors

Limits rodent and bird contact with the refuse

As an operational requirement for vehicle access to the active face

Helps prevent fires

Decreases the unsightliness of the facility.

Fig. 15.5 Mulched leaves used at Keele Valley Landfill, Vaughan, Canada, for daily cover.

Another alternative uses shredded brush and vegetation as a daily cover soil, avoiding the need for temporary storage. Figure 15.6 shows a hammermill being used to shred the wastes. In this particular application, when wooden pallets are shredded, the material goes to an energy-from-waste facility. When green vegetation is shredded, the material is used as a daily cover.

Consideration of alternative daily cover materials is an active area of investigation. Bypassing cover material acquisition and avoiding the use of site volume for placement of materials other than refuse, represent very substantial overall benefits. For example, a 4:1 ratio of refuse to daily cover volume is typically employed. Adjustments to this 20 percent of refuse space currently being

Fig. 15.6 Hammermill being utilized to shred wood and
vegetation partly for energy-from-waste facility and partly for
use as daily cover.

filled with soil as daily cover presents an important opportunity for increasing
site life. Nevertheless, the effectiveness and appropriateness of individual materials as artificial covers is weather and site-dependent.

The largest quantity of water that enters a landfill and ultimately becomes
leachate, enters during the period when the landfill is being filled. Thus, in some
landfill operations, a very thick layer of soil (1 to 2 m or 3 to 6 feet) is placed temporarily over a completed cell. Rainfall that infiltrates this cover is retained in
the cover by virtue of its field capacity. Before more wastes are placed, the soil
may be removed and stockpiled. By using this operating technique of temporarily storing additional cover material over a completed cell, the amount of
water entering the landfill can be significantly decreased (and also serves to
increase density and reduce future settlement).

Ensuring the availability of the daily cover material requires long-term
planning due to the substantial quantities involved. As a typical example as
indicated above, a 4:1 waste to daily cover ratio (and possibly interim cover) is
typically required. When the additional requirements for intermediate and final
cover are added (as discussed below), the overall requirements for cover material
are considerable. In some situations, the cover material can be obtained from
the excavations needed for future waste disposal - obviously this is an ideal situation but the practicality of timing and volumes of materials needed, may be limited. The materials' sources must have adequate capacity to meet the full
demand of the daily cover and most of the final cover materials needed to

develop, operate and close the landfill site. The soil balance must be carefully monitored and as suitable clean fill becomes available from local construction, it may be accepted to meet any shortfall. As well, top soils can be stripped and stockpiled for use in final restoration of the disposal cells. For the short term, stockpiling of cover material may be practiced. An indication of the basic considerations applicable to stockpiling operations is provided in Table 15.2.

Table 15.2 LIST OF CONSIDERATIONS APPLICABLE TO STOCKPILING OF COVER MATERIAL

Visual	Visual impacts must be considered when siting locations for stockpiles
Dust	Stockpiles have the potential for dust generation during both the transport to, and from the stockpile itself, and from wind erosion of its surface
Noise	The associated equipment traffic generates noise
Economics	Use of stockpiles implies increased costs associated with double-handling of materials resulting in a practical economic consideration to minimize stockpiling, where possible.

Table 15.3 PURPOSES OF FINAL SOIL COVER

To encourage surface runoff while discouraging erosion of the cover itself; the intent is to provide an effective low-permeability cap to the landfill site that serves to control the infiltration of surface water into the waste and limit the generation of leachate

To retain moisture for vegetative root growth

To reduce or enhance landfill gas migration (depending on the design objectives)

To provide the final shaping and contouring of the site in accordance with the end use objectives

To provide a base for the establishment of a suitable ground cover.

Intermediate Cover   Intermediate cover provides the same general functions as daily cover, with its primary objective being the reduction of moisture entering the refuse. Intermediate cover is normally 30 cm (12 in.) thick and is applied to areas of the fill that are inactive, or proposed to be inactive, for lengthy periods of time (e.g., one year).

Composted materials are also increasingly being used for intermediate cover. The compost need not be fully cured before being used as an intermediate cover material (although odors may be a problem). The compost placed on the refuse should serve as an "odor filter" for the underlying refuse. The use of composted refuse for intermediate cover is expected to increase significantly in the coming years, as the conservation of landfill capacity becomes more crucial.

Final Cover   To be effective, the final cover must be compacted (except the top soil in which vegetative growth is desired), uniformly applied, and sloped to enhance surface runoff as opposed to allowing infiltration. Table 15.3 lists the functions of the final cover. The depth of final cover, the specifics of the composition (e.g., low-permeability soil, geomembrane (if utilized), vegetative cover, and so forth), and the design requirements depend on applicable governmental regulations. The intent is that the landfill be closed in a manner that minimizes the need for further maintenance.

The volume requirements and timing of availability of the various cover materials are functions of the individual covers, as detailed previously. It is important to restate, however, that the total volume requirements are considerable and involve long-term planning to ensure their availability. Table 15.4 indicates the quantities of soil required for a moderate-size landfill site.

Table 15.4   SOIL QUANTITY REQUIREMENTS FOR DAILY AND FINAL COVER

	Estimated Refuse Volume (m^3)	Daily Cover Soil Volume (m^3)	Final Cover Soil Volume (m^3)	Topsoil Volume (m^3)
Landfill Disposal Area				
Stage 1	74,700	18,700	8,500	1,400
Stage 2	87,400	21,800	7,600	1,270
Stage 3	118,000	29,500	10,900	1,840
Stage 4	105,000	26,200	13,000	2,160
Subtotal	385,100	96,200	40,000	6,670
Additional Areas				
Landfill site berm	0	0	3,000	0
Roadway Materials	0	4,000	0	0
Screening berm	0	36,800	0	0
Drainage ditch	0	0	0	1,900
Subtotal	0	40,800	3,000	1,900
Total Requirements	385,100	137,000	43,000	8,570

Source: From Conestoga-Rovers & Associates, (1990).

Note: 4:1 ratio of refuse to daily cover used for projections.

Fig. 15.7 Operations at Keele Valley Landfill. In the foreground, the equipment is preparing the clay liner. In the background, note the berm preventing run-on of rainfall from the region of liner being prepared.

15.1.3 Prevention of Precipitation Run-On

If the objective is to minimize leachate production, consideration must be given to precipitation falling on the open portions of the landfill, prevention of standing water (rainwater) from remaining in depressions in the landfill cover, and run-on of rainfall from adjacent areas, as contributions to the moisture content of the solid waste in a landfill. Note the use of the berms at the toe of the placed refuse in Figure 15.7. Hence, good operation and design must focus on minimizing these sources of introduction of moisture. Strategies to minimize run-on include the following:

- Maintain as small a working face as possible.
- Continuously place and grade the daily cover.
- Maintain daily, intermediate, and final cover to avoid depressions.
- Design and maintain a surface water control system to minimize run-on and maximize run-off.

15.2   OPERATIONAL CONTROL CONSIDERATIONS

15.2.1 Elements of Landfill Cost

As landfill disposal practices have become more complex, the total cost of providing this management option has increased dramatically. Capital costs of developing an up-to-date site with leachate and gas control systems have greatly contributed to these increases. Landfill costs are itemized in Table 15.5. Many of these costs are incurred after site closure and thus after receipt of revenue from waste tipping fees. Consequently, these costs must be funded through finances deposited in a funded trust built up over the active life of the site.

The owner or operator of the landfill, including local governments, must demonstrate the financial capability to perform any needed closure and postclosure activities or any necessary remedial actions. Thus, it is imperative to develop a funded trust during the active life of the site to cover any of these potential costs.

15.2.2 Site Life Projections

Accurate projections of the quantities and composition of solid wastes are essential for landfill site capacity calculations. Normal projections are based on a per capita generation rate of refuse and projected contributing populations. There is increasing concern about the effect on these calculations of, for example, consumption patterns, economic business cycles, and modifications to packaging techniques. There are also uncertainties associated with the impacts of recycling and composting, which will change the amount and nature of material being sent to the landfill.

Accurate projections of the quantities of solid wastes are essential for determining the type, size, design, and location of facilities, the transportation routes from the waste sources to the facilities, personnel needs, and equipment requirements.

15.2.3 Degree of Compaction

Municipal solid wastes contain a heterogeneous mixture of materials. This material can be compressed and crushed under relatively low pressure, but in a landfill these hollow and compressible items are incorporated within the mass of solid waste, which acts as a cushion. Spreading and compacting the waste in layers approximately 0.7 m (2 ft) thick and then compacting the waste by making two to five coverage passes reduces the problem.

Table 15.5 COMPONENTS OF LANDFILL COST

Predevelopment Costs

  Site characterization

  Environmental assessment

  Engineering design

  Hydrogeologic investigation

  Professional service fees—design/approvals

  Legal consultation

Construction Costs

  Land clearing

  Excavation

  Liner and leachate collection system installation

  Leachate management—pumping station and/or treatment systems

  Surface water control and final cover construction

  Gas management system

  Groundwater monitoring systems

  Site structures

Operations

  Equipment and personnel

  Leachate and landfill gas management

  Environmental monitoring costs

  Community relations

  Impact management—dust, odors, birds

Closure Costs

  Cap/final cover

  Seeding

  Runoff control

Long-term Care Costs

  End use plan costs, including trees and shrubs

  Site inspections

  Land service care

  Leachate and gas management

  Environmental monitoring

  Insurance

If the waste is properly spread and compacted, a field density (wet weight) of $475kg/m^3$ (800 lb/yd^3) (of refuse alone) is possible. Some of the benefits of high in-place density of solid waste obtained by good refuse compaction are listed in Table 15.6.

The number of passes a compactor machine makes over the solid waste also affects density. A machine typically makes three to five passes to achieve the best results. More than five passes does not achieve sufficient additional compaction to warrant the effort. The slope of the working face should be kept steep; a 3:1 to 5:1 slope is best, depending upon the type of compaction equipment used and the size of the landfill operation.

Landfill density can also be increased by routing traffic over areas previously filled. Vibration and weight help to consolidate and compress the waste, and to eliminate voids. Surcharging filled area with cover soil stockpiles also aids in compaction due to their weight and the vibration caused by the earth-moving activity. Utilization of deep fills (greater than 15 m) also compresses the lower layers of waste, increasing the rate at which density is increased (Bleiker et al., 1993).

Table 15.6 BENEFITS OF GOOD REFUSE COMPACTION

Extends the life of the site

Decreases settlement of the refuse in the future, thereby minimizing ponding and surface repair problems

Reduces voids in the refuse

Reduces windblown litter

Discourages insects and rodents

Reduces the possibility that waste will wash away during heavy precipitation events

Reduces the needed total volumes of daily cover

Provides a more solid roadbed, reducing maintenance and repair of on-site roads during site operation.

15.2.4 Weather Considerations

Operations will presumably function well under good weather conditions, but most operational problems are caused by weather conditions and the type of wastes being handled. For example, daily cover material with fine-textured soil (e.g., clay) is desirable for preventing some problems but creates operational problems during wet weather. Plans for site access during wet weather (e.g., haul roads must be passable in all weather conditions) and consideration of drainage patterns to minimize problems are essential.

In more northerly climate conditions, planning for winter conditions is also essential. Frost creates considerable wear on equipment, difficulty in handling cover material, and potential accident hazards. Other problems include equipment reaction to cold weather, frostbite, snow removal, and fires caused by cinders or ashes buried with the solid waste. Table 15.7 gives some suggestions for possible approaches to utilize during winter operations.

Table 15.7 SUGGESTIONS FOR WINTER OPERATIONS

Use coarse-textured soil.

Excavate from sidewalls of borrow pit, where frost penetration is less.

Stockpile soil for winter use.

Protect soil with leaves, plastics, or other insulating materials.

Treat stockpiled soil with calcium chloride to reduce freezing.

Use "frost ripper" to rip cover material.

Minimize traffic over borrow areas to reduce compaction of the surface material and, ultimately, frost penetration.

Use plug-in heaters for equipment and/or protected storage.

Monitor incoming solid waste closely, providing separate areas to discharge "hot" loads, and/or temporarily storing material.

15.2.5 Handling of Special Wastes

Provisions must be made for the handling of special wastes, including:

- *Bulky wastes* such as appliances (e.g., "white goods"), furniture, and tree stumps. Disposal of these wastes may involve:
    - Disposal in a separate section of the landfill
    - The dumping at the toe of the slope and compaction of other solid wastes around them
    - Crushing of objects on solid ground to increase compaction obtainable
    - Salvaging and reselling the material for scrap.

- *Construction/demolition wastes.* Some of these materials may be usable for roadways directly or after mulching and shredding. Others can be incorporated with normal wastes or utilized as part of the daily cover.
- *Biomedical wastes.* These come from health care facilities, nursing homes, home health care, veterinary care facilities, and laboratories. The disposal of biomedical waste usually requires special approval. Any material disposed of should be covered immediately to reduce occupational exposures, whereas some should be incinerated or sterilized prior to disposal to kill entrained pathogens.
- *Pesticide waste containers.* These should be disposed of only after they have been rinsed and holes have been knocked into them.

- *Asbestos wastes.* These wastes require special handling procedures. Disposal involves the isolation of asbestos material to prevent fiber release to air or water. Landfilling is recommended as an environmentally sound isolation method because asbestos fibers are virtually immobile once buried. However, some governmental restrictions exist on disposal of asbestos. Disposal techniques such as incineration or chemical treatment are not appropriate due to the properties of asbestos. EPA has established asbestos disposal requirements under National Emission Standard For Hazardous Air Pollutants (NESHAPs) (40 CFR Part 61, Subpart M) and specified federal requirements for solid waste disposal under RCRA (40 CFR Part 257). Advance EPA notification of the intended disposal site is required by NESHAPs.

  Required disposal procedures include the following:
  - Asbestos should be put in closed containers or bags that can be easily inspected.
  - Material in bulk must be dampened and covered before shipment to the landfill.
  - Landfill managers must be notified of asbestos deliveries in advance so cover material will be available.
  - Wind velocities should not exceed 10 m/h during disposal operations.
  - Employees handling asbestos should wear protective clothing and dust masks.
  - Asbestos should be placed at the toe of the working face and covered immediately.

- *Automobile tires.* Tires create unique problems in landfill operations and are generally not accepted. If they are accepted, the following procedures are usually adhered to:
  - Place tires at the bottom of the working face and compact solid wastes on top.
  - Separate stockpiles from the active fill area and remove frequently if salvage is being practiced.
  - Use a tire splitter or shredder.
  - Consider special charges to cover the cost of disposal of the tires.

- *Brush, plastics and lightweight materials.* If not handled correctly, these materials can result in considerable litter. The following procedures should be practiced:
  - Cover with other solid waste material as soon as possible.
  - Use caution when compacting, since these materials can plug the air intakes or radiators of equipment.
  - Chip the brush and use for mulch if at all possible.
  - Pick up all the litter on the site daily.

- *Wastes of unknown characteristics*. Liquids, sludges, ashes, and chemicals should not be accepted at the site unless the facility is specifically designed and permitted to receive these types of wastes. If unknown wastes appear, do not place in the landfill until information is obtained from the hauler, and regulatory agency approval is obtained. Municipal solid waste landfills are now forbidden from accepting hazardous wastes from business such as automotive service centers, dry cleaners, printing shops, and electroplaters.

15.3 SITE OPERATIONS

15.3.1 Noise Control

Noise may create problems by increasing stress, causing hearing impairment (primarily for on-site personnel) and being a nuisance to adjoining residents. Noise is caused by routine travel of trucks and landfill equipment and acceleration of engines while working. Noise control can be accomplished by a number of methods, all of which are highly recommended, since noise control is an important ongoing activity at a landfill site. These methods include:
- Use of trees and berms to attenuate noise
- Proper phasing of operations to create a buffer zone or barrier between the source and nearby residents
- Appropriate equipment maintenance, including muffler repair as needed
- Regulation of hours of operation to reflect adjacent land uses
- Maximizing the separation distance between the active site face and residents.

15.3.2 Odor Control

Odors are normally a seasonal problem and hence can be incorporated into planning. The problems with odors include community complaints, public perceptions, and nuisance conditions for those affected. Odors may be from toxic or irritating gases that could have adverse health effects.

The sources of odors are putrescible wastes, carcasses, sludges, landfill gas, sulfides and mercaptans, and leachate seeps. Control of the odors can be accomplished by immediately covering the solid waste that has reached stages of decomposition. If the solid wastes require special handling, tipping fees should reflect this. "Masking" chemicals for special problems can be employed, but these chemicals are generally costly and not always effective. The placement and utilization of a horizontal gas collection system as the site is built can greatly assist in minimizing odors. Proper venting and flaring of landfill gas will also normally be required, as discussed in Chapters 13 and 14. The collection or prevention of leachate escape will minimize odors, but odors associated with the leachate treatment itself, as discussed in Chapter 11, must be considered.

Fig. 15.8   Movable fencing for litter control.

15.3.3 Litter Control

Problems that are created by litter include unacceptable appearance, a food source for vermin, blight, and fires. Litter is the main source of many complaints from nearby residents, since it is something that can easily be seen by the public. This results in a public relations problem, as people perceive that the facility is being operated improperly. Litter is caused by open loads, windy-day operation, operational techniques, and dumping practices.

The following procedures will help litter control.

- Require open loads to be covered.
- Move operations to protected areas during windy conditions, plant some windbreaks, or operate the working face so that it is opposite to the direction of prevailing winds. Alternatively, use movable fencing, such as depicted in Figure 15.8.
- Unload solid waste at the bottom of the slope and push materials up the slope.
- Compact and cover the refuse more frequently during the day.
- Confine the dumping area to a minimal surface area.
- Collect the litter on an ongoing basis.
- Stop operations in excessively windy conditions.

It is much easier to control litter at the source. If not, it can be controlled at the periphery of the site, as indicated in Figure 15.9.

Fig. 15.9   Litter removal from peripheral fencing.

15.3.4 Dust Control

The potential problems with dust include allergic reactions, nuisances, accidents, increases in equipment problems, and property depreciation for nearby lands. Dust is caused by traffic on roadways, earthmoving activities, dumping and compaction activities, and wind. Dust control measures include the following:

- Consider using all-weather roads (asphalt) with proper maintenance. Utilize speed bumps as appropriate.
- On dirt roads, apply calcium chloride (if the humidity is over 30 percent) at the rate of 0.2 to 0.4 kg/m^2 (0.4 to 0.8 lb/yd^2); apply water from trucks as necessary.
- Time earthmoving activities to reduce the generation rate of dust.
- Move dumping and compaction areas to reduce the impact on neighbors during dusty, dry, or windy conditions.
- Use vegetation and windbreaks to reduce wind speed at ground level.

15.3.5 Bird Control

Seagulls often use landfills as their "restaurants." Problems caused by the seagulls include danger to aircraft and contamination of surface water. Nuisance conditions can also occur when the birds pick up refuse and drop it in nearby fields (which causes choking of farm animals).

Bird control is very difficult to achieve. In the United States, seagulls are a protected species, and approval is needed to employ an elimination program. Seagulls have been successfully controlled using elevated wires and/or string netting over a fill—the nets cause interference with the seagulls' "radar" system. Obviously, the use of netting has ramifications in terms of site operation.

Other bird control measures that have been tried with varying degrees of success include intermittent killing, loud noises, chemical poisons, tape recordings of birds in distress played over loudspeakers, and bird models illustrating death or distress. Excellent daily operations help a great deal in minimizing the problem.

15.3.6 Rodent Control

Rodent populations occur because they are brought to the site in loads or migrate to the site. They remain if there is available food and shelter. Shelter areas are created by salvage storage areas or voids in bulky or demolition wastes. Rodents are usually not visible during normal working hours. If they are observed, it is an indication that a severe infestation is likely present.

Rodents can spread disease (rat bite fever, rabies, typhus, and the like), destroy property, and contaminate food. Rodents are difficult to eliminate. The program must be total elimination; the elimination of food will simply cause the problem to migrate elsewhere.

15.3.7 Insect Control

Insects of concern include flies and mosquitoes. The potential problems from insects include disease transmission and nuisance. Flies are capable of transmitting diseases such as salmonella. Mosquitoes transmit encephalitis and other arthropod-borne viral diseases. Insects are attracted by food sources, shelter, and breeding areas, and they are brought in with the waste stream. They can be controlled by covering the solid waste to eliminate food, shelter, and breeding areas and by extermination.

15.4   SITE OPERATIONS FACILITIES

15.4.1 Record-keeping

For protection from liability and considering possible future requirements for notification on disposal site deeds, a landfill owner must maintain documentation of disposal materials. The stringency of the record-keeping requirements has increased dramatically in recent years. In particular, the estimated depth of the waste below the surface should be recorded whenever a landfill section is closed.

15.4.2 Perimeter Fencing

Fencing must be periodically inspected and repaired in the regular course of site maintenance activities. All other portions of the site should be delineated with a post and wire fence to limit access by people and animals.

15.4.3 Signs

To improve site operation and for public safety reasons, a number of signs and notices must be placed and maintained at various locations at the site. The minimum list of required signs includes the following:
- Speed limit signs on all site access roads
- Warning signs at all manholes and valve chambers where combustible gas may be present
- No Entry signs at all maintenance and monitoring access roads
- Appropriate sign at the main site entrance indicating the operating hours
- Stop Ahead and notification signs for any public drop-off area indicating that all public traffic must stop at the attendant's trailer for direction regarding unloading wastes and/or recyclables
- Appropriate signs indicating current tipping fees and charge rates at the public drop-off area
- Directional signs directing people to the disposal face/area.

15.4.4 Safety and Security

The manager of a landfill is responsible for both the security of the site and the safety of users, employees, and the general public. In recognition of this responsibility, the manager must understand and support the security measures planned for the facility.

Commonly, a landfill is set off from surrounding properties by fences or other barriers (e.g., ditches, bodies of water, or extensive open space). Site security includes controlling access to the site and supervising all activities of all persons on the site, including identifying all substances to be placed in the disposal site. Thus, site security includes
- The employment of appropriately trained staff to control access to the site by vehicular and foot traffic
- The maintenance of physical features and components such as gates, fences, bridges, moats, and streams
- The examination of the solid waste presented for disposal to minimize the potential for receiving solid wastes that are not acceptable, such as hazardous and toxic materials
- The operation of an electronic weigh scale and accessories such as traffic lights
- Maintenance of site-operation vehicles.

Table 15.8 lists the array of vehicles required for site operation.

Fig. 15.10   Dropoff area for public access for recycling
(Yolo County Landfill, California.

Table 15.8   EQUIPMENT NEEDED FOR SITE OPERATION

Equipment to maintain roads (road graders) for access roads

Equipment for transporting and spreading daily and final cover materials

Tanker truck (if needed) to transport liquid from the leachate collection system to the treatment plant

Compactors to level and compact the waste and daily cover. A crawler or rubber-tired tractor can be used with a dozer blade, trash blade, or a front-end loader. Bulldozers are used for miscellaneous operations and construction including berm construction, refuse placement and leveling, and daily cover leveling. Other equipment used normally only at large landfills includes scrapers, compactors, draglines, rippers, and graders.

Road watering truck

Pickup trucks for monitoring and maintenance activities.

15.4.5 Public Drop-off Area

The public drop-off area utilizes a ready-access point for drop-off by the public for goods for recycling. The area may also be combined with a transfer station. The facility must be amenable to drop-off from small vehicles such as cars, half-ton trucks, and from larger industrial and commercial vehicles.

It may be appropriate to construct a berm (e.g., 2 m) to visually screen the bin storage area and to serve as a windbreak to limit blowing litter.

An example of a drop-off facility is indicated in Figure 15.10.

References

Bleiker, D., McBean, E., and Farquhar, G. 1993. Refuse Sampling and Permeability Testing at the Brock West and Keele Valley Landfills. Paper presented at *Sixteenth International Madison Waste Conference.* Madison, Wisconsin, September.

Conestoga-Rovers & Associates 1990. Design and Operation Report: Mid-Huron Landfill Site. Waterloo, Ontario, Canada, October.

MONITORING PROGRAM CONCERNS

16.1 COMPONENTS OF A MONITORING PROGRAM

The intent of a monitoring program is to determine the degree to which a landfill and any associated containment system is functioning in accord with the design objectives. As noted at numerous locations within the previous chapters, concern is with the excursion of contaminants off the site as leachate or as landfill gas or both. It is essential, therefore, that the monitoring program identify the extent to which any excursions are occurring.

The monitoring activities should include:

- A perimeter monitoring well system for detection of off-site migration of contaminants
- Lysimeters placed under the liner of each section of the landfill
- A perimeter storm-water ditching system around the active sections of the landfill to direct runoff to detention ponds. The storm water is then monitored prior to controlled release or infiltration.

Regulatory standards typically specify minimum levels of monitoring. However, regulatory standards are continuously evolving and differ from one governmental jurisdiction to another. Therefore, we focus on the principles and some of the details associated with such features as the design components of a monitoring well. The reader interested in the specifics of the existing regulatory standards must therefore contact the local authority for more specific guidance.

457

Some of the features that are monitored in and around the landfill include

- Leachate head within the refuse
- Hydraulic head in the de-watering system
- Leachate leakage quantities through the landfill liner
- Gas migration through the landfill liner
- Groundwater quality around the site
- Gas characteristics in the refuse (pressure, temperature, methane content);
- Gas in the soil and the atmosphere around the landfill
- Head and quality of the leachates in the leachate collection tank (if it exists)
- Visual examples of leachate seeps through side slopes
- Visual examples of extensive sedimentation onto adjacent vegetation.

The preceding list is not purported to be exhaustive but is intended to indicate the considerable extent of monitoring program requirements.

Groundwater monitoring wells are a principal means for characterizing the soil stratigraphy and downward movement of leachate to the groundwater. In lined landfills, the presence of leachate in the saturated zone below or beyond the landfill is a clear indication of failure of the containment system. Surface water contamination is not normally a major concern for toxics, and visual monitoring of the slopes will probably indicate any horizontal migration of leachate out the sides of the landfill. However, sediments and debris are frequently surface water-related problems. Table 16.1 lists examples of monitoring program requirements.

Table 16.1 EXAMPLES OF MONITORING PROGRAM REQUIREMENTS

	Location	Parameter	Frequency of Monitoring
1.	*Groundwater*		
	On-site obervation wells	Water Quality*	Monthly
			Quarterly
			Tri-annually
			Semi-annually
		VOCs	Annually
		Water Quality*	Monthly
			Quarterly
			Tri-annually
			Semi-annually
		Total Well Inspection	Semi-annually
	Off-site wells	Quality	Semi-annually

Table 16.1 EXAMPLES OF MONITORING PROGRAM REQUIREMENTS

2. *Liner Performance*

Lysimeters	Quantity	Quarterly
	Quality	Quarterly
Conductivity sensor array (if installed)	Conductivity	Continuous
Vibrating wire piezometers (if installed)	Leachate head on liner	Continuous

3. *Landfill Gas*

Collection trenches	Static pressure, flow, temperature, methane concentration	Weekly
Collection wells	Static pressure, flow, temperature, methane concentration	Weekly
LFGflare station	Static pressure, flow, temperature, methane concentration	Daily
Gas migration probes	Pressure, combustible gas concentration	Monthly
Condensate	Quantity	Monthly
	Quality	Monthly

4. *Landfill Leachate*

Leachate	Quantity	Monthly
	Quality	Monthly
Pumping stations	Floats and level alarms	Continuous
Discharge to sanitary sewer	Quantity	Continuous
	Quality	Monthly
Leachate collection system cleanouts	Sediment buildup	Annually
Leachate collection pipes	Sediment buildup, integrity	Annually
Sanitary sewer	Sediment buildup, integrity	Annually

5. *Surface Water*

Swales	Sediment buildup, ground cover	Major storm events & quarterly

Table 16.1 EXAMPLES OF MONITORING PROGRAM REQUIREMENTS

	Silt curtains	Sediment buildup, damage	Major storm events & semi-annually
	Ditch inlets and manholes	Blockage	Semi-annually
	Solid waste area storm basins	Sediment buildup	Major storm events & monthly
	Off-site	Flow	Continuous
		Quality	Tri-annually
6.	*Climatic Conditions*		
	Site office	Wind speed, wind direction, barometric pressure, temperature	Continuous
7.	*Site Restoration*		
	Final cover	Erosion, surface seeps, groundwater discoloration, cracking, ponding	Continuous
	Final cover	Differential settlement	Continuous/Semi-annually
	Odors	Surface seeps	Continuous
9.	*Access Roads and Services*		
	Surfaces	Clean Asphalt, regrading granular surface	Continuous
	Perimeter fencing	Damage, maintenance	Continuous
	Road and warning signs	Damage, maintenance	Continuous

* The frequency of monitoring varies for specific observation wells.

A groundwater monitoring program must include sampling and analytical methods that (1) are appropriate for groundwater sampling and (2) accurately measure constituents. Sampling procedures and sampling frequency therefore have to be designed to protect human health and the environment. The sampling procedure has to ensure that the statistical procedure used to evaluate samples has an acceptably high probability of detecting contamination, if it is present.

The monitoring effort associated with a landfill is substantial. Questions that must be addressed include what equipment to use, where to install the equipment, when the equipment should be dedicated to that use, and what frequency of monitoring is required. We will examine an array of features associated with monitoring and the implications to the design of the monitoring program.

16.2   ACCESS FEATURES FOR MONITORING

16.2.1 Groundwater Monitoring

For landfill investigations, groundwater monitoring wells are installed (1) to measure groundwater level elevations (to determine flow direction), (2) to test the physical hydraulic parameters of the hydrogeologic units (to determine permeabilities), and (3) to collect groundwater samples for chemical analyses (to determine if groundwater contamination is occurring). The selection of locations for monitoring wells is directed by all three objectives.

The monitoring program for groundwater must include consistent sampling and analysis procedures designed to ensure monitoring results of the groundwater quality at a number of sampling locations. A groundwater monitoring program must afford the owner/operator and the regulatory agency the opportunity to evaluate the following:

- The performance of technical design features such as liners and leachate collection systems
- The occurrence of groundwater contamination, including the degree and significance of groundwater contamination
- Interference from non-contaminant sources, such as elevated chloride levels from road salting.

Although seemingly a simple task, the task of collecting samples representative of aquifer chemical conditions is a challenge. Numerous events may impact the water quality levels, including:

- loss of volatile constituents as a result of changes in pressure or due to agitation
- a change in reactive constituents due to use of improper preservatives
- contamination of the sample by dirty equipment and/or laboratory analytical equipment
- delay prior to analysis at the laboratory
- interference from noncontaminant sources, such as elevated chloride levels from road salting.

The sample and the laboratory analyses must be obtained and documented in a manner that demonstrates technical and legal validity. Thus, the sampling plan must carefully delineate the procedures and techniques for sample collection, sample preservation and shipment, laboratory analytical procedures, and chain-of-custody control.

Measurement should include depth to standing water, with measurements to 0.01 ft. However, the water standing in a well prior to sampling may not be representative of in situ groundwater quality. Therefore, the standing water in the well and filter pack should be removed so that the water in the surrounding soil and refuse formation can replace the stagnant water. Typically, three well volumes of water are removed prior to sample collections.

The procedure used for well evacuation depends on the hydraulic yield characterization of the well. Low-yield wells (wells incapable of yielding three well volumes) should be evacuated to dryness.

Well Locations   Wells must be installed at the closest practicable distance from the waste management unit boundary. The set of wells must consist of a sufficient number of wells installed at appropriate locations and depths to yield groundwater samples sufficient to characterize (1) the quality of background levels (not affected by leakage from the landfill), (2) downgradient levels (regions that may have been impacted by the landfill), and (3) a region between the site and any other possible sources of contamination.

Screened Interval   The sample being recovered in the monitoring is intended to be representative of the specified vertical interval over which the monitoring well screen is situated.

Well Installation and Construction Details   Specifics on typical monitoring well construction details are described in Section 16.3. At this point it is sufficient to indicate that monitoring wells must be cased in a manner that maintains the integrity of the monitoring borehole. The casing must be screened or perforated and packed with sand or gravel. The annular space (i.e., the space between the borehole and the well casing above the sample depth) must be sealed to prevent vertical migration of contamination from overlying intervals and the groundwater. The drilling technique and monitoring well design, if not selected appropriately, may alter the chemistry of the groundwater samples, which might be mistaken for contamination from the landfill. Additionally, drilling equipment must be cleaned prior to on-site use, and drilling should proceed (where possible) from areas of least contamination to areas of greatest contamination.

The most common casing materials in use are stainless steel, Teflon, and PVC. Casing and screens should be constructed out of stainless steel to eliminate interference from elevated iron levels leached from the well casing. Leaching of organics from PVC glues is a concern. When PVC is used, threaded joints should be employed.

Additional guidance on well installation and construction details is provided in numerous sources, including Gilham et al. (1983).

Sampling Parameters   There are potentially many individual parameters to be monitored. However, the general practice is to include a number of general parameters in the sampling set and then, where the local situation merits, to add additional parameters to the set of general parameters.

The selection of parameters to be analyzed is based on the chemical characteristics of the unimpacted water in the area, the type of landfilled waste, and

indicator   parameters   known   to   signal   the   advancement   of   a
leachate-contaminated plume of water.

The general set of parameters frequently utilized include:

- Water levels or groundwater elevations measured in each well immediately
  prior to water quality sampling, to determine the rate and directions (horizontal and vertical) of the groundwater flow. The groundwater flow direction dictates directions of advective contaminant transport.
- Total dissolved solids (TDS), total organic carbon (TOC), chemical oxygen
  demand (COD), hardness, temperature (°C), and electrical conductivity.
  Specific conductance is a measure of the ability of a substance to transmit
  an electrical current. High values of specific conductance indicate the presence of inorganic materials dissolved in the water.

Additional parameters might include:

- Inorganics (metals, alkalinity, ions)
- Organics (volatile organics)
- Surrogates/site-specific constituents (e.g., total phenols).

If there is little or no contamination evident in existing monitoring records,
monitoring requirements are frequently limited to indicator parameters only, to
signal the advance of any leachate-contaminated plume.

Table 16.2 MONITORING PARAMETERS FOR SAMPLING

Ammonia (as N)	Total organic carbon (TOC)
Bicarbonate (HCO_3^-)	pH
Calcium	Arsenic
Chloride	Barium
Iron	Cadmium
Magnesium	Chromium
Manganese (dissolved)	Cyanide
Nitrate (as N)	Lead
Potassium	Mercury
Sodium	Selenium
Sulfate (SO_4^-)	Silver
Chemical oxygen demand (COD)	VOCs
Total dissolved solids (TDS)	

Table 16.2 lists monitoring parameters or constituents as indicated in the Federal Register (1988).

Selection of which parameters to monitor must consider a number of features that are specific to individual constituents.

- Chloride is an important parameter for analysis. It is present in abundance from both municipal and industrial refuse and is essentially unretained by soil mechanisms because it is nonreactive both physically and biologically. It diffuses quickly and often signals the advancement of a plume of contaminated water.

- Ammonia, and other forms of nitrogen in water, principally nitrate, are often indicators of sewage, fertilizers, nitrogenous aerosols, plastics, and drugs. Nitrate is very mobile and is closely monitored for health reasons. It may also indicate whether or not the landfill has entered its anaerobic stage. In the anaerobic stage, anaerobic decomposition dominates and the entire landfill is in a chemically reducing state that results in more ammonia than nitrate or nitrite. Nitrite is an indicator of active biological activity.

- Sodium is the principal alkali metal and tends to remain in solution and not be subject to attenuation. Its sources in leachate may stem from the extensive use of sodium salts in industry and domestic activities (paper, soap, borax, and baking).

- Sulfate is the most common form of sulfur in landfills. It is quite mobile and is helpful for monitoring leachate movement. It is easily reduced to sulfides, which complex readily with metals.

- Potassium is part of the parameter list because it is essential for plant and animal nutrition. It can also correlate with the release of oxygen-demanding organic materials during refuse decomposition, since it is part of the organic materials in plants.

- Magnesium in landfills is most often due to cosmetics, cement, and textiles.

- Copper is sometimes tested for because of health guidelines but is not very mobile in soil and is normally of limited value in monitoring programs.

- Lead comes from batteries, photography, older lead-based paints and lead pipe disposed of in the landfill. It is toxic to all forms of life at certain levels and is tested for to ensure that health guidelines are not exceeded. Acidity in leachate causes lead to be released from refuse.

- Zinc is present in the landfill due to batteries, solder, and fluorescent lights. It is of concern with regard to plants and aquatic life.

- Iron, due to the corrosion of iron and steel, may also exist at elevated levels. Guidelines restricting the concentration of iron in water are due to taste and discoloration factors.

- Biological oxygen demand (BOD) is a measure of the oxygen required to oxidize organic matter, by aerobic microbial decomposition, to stable inorganic forms.

- Chemical oxygen demand (COD) is the oxygen required to chemically oxidize organic matter (COD>BOD).
- Suspended solids in leachate arise from solids carried out of the refuse. They alter both the physical and chemical characteristics of water.
- Hardness is the sum of the calcium and magnesium present in the water. It is reported in terms of equivalents of calcium carbonate.
- If the pH is less than 2 or greater than 11, the water is considered corrosive. The toxicity and complex forms of metals are strongly influenced by pH. The acceptable range of pH for water is 6.5 to 8.5. Conductivity is sensitive to variations in dissolved solids, with which it has almost a direct mathematical relationship, and mineral content, both of which are elevated in leachate due to contact with refuse. The strength of the leachate is sometimes measured in terms of conductance; greater conductance signifies stronger leachate. Turbidity results from suspended particles.
- Metals, VOCs, pesticides, PCBs, and other trace organics are associated with hazardous waste dumping. They are sometimes analyzed for in the early stages of a monitoring program if it is not known if hazardous waste was disposed of at a municipal solid waste disposal site. Until this fact is established, these parameters are tested for approximately once a year. If hazardous wastes are not present in a municipal landfill, they are unlikely to be analyzed for except in the first stages of monitoring.

Sampling Frequency   Sampling frequency selection is a function of hydrogeologic considerations. Regulatory requirements are changing, but criteria reflected in the selection of the sampling frequency consider the aquifer flow rate and the resource value of the groundwater. Groundwater may be monitored quarterly, biannually, or annually, depending on the type of waste, size and design of the landfill, and aquifer material. If the site is contaminated, it is tested often during the year, probably four times, and possibly six. If the site does not show signs of contaminating the surrounding area, the site may be tested only once or twice yearly. All sites should be tested at least quarterly for a year until the site has been properly characterized and long-term monitoring plans have been made. If contamination is detected, monitoring may become part of an enforcement action for the protection of groundwater users and the environment.

Sampling Techniques   The sampling technique must ensure that the groundwater sample is representative of the groundwater. Thus, it is essential to minimize physical alteration and chemical contamination of the sample, for example, decontaminating any purging equipment and using dedicated samplers and tubing or thoroughly decontaminating them. Thus, specification of sampling techniques require determination of the need for such features as dedicated versus nondedicated samplers; decontamination procedures; purging—the water contained within each monitoring well should be purged by removing well volumes of water using a bailer or a pump prior to sampling to obtain a fresh sample; or quality assurance samples and blanks. Surging, bailing, pumping and air lifting are methods of developing wells (Scalf et al., 1981). Table 16.3 describes the advantages and disadvantages of different types of sampling devices for water quality.

Data Handling and Reporting Protocol All data must be quality assured and controlled (QA/QC) to ensure that the results can be quoted with confidence. The range of QA/QC concerns is very broad. They include the documentation in the operating record of the design, installation, and decommissioning of any monitoring wells, piezometers and other measurements, sampling, and analytical devices.

Table 16.3 ADVANTAGES AND DISADVANTAGES OF VARIOUS WATER QUALITY SAMPLING DEVICES FOR MONITORING WELLS

Type	Advantages	Disadvantages
Bailer	Can be constructed in a wide variety of diameters	Sampling procedure is time consuming; it is sometimes impractical to properly evacuate casing before taking samples
	Can be constructed from a wide variety of materials	Aeration may result when transferring water to the sample bottle
	No external power source required	
	Extremely portable	
	Low surface-area-to-volume ratio, resulting in a very small amount of outgassing of volatile organics while sample is contained in bailer	
	Easy to clean; readily available; inexpensive	
Suction-lift pump	Relatively portable	Sampling is limited to situations where water levels are within about 20 ft of the ground surface
	Readily available	Vacuum effect can cause the water to lose some dissolved gases
	Inexpensive	
Air-lift samplers	Relatively portable	Procedure causes changes in carbon dioxide concentrations; therefore, this method is unsuitable for sampling for pH-sensitive parameters
	Readily available	In general, this method is not an appropriate method for acquisition of water samples for detailed chemical analyses because of degassing effect on sample

Table 16.3 ADVANTAGES AND DISADVANTAGES OF VARIOUS WATER QUALITY SAMPLING DEVICES FOR MONITORING WELLS

	Inexpensive	Oxygen is impossible to avoid unless elaborate precautions are taken
	Suitable for well development	
Gas operated pump	Can be constructed in diameters as small as 25.4 mm (1 inch)	A gas source is required
	Can be constructed from a wide variety of materials	Large gas volumes and long cycles are necessary when pumping from deep wells
	Relatively portable	Pumping rates are lower than those of suction or jet pumps
	Reasonable range of pumping rates available	Commercial units are relatively expensive
	Driving gas does not contact water sample, eliminating possible contamination or gas stripping	
Submersible pump	Wide range of diameters	With one exception, submersible pumps are too large for 51 mm (2 in.) diameter well casing
	Constructed from various materials	Conventional units are unable to pump sediment-laden water without damaging the pump
	12V pump is highly portable; other units are fairly portable	44.5 mm (1-3/4 in.) pump delivers low pumping rates at high heads
	Depending on size of pump and pumping depths, relatively large pumping rates are possible for wells larger than 51 mm (2 in.) diameter	Smallest diameter pump is relatively expensive
	Readily available	
	44.5 mm (1-3/4 in.) helical screw pump has rotor and stator construction that permits pumping fine-grained materials without damage to the pump	

Source: Driscoll, 1986.

Inaccurate results cost as much as accurate results initially, but the inaccurate results may later come back to cause more significant trouble. The best approach is to carry out the sampling properly from the outset.

Data handling and reporting protocols require numerous decisions including the following:

- Chain-of-custody requirements — collection, preservation, and analysis of the groundwater samples are essential to obtain representative data
- Continuous versus non-continuous monitoring
- Digital versus analog recorders
- Remote versus accessible monitoring capabilities
- Database design
- Reporting requirements to establish the appropriate scope and frequency.

Factors that may compromise sample integrity include:

- Contamination of samples by dirty equipment or dirty analytical equipment
- Loss of volatile constituents as a result of changes in pressure and/or agitation
- Improper sample preservation.

The sampling and analysis plan must include procedures and techniques for sample collection, sample preservation and shipment, analytical procedures, and chain-of-custody control.

Field blanks, trip blanks (for VOCs), and bailer rinses form part of a comprehensive QA/QC plan. *Field blanks* are samples of deionized water that are taken throughout the entire sampling and analysis procedure. They are used to identify contamination due to ambient conditions on site, bottle contamination, or laboratory practices. *Trip blanks* are generally used only for VOC analysis. They are lab-prepared samples of organic-free deionized water that travel with the sample containers from the lab to the field and back to the lab to indicate any contamination from leakage due to other sample bottles. They are not opened until they are returned to the lab. A bailer rinse specifically tests cross-contamination from sampling equipment. Duplicate samples are also taken as a check on field sampling procedures and laboratory testing methods. The personnel in the laboratory are not informed as to which samples are blanks or duplicates to prevent bias during analyses.

Data Interpretation   The statistical procedures used to evaluate samples must have an acceptably low probability of failing to identify contamination. Statistical procedures for examining these features are described in Section 16.7.

16.2.2 Monitoring for Leakage

Lysimeters are constructed in or below the liner of a landfill to obtain samples for the purpose of leak detection. Suction lysimeters are designed to collect soil moisture before it reaches the groundwater.

Additional means of monitoring are provided by instruments such as thermocouple psychrometers, which are meters used to detect changes in moisture content. Gamma ray attenuation probes are also used to detect changes in moisture content. They function on the basis of the degree of attenuation of gamma rays in response to bulk density changes and water content of the medium.

16.2.3 Surface Water Monitoring

Surface water may be characterized by the same methods as used for groundwater, but surface water in any given area may be more varied than groundwater because more diverse factors may be influencing it.

Surface water locations are sampled by submersing the sampling container at the midpoint in the body of the water. Stagnant water should not be sampled, if this can be avoided. If several samples are to be taken, the sampler should start at a downstream position and move upstream. Care must be taken to not unduly agitate the water.

16.2.4 Field Analyses

Some preparation and analyses of samples are frequently performed in the field. Qualitatively, the odor, color, and turbidity are reported. The temperature (°C), pH (standard units), and conductivity (μmhos/cm) are also recorded. The temperature is needed to correct conductivity measurements. The pH of the samples is measured in the field because it is affected by dissolved carbon dioxide in the sample. As the sample is transported from the field to the lab, CO_2 exits the sample, causing the pH to increase. Conductivity is measured in the field, since it is temperature sensitive. Further, it is also dependent on suspended solids, which are removed during filtering.

Surface water is never filtered, but some groundwater is, depending on the parameter(s) being tested. Currently, there is considerable discussion regarding the need for field filtration, and different policies may be followed. If the dissolved form of the parameter is required (total dissolved solids), the sample is filtered. For example, organic parameters are not normally filtered. Inorganic parameters are filtered unless a total parameter concentration is needed (e.g., total heavy metals). The filtration procedure, when and if it is employed, may involve vacuum filtration through a 1.2 μ glass fiber filter and then through a 0.45 μ filter. The initial filtration is optional and is performed merely to save time in the second step by removing large particulates. The 0.45 μ filter removes suspended solids. Preservatives in the form of nitric or sulfuric acid may then be added, again depending on the desired parameter. The amount of acid added is generally around 1 mL. The object is to cause the pH of the sample to drop to approximately 2. Preservation slows microbial activity so that COD, BOD, and organic nitrogen parameters are not substantially changed before analyses. Acid preservation is also used to keep metals in solution. However, in unfiltered samples, this may put into solution metals adhered to or part of the suspended solids.

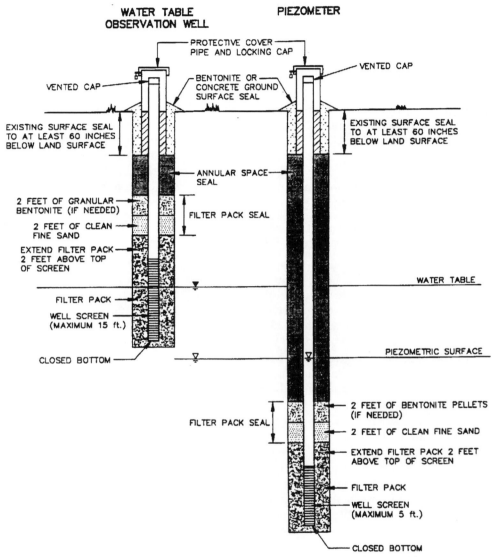

Fig. 16.1 Typical water table observation well and piezometer
construction details. [CR, Register, January 1990, (No. 409, EFF. 2-1-90)]

16.3   MONITORING WELL CONSTRUCTION DETAILS

16.3.1 Construction Materials for Groundwater Monitoring

Typical monitoring well construction details are depicted in Figures 16.1
and 16.2. The monitoring well construction must allow for collection of data on
water levels and access to obtain a sample of water for analyses. Additional fea-

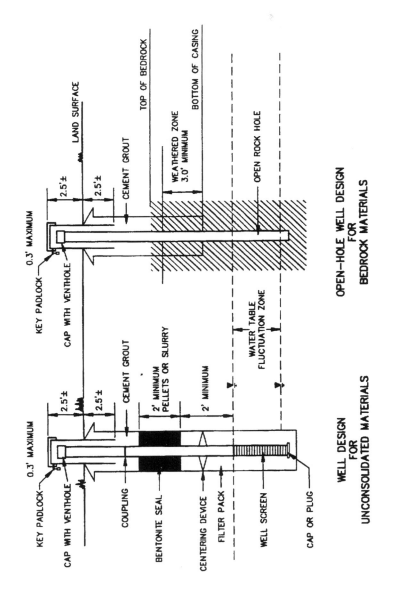

Fig. 16.2 Typical monitoring well designs for unconsolidated and bedrock materials.

tures of monitoring wells involve consideration of the composition of the materials being employed. Materials used in well installation/construction include PVC, Teflon, and galvanized and stainless steel. Selection of well construction materials must consider inertness, strength, ease of installation, cost, and geophysical characteristics consistent with the designed use of the monitoring well.

Prior to 1985, polyvinyl chloride (PVC) was the most commonly used casing material for groundwater monitoring. However, the concern was that the materials used in the monitoring well and sample collection might affect the integrity of the samples. These samples might sorb or leach constituents, thereby affecting the quality of the groundwater samples, or the casings might not have the long-term structural characteristics.

In recent years, a number of investigations have reported on these types of concerns. Sorption of metals by plastic and glass containers can be significant (Eicholz et al., 1965; Robertson, 1968; Bateley and Gardner, 1977; and Masse et al., 1981). Houghton and Berger (1984) found that a steel-cased well appeared to leach a number of metal species, including Fe, Cd, Cr, Cu, Mn, Mo, Se, and Zn. Reynolds and Gilham (1986) tested both PVC and PTFE for sorption of trace levels (ppb) of five halogenated organics: 1,1,1-trichloroethane, 1,1,2,2-tetrachloroethane, bromoform, tetrachloroethylene, and hexachloroethane.

Organic substances may also leach from casing materials. One would not expect to find much in the way of organics leaching from metal casings with the exception of those that might be derived from lubricants.

Components of PVC that can leach include vinyl chloride monomer, plasticizers, thermal stabilizers, pigments, lubricants, fillers, impurities, and transformation products (Parker et al., 1988). Parker et al. found that PTFE is superior to PVC when samples must be collected for trace levels of metals, but PTFE was reported as having significant losses of a series of chlorinated compounds.

Table 16.4 lists the advantages and disadvantages of various well casing and screen materials.

Monitoring Well Nest Well nests (clusters) are designed to monitor different levels within an aquifer or different aquifers at the same location. Well nests consist of two or more wells very close together but having different screened depths. Well nests are essential at many sites because contaminants can move downward as they move away from sites. Thus, using vertical nests provides a vertical characterization of water quality.

Groundwater Monitoring Well Development After installation, a well must be developed (or cleaned) to remove fines accumulated in the borehole during installation and to restore the natural permeability of the surrounding soil. The best development method is one that causes water to flow rapidly both into and out of the well screen.

A well should be developed until the water is clear or until the conductivity, pH, and temperature of the pumped water remain constant for a period of time. Well installation must be fully documented in the well installation log.

Factors Impacting Groundwater Quality   Numerous factors must be considered in obtaining representative water quality measurements. A number of publications examine the implications of various factors in terms of their influence (e.g., Nielsen, 1989). We will focus on the list in Table 16.5 which is a summary of the factors affecting groundwater quality measurements.

Table 16.4   WELL CASING AND SCREEN MATERIALS

Type	Advantages	Disadvantages
PVC (Polyvinyl chloride)	Lightweight	Weaker, less rigid, and more temperature sensative than metallic materials
	Excellent chemical resistance to weak alkalies, alcohols, aliphatic hydrocarbons, and oils	May adsorb some constituents from groundwater
	Good chemical resistance to strong mineral acids, concentrated oxidizing acids, and strong alkalies	May react with and leach some constituents from groundwater
	Readily available	Poor chemical resistance to ketones, esters, and aromatic hydrocarbons
	Low priced compared with stainless and teflon	
Polypropylene	Lightweight	Weaker, less rigid, and more temperature sensative than metallic materials
	Excellent chemical resistance to mineral acids	May react with and leach some constituents into groundwater
	Good to excellent chemical resistance to alkalies, alcohols, ketones, and esters	Poor machinability—it cannot be slotted because it melts rather than cuts
	Good chemical resistance to oils	
	Fair chemical resistance to concentrated oxidizing acids, aliphatic hydrocarbons, and aromatic hydrocarbons	
	Low priced compared with stainless steel and teflon	

Table 16.4   WELL CASING AND SCREEN MATERIALS

Teflon	Lightweight	Tensile strength and wear resistance low compared with other engineering plastics
	High impact strength	Expensive relative to other plastics and stainless steel
	Outstanding resistance to chemical attack; insoluble in all organics except a few exotic flourinated solvents	
Kynar	Greater strength and water resistance than Teflon	Not readily available
	Resistance to most chemicals and solvents	Poor chemical resistance to ketones
	Lower priced than Teflon	
Mild steel	Strong, rigid; temperature sensitivity not a problem	Heavier than plastics
	Readily available	May react with and leach some constituents into groundwater
	Low priced relative to stainless steel and Teflon	Not as chemically resistant as stainless steel
Stainless steel	High strength at a great range of temperatures	Heavier than plastics
	Excellent resistance to corrosion and oxidation	May corrode and leach some chromium into highly acidic waters
	Readily available	May act as a catalyst in some organic reactions
	Moderate price for casing	Screens are higher priced than plastic screens

Source: Driscoll, 1986

16.3.2 Landfill Gas Monitoring

Landfill gas monitoring must be carried out within the soil profile and in the atmosphere within the vicinity of the landfill. Monitoring within the soil profile is frequently accomplished using gas monitoring wells. Well nests are frequently used, in a manner and for purposes similar to those for groundwater sampling, as discussed in Section 16.3.1. The need for well nests is particularly relevant for gas monitoring, since, as discussed in Chapter 13 gas migration is dramatically affected by the existence of permeable strata.

In some respects, gas sampling is simpler than monitoring groundwater quality in that, at least for CO_2 and CH_4, the equipment is portable. Thus, explosion meters are used to detect the presence of methane gas. More sophisticated (accurate) analyses for trace gas constituents must involve laboratory analyses.

It is noteworthy that groundwater testing for pH, alkalinity and hardness may give an indication of landfill gas generation. These values are affected by carbon dioxide generation and migration. Carbon dioxide is highly soluble and if present in sufficient quantity, will alter the water quality parameters.

Frequency for landfill gas monitoring must be:

- Utilized on a systematic basis (premised on generation rates)
- Based on soil types (profile characteristics)
- Based on proximity of structures
- Increased if landfill gas is found.

Table 16.5   FACTORS IMPACTING GROUNDWATER QUALITY MEASUREMENTS

Factor
Type of drilling procedure and fluid used
Well development and effectiveness
Type of sampling pump (e.g., a special pump is used for VOC sampling)
Type of materials present in the sampling train
Length of time and rate at which the well is pumped
Location of the monitoring well
Length of the well screen
Selection of materials used in the casing and the screen
Type of sealant used (e.g., bentonite or cement)

16.4   STATISTICAL ANALYSES OF MONITORING DATA

16.4.1   Overview

Background water quality is typically obtained from a hydraulically upgradient well for each of the monitoring parameters or constituents required. The well does not have to be upgradient but must be representative of concentrations in the absence of the site.

The need for statistical analyses of monitoring data arises because a number of sources of uncertainty affect groundwater quality measurements. These sources of uncertainty include sampling uncertainty, laboratory error uncertainty, and the vagaries of nature. The result is that upgradient and downgradient sampling data must be examined to determine if there has indeed been a statistically significant change in water quality or whether quantified differences could have arisen purely by chance. The examination involves utilization of a statistical significance test.

A number of statistical discrimination models exist. However, the assumptions within the available models are frequently violated due to the characteristics of the data, including the following:

1. Typically, the number of samples is small.
2. Frequently, the data at two locations being tested have unequal variances (e.g., see McBean and Rovers, 1985).
3. Frequently, the data are skewed in distribution (i.e., they are not Gaussian or normally distributed). Sometimes this difficulty can be corrected by a data transformation (e.g., see Rovers and McBean, 1981).
4. The data sets include "less than" detection limits.

Use of the less-than value as equal to the detection value will maximize the estimated value of this mean and minimize the standard deviation over the true (but unknown) value. A number of different approaches have been presented in the technical literature, and the interested reader is referred to, for example, McBean and Rovers, 1984, for further information.

Because of the nature of the data, frequently there is considerable difficulty in identifying the most appropriate statistical discrimination test to apply in a particular application. We will focus on relatively "well balanced" situations where there are not severe problems with utilizing the statistical tests. The interested reader is referred to Box et al., 1978; Moroney,1975; and Mezei,1990, as examples for the more complex aspects of data analyses.

16.4.2 The Standard T-Test

A result is referred to as being statistically significant when a finding is established that is very unlikely to have arisen by chance (e.g., the difference between background groundwater quality and contaminated groundwater is difficult to ascribe to chance, and the difference must, in all common sense, be accepted as a real difference).

As a means of reflecting the probability of significance, a series of discrimination procedures have evolved, with the Student's t-test (and approximations therefrom) being the most commonly used.

Mathematically, the procedure tests whether the means from two sets of measurements, say X (where elements of X are x_i where $i = 1... m$) and Y (where elements of Y are y_j; where $j = 1... n$) are the same. The means of each of X and Y are calculated from

$$\bar{x} = \frac{\sum\limits_{i=1}^{m} x_i}{m} \qquad (16.1)$$

and

$$\bar{y} = \frac{\sum\limits_{j=1}^{n} y_j}{n} \qquad (16.2)$$

Similarly, the standard deviations of X and Y are calculated from

$$S_x = \sqrt{\frac{\sum\limits_{i=1}^{m} (x_i - \bar{x})^2}{m-1}} \qquad (16.3)$$

and

$$S_y = \sqrt{\frac{\sum\limits_{j=1}^{n} (y_j - \bar{y})^2}{n-1}} \qquad (16.4)$$

Assuming that X and Y are normally distributed with the same variance but that their population means μ_x and μ_y are different, then the difference between the means $\bar{x} - \bar{y}$ will be normally distributed with mean $(\mu_x - \mu_y)$ and variance $\sigma^2 (1/m) + (1/n)$. Then

$$t' = \frac{|\bar{x} - \bar{y}|}{\sqrt{\dfrac{(m-1) S_x^2 + (n-1) S_y^2}{m+n-2}}} \cdot \frac{1}{\sqrt{\dfrac{1}{m} + \dfrac{1}{n}}} \qquad (16.5)$$

where:

$| \ |$ denotes absolute value.

The t follows a t-distribution with $m + n - 2$ degrees of freedom. The portion of the denominator within the first square root sign in Equation 16.5 is referred to as the *pooled variance* and is calculated when the sample variances are deemed sufficiently alike to be independent estimates of the same population variance.

Table 16.6   5% LEVEL OF VARIANCE RATIO ASSOCIATED WITH EQUATION 16.6

	Number of Degrees of Freedom in the Greater Variance Estimate							
Number of Degrees of Freedom in Lesser Variance Estimate	**1**	**2**	**3**	**4**	**5**	**10**	**20**	**∞**
1	161	200	216	225	230	242	248	254
2	18.5	19.0	19.2	19.2	19.3	19.4	19.4	19.5
3	10.1	9.6	9.3	9.1	9.0	8.8	8.7	8.5
4	7.7	6.9	6.6	6.4	6.3	6.0	5.8	5.6
5	6.6	5.8	5.4	5.2	5.0	4.7	4.6	4.4
10	5.0	4.1	3.7	3.5	3.3	3.0	2.8	2.5
20	4.3	3.5	3.1	2.9	2.7	2.3	2.1	1.8
∞	3.8	3.0	2.6	2.4	2.2	1.8	1.6	1.0

Source: Moroney, 1975

The significance of the difference between sample variances can be tested using the variance ratio test. The variance ratio test is defined as

$$F = \frac{\text{greater estimate of the variance of the population}}{\text{lesser estimate of the variance of the population}} \tag{16.6}$$

It follows that the larger F is, the less likely it is that the two samples are from the same population. However, with small sample sizes (m and n small) it is possible that F can be different from unity. Table 16.6 gives the value of F that will be exceeded with a given degree of probability for various sample sizes. (Note that the number of degrees of freedom for a sample of n items is equal to n-1).

In the event that there is a significant difference between the variances (i.e., that ratio of variances estimated from the samples exceeds the corresponding values in Table 16.6), the assumptions of the t-test are violated, and erroneous conclusions may be determined if the test is used. In this situation, a possible alternative test to employ is Cochran's approximation test, described next.

Assuming no failure of the variance test, the t-statistic, t', is compared to a critical value from the t-distribution, t_c, determined using degrees of freedom of $n + m$ - 2 at a selected level of significance of $\propto$. Decisions on whether $\bar{x}$ and $\bar{y}$ are statistically significantly different are based on whether $t' \geq t_c$ or $t' < t_c$.

16.4.3 Cochran's Approximation Test

Cochran's test computes the t' statistic from

$$t' = \frac{|\bar{x} - \bar{y}|}{\sqrt{\dfrac{S_x^2}{m} + \dfrac{S_y^2}{n}}}$$

(16.7)

The critical t-statistic, t_c, is computed from

$$t_c = \frac{\dfrac{S_x^2}{m}t_x + \dfrac{S_y^2}{n}t_y}{\dfrac{S_x^2}{m} + \dfrac{S_y^2}{n}}$$

(16.8)

where:

t_x and t_y are determined from t-tables for degrees of freedom of m - 1 and n - 1, respectively, for a specified level of significance, $\propto$.

Noteworthy about Cochran's test is the relaxation of the requirement that the variance of the data at both sampling locations be of the same magnitude, as specified as essential to the standard t-test.

To decide whether $\bar{x}$ - $\bar{y}$ are statistically significantly different, compare t' with t_c:

If $t' \geq t_c$, conclude there is most likely a significant difference.

If $t' < t_c$, conclude there is most likely not a significant difference.

For further discussion on Cochran's Approximation, see Iman and Canover (1983) and McBean et al., 1988.

16.4.4 Alternative Testing

The preceding subsections indicate just two of the possible tests used for statistical significance testing. Numerous others exist, and the interested reader is referred to the technical literature for these (e.g., see McBean and Rovers, 1989).

References

Bately, G., and Gardner, D. 1977. Sampling and Storage of Natural Waters for Trace Metal Analysis. *Water Research* 11:745–756.

Box, G. E. P., Hunter, W. G., and Hunter, J. S. 1978. *Statistics for Experimenters.* Wiley Interscience, New York.

Eicholz, G., Nagel, A., and Hughes, R. 1965. Adsorption of Ions in Dilute Aqueous Solutions on Glass and Plastic Surfaces. *Analytical Chemistry* 37:863–868.

Driscoll, F. 1986. *Groundwater and Wells*. Report of Johnson Division, St. Paul, Minn.

Federal Register. 1988. 53, no. 168 (30 August):33413.

Gilham, R., Robin, M., Barker, J., and Cherry, J. 1983. Groundwater Monitoring and Sample Bias. *American Petroleum Institute*. Publication 4367, 206 pp.

Houghton, R., and Berger, M. 1984. "Effects of Well-Casing Composition and Sampling Method on Apparent Quality of Groundwater." *Fourth National Symposium on Aquifer Restoration and Groundwater Monitoring*, Las Vegas, May pp. 203–213.

Iman, R. A., and Canover, W. 1983. *A Modern Approach to Statistics*. John Wiley, New York.

Masse, R., Maessen, F., and DeGoeij, J. 1981. Loss of Silver, Arsenic, Cadmium, Selenium and Zinc Traces from Distilled Water and Artificial Sea-Water by Sorption on Various Container Surfaces. *Analytica Chimica Acta* 127:191–193.

McBean, E., and Rovers, F. 1984. Alternatives for Handling Detection Limit Data in Impact Assessments. *Groundwater Monitoring Review* Spring:42–44.

McBean, E., and Rovers, F. 1985. Analysis of Variances as Determined from Replicates Versus Successive Samplings. *Groundwater Monitoring Review* Summer: 61–64.

McBean, E., and Rovers, F. 1989. "Flexible Selection of Statistical Discrimination Tests for Field-Monitored Data." In *ASTM Standards Development for Groundwater and Vadose Monitoring*, ed. D. Nielsen. Philadelphia, Pa.

McBean, E., Kompter, M., and Rovers, F. 1988. A Critical Examination of Approximations Implicit in Cochran's Procedure. *Groundwater Monitoring Review* Winter: 83–87.

Mezei, L. M. 1990. *Practical Spreadsheet Statistics and Curve-fitting for Scientists and Engineers*. Prentice Hall, Englewood Cliffs, N.J.

Moroney, M. J. 1975. *Facts from Figures*. Pelican Books, Middlesex, England.

Nielsen, D. ed. 1989. *ASTM Standards Development for Groundwater and Vadose Zone Monitoring*. Philadelphia, Pa.

Parker, L. V., Hewitt, A. D., and Jenkins, T. F. 1988. "Influence of Well Casing Materials on Chemical Species in Groundwater." Presented at 13th Annual Environmental Quality R & D Symposium, U.S. Army Toxic and Hazardous Materials Agency, Aberdeen Proving Ground, Maryland, November 15–17.

Reynolds, G., and Gilham, R. 1986. "Adsorption of Halogenated Organic Compounds by Polymer Materials Commonly Used in Groundwater Materials." In *Proceedings of the Second Canadian/American Conference on Hydrogeology*, Banff, Alberta, June 25–29.

Robertson, D. 1968. The Adsorption of Trace Elements in Sea Water on Various Container Surfaces. *Analytica Chimica Acta* 42:533–536.

Rovers, F., and McBean, E. 1981. Significance Testing for Impact Evaluation. *Groundwater Monitoring Review* Summer:39–43.

Scalf, M. R., McNabb, J. F., Dunlap, J. Cosby, R. L., and Fryberger, J. 1981. *Manual of Groundwater Quality Sampling Procedures*. Joint Report of the National Water Well Association and the R. S. Kerr Environmental Research Laboratory. U.S. EPA. 93 pp.

16.5 PROBLEMS

The following upgradient and downgradient data were obtained from sampling an aquifer beneath a landfill. Determine whether the landfill is having a statistically significant impact on groundwater.

Station 1				
Date	TOC	COD	Arsenic	pH
March 83	52.3	425	23	8.5
September 83	56.6	435	45	8.6
December 83	46.0	634	33	7.9
March 84	49.6	555	19.9	8.3
September 84	51.3	479	23	8.4
December 84	39.9	532	34	8.1
March 85	43.2	439	41	8.1
September 85	43.9	479	38	8.2
December 85	49.0	434	39	7.9
March 86	52.8	485	41	7.8
September 86	42.9	523	38	7.9
March 87	56.9	510	39	8.1
Station 2 (downgradient from 1)				
March 83	61.9	530	41	7.2
September 83	58.9	510	38	7.1
December 83	61.5	489	42	7.0
March 84	62.2	499	44	7.3
September 84	58.9	519	no sample	7.4
December 84	62.9	539	57	7.7
March 85	61.8	539	47	7.4
September 85	46.9	569	52	7.1
December 85	57.8	589	51	7.8
March 86	62.7	458	49	7.3
March 87	58.6	555	55	7.4

Table A.1: CONVERSION FACTORS

	Units				Multiplier
Length	cm	$\times$	10^{-2}	=	m
		$\times$	0.3937	=	in.
		$\times$	3.281×10^{-2}	=	ft
	mils	$\times$	.0254	=	mm
	mm	$\times$	0.03937	=	in.
	in.	$\times$	2.54	=	cm
	ft	$\times$	0.3048	=	m
	mi	$\times$	1.609	=	km
Area	m^2	$\times$	10.76	=	ft^2
	ha	$\times$	2.471	=	acres
	ft^2	$\times$	9.29×10^{-2}	=	m^2
	yd^2	$\times$	0.8361	=	m^2
	acre	$\times$	43,560	=	ft^2
	mi^2	$\times$	640	=	acres
	acre	$\times$	4046.80	=	m^2
Volume	m^3	$\times$	1000	=	L
	m^3	$\times$	35.31	=	ft^3
	L	$\times$	3.531×10^{-2}	=	ft^3
	acre-ft	$\times$	1.2335×10^3	=	m^3
	ft^3	$\times$	28.317	=	L
	yd^3	$\times$	0.7645	=	m^3
	gal (U.S.)	$\times$	3.7854×10^{-3}	=	m^3
	gal (U.S.)	$\times$	3.7854	=	L
Mass	g	$\times$	6.852×10^{-5}	=	slug

Table A.1: CONVERSION FACTORS

	Units		Multiplier		
		$\times$	2.205×10^{-3}	=	lb
	ton	$\times$	0.9072	=	tonne
	kg	$\times$	2.2046	=	lb
Density	slug/ft^3	$\times$	515.4	=	kg/m^3
		$\times$	32.17	=	lb/ft^3
	g/cm^3	$\times$	1.940	=	slug/ft^3
Force	dyne	$\times$	10^{-5}	=	N
		$\times$	2.248×10^{-6}	=	lb
	N	$\times$	0.2248	=	lb
Pressure	atm	$\times$	1.013×10^6	=	dyne/cm^2
		$\times$	406.8	=	in. of water
		$\times$	76	=	cm Hg
		$\times$	1.013×10^5	=	P
		$\times$	14.70	=	lb/in.^{2}
	pascal or	$\times$	9.869×10^{-6}	=	atm
	(N/m^2)	$\times$	4.015×10^{-3}	=	in. of water
		$\times$	1.450×10^{-4}	=	lb/in.^{2}
Energy, Work, Heat	Btu	$\times$	777.9	=	ft.-lb
		$\times$	1055	=	J
		$\times$	252	=	cal
		$\times$	2.93×10^{-4}	=	kWh
	J	$\times$	9.481×10^{-4}	=	Btu
Temperature	°F	$\times$	0.555 (°F − 32)	=	°C
	°F	$\times$	0.555 (°F + 459.67)	=	K

Table A.2: SOME PHYSICAL CONSTANTS

Universal gas content	R	8.31 J/mol °K
Acceleration due to gravity	g	32.17 ft/s^2
		9.807 m/s^2
Specific weight of water (68°F)		62.31 lb/ft^3
Standard atmosphere		14.7 lb/in.^{2}
		101.33 kPa
		33.9 ft of water
		10.33 m of water
		760 mm Hg
Temperature		
Rankin	°R	459.6 + °F
Kelvin	°K	273.0 + °C

Table A.3: SI PREFIXES

Factor	Prefix	Symbol	Factor	Prefix	Symbol
10^{18}	exa	E	10^{-1}	deci	d
10^{15}	peta	P	10^{-2}	centi	c
10^{12}	tera	T	10^{-3}	milli	m
10^{9}	giga	G	10^{-6}	micro	μ
10^{6}	mega	M	10^{-9}	nano	n
10^{3}	kilo	k	10^{-12}	pico	p
10^{2}	hecto	h	10^{-15}	femto	f
10^{1}	deka	da	10^{-18}	atto	a

Table A.4: VALUES OF THE ERROR FUNCTION erf (ψ) AND COMPLEMENTARY ERROR FUNCTION erf (ψ) FOR VALUES OF ψ VARYING FROM 0 TO 3.0

ψ	erf(ψ)	erfc(ψ)
0	0	1.0
0.05	0.056372	0.943628
0.1	0.112463	0.887537
0.15	0.167996	0.832004
0.2	0.333703	0.777297
0.25	0.276326	0.723674
0.3	0.328627	0.671373
0.35	0.379382	0.620618
0.4	0.428392	0.571608
0.45	0.475482	0.524518
0.5	0.520500	0.479500
0.55	0.563323	0.436677
0.6	0.603856	0.396144
0.65	0.642029	0.357971
0.7	0.677801	0.322199
0.75	0.711156	0.288844
0.8	0.742101	0.257899
0.85	0.770669	0.229332
0.9	0.796908	0.203092
0.95	0.820891	0.179109
1.0	0.842701	0.157299
1.1	0.880205	0.119795
1.2	1.910314	0.089686
1.3	0.934008	0.065992
1.4	0.952285	0.047715
1.5	0.966105	0.033895
1.6	0.976348	0.023652
1.7	0.983790	0.016210
1.8	0.989091	0.010909
1.9	0.992790	0.007210

Table A.4: VALUES OF THE ERROR FUNCTION erf (ψ) AND COMPLEMENTARY ERROR FUNCTION erf (ψ) FOR VALUES OF ψVARYING FROM 0 TO 3.0

ψ	erf(ψ)	erfc(ψ)
2.0	0.995322	0.004678
2.1	0.997021	0.002979
2.2	0.998137	0.001863
2.3	0.998857	0.001143
2.4	0.999311	0.000689
2.5	0.999593	0.000407
2.6	0.999764	0.000236
2.7	0.999866	0.000134
2.8	0.999925	0.000075
2.9	0.999959	0.000041
3.0	0.999978	0.000022

$$\text{erf}(\psi) = \frac{2}{\pi} \int_{O}^{\beta} \varepsilon^{-\varepsilon^2} d\varepsilon$$

$$\text{erf}(-\psi) = -\text{erf}\,\psi$$

$$\text{erf}(\psi) = 1 - \text{erf}(\psi)$$

Table A.5: STUDENT'S t DISTRIBUTION

The first column lists the number of degrees of freedom (υ). The headings of the other columns give probabilities (P) for t to exceed the entry value. Use symmetry for negative t values.

υ	.10	.05	.025	.01	.005
1	3.078	6.314	12.706	31.821	63.657
2	1.886	2.920	4.303	6.965	9.925
3	1.638	2.353	3.182	4.541	5.841
4	1.533	2.132	2.776	3.747	4.604
5	1.476	2.015	2.571	3.365	4.032
6	1.440	1.943	2.447	3.143	3.707
7	1.415	1.895	2.365	2.998	3.499
8	1.397	1.860	2.306	2.896	3.355

Table A.5: STUDENT'S t DISTRIBUTION

υ	.10	.05	.025	.01	.005
9	1.383	1.833	2.262	2.821	3.250
10	1.372	1.812	2.228	2.764	3.169
11	1.363	1.796	2.201	2.718	3.106
12	1.356	1.782	2.179	2.681	3.055
13	1.350	1.771	2.160	2.650	3.012
14	1.345	1.761	2.145	2.624	2.977
15	1.341	1.753	2.131	2.602	2.947
16	1.337	1.746	2.120	2.583	2.921
17	1.333	1.740	2.110	2.567	2.898
18	1.330	1.734	2.101	2.552	2.878
19	1.328	1.729	2.093	2.539	2.861
20	1.325	1.725	2.086	2.528	2.845
21	1.323	1.721	2.080	2.518	2.831
22	1.321	1.717	2.074	2.508	2.819
23	1.319	1.714	2.069	2.500	2.807
24	1.318	1.711	2.064	2.492	2.797
25	1.316	1.708	2.060	2.485	2.787
26	1.315	1.706	2.056	2.479	2.779
27	1.314	1.703	2.052	2.473	2.771
28	1.313	1.701	2.048	2.467	2.763
29	1.311	1.699	2.045	2.462	2.756
30	1.310	1.697	2.042	2.457	2.750
40	1.303	1.684	2.021	2.423	2.704
60	1.296	1.671	2.000	2.390	2.660
120	1.289	1.658	1.980	2.358	2.617
∞	1.282	1.645	1.960	2.326	2.576

Table A.6: AREAS OF A STANDARD NORMAL DISTRIBUTION

Table entry's give the proportion under the entire curve that is between z = 0 and a positive value of z. Areas for negative values of z are obtained by symmetry.

z	.00	.01	.02	.03	.04	.05	.06	.07	.08	.09
0.0	.000	.0040	.0080	.0120	.0160	.0199	.0239	.0279	.0319	.0359
0.1	.0398	.0438	.0478	.0517	.0557	.0596	.0636	.0675	.0714	.0753
0.2	.0793	.0832	.0871	.0910	.0948	.0987	.1026	.1064	.1103	.1141
0.3	.1179	.1217	.1255	.1293	.1331	.1368	.1406	.1443	.1480	.1517
0.4	.1554	.1591	.1628	.1664	.1700	.1736	.1772	.1808	.1844	.1879
0.5	.1915	.1950	.1985	.2019	.2054	.2088	.2123	.2157	.2190	.2224
0.6	.2257	.2291	.2324	.2357	.2389	.2422	.2454	.2486	.2517	.2549
0.7	.2580	.2611	.2642	.2673	.2703	.2734	.2764	.2794	.2823	.2852
0.8	.2881	.2910	.2939	.2967	.2995	.3023	.3051	.3078	.3106	.3133
0.9	.3159	.3186	.3212	.3238	.3264	.3289	.3315	.3340	.3365	.3389
1.0	.3413	.3438	.3461	.3485	.3508	.3531	.3554	.3577	.3599	.3621
1.1	.3643	.3665	.3686	.3708	.3729	.3749	.3770	.3790	.3810	.3830
1.2	.3849	.3869	.3888	.3907	.3925	.3944	.3962	.3980	.3997	.4015
1.3	.4032	.4049	.4066	.4083	.4099	.4115	.4131	.4147	.4162	.4177
1.4	.4192	.4207	.4222	.4236	.4251	.4265	.4279	.4292	.4306	.4319
1.5	.4332	.4345	.4357	.4370	.4382	.4394	.4406	.4418	.4429	.4441
1.6	.4452	.4463	.4474	.4484	.4495	.4505	.4515	.4525	.4535	.4545
1.7	.4554	.4564	.4573	.4582	.4591	.4599	.4608	.4616	.4625	.4633
1.8	.4641	.4649	.4656	.4664	.4671	.4678	.4686	.4693	.4699	.4706

Table A.6: AREAS OF A STANDARD NORMAL DISTRIBUTION

Table entry's give the proportion under the entire curve that is between z = 0 and a positive value of z. Areas for negative values of z are obtained by symmetry.

z	.00	.01	.02	.03	.04	.05	.06	.07	.08	.09
1.9	.4713	.4719	.4726	.4732	.4738	.4744	.4750	.4756	.4761	.4767
2.0	.4772	.4778	.4783	.4788	.4793	.4798	.4803	.4808	.4812	.4817
2.1	.4821	.4826	.4830	.4834	.4838	.4842	.4846	.4850	.4854	.4857
2.2	.4861	.4864	.4868	.4871	.4875	.4878	.4881	.4884	.4887	.4890
2.3	.4893	.4896	.4898	.4901	.4904	.4906	.4909	.4911	.4913	.4916
2.4	.4918	.4920	.4922	.4925	.4927	.4929	.4931	.4932	.4934	.4936
2.5	.4938	.4940	.4941	.4943	.4945	.4946	.4948	.4949	.4951	.4952
2.6	.4953	.4955	.4956	.4957	.4959	.4960	.4961	.4962	.4963	.4964
2.7	.4965	.4966	.4967	.4968	.4969	.4970	.4971	.4972	.4973	.4974
2.8	.4974	.4975	.4976	.4977	.4977	.4978	.4979	.4979	.4980	.4981
2.9	.4981	.4982	.4982	.4983	.4984	.4984	.4985	.4985	.4986	.4986
3.0	.4987	.4987	.4988	.4988	.4989	.4989	.4989	.4989	.4990	.4990

APPENDIX B

Table B.1: PROPERTIES OF METHANE (CH₄)

Molecular weight	
Visocity (gas @ 1 atm)	
@ 4.4 °C	0.0106 cP
@37.8 °C	0.0116 cP
Flammable limits in air	5.3–14% (by volume)
Density (gas @ 0 °C and 1 atm)	0.72 g/L
Specific heat (gas @ 60 °F and 1 atm)	
C_p	0.5271 cal/(g · °C)
C_v	0.4032 cal/(g · °C)

Table B.2: PROPERTIES OF CARBON DIOXIDE (CO₂)

Molecular weight	44.01
Visocity (gas @ 1 atm)	
@70 °C	0.0148 cP
Density (gas @ 0 °C and 1 atm)	1.521
Specific heat (gas @ 25 °F and 1 atm)	
C_p	0.205 cal/(g · °C)
C_v	.1565 cal/(g · °C)

Table B.3: HENRY'S LAW CONSTANTS FOR SELECTED GASES

	H (atm/mol fraction)			
T (°c)	CO_2	H_2S	CH_4	O_2
0	0.728×10^4	0.0268×10^4	2.24×10^4	2.55×10^4
10	0.104	0.0367	2.97	3.27
20	0.142	0.0483	3.76	4.01
30	0.186	0.0609	4.49	4.75
40	0.233	0.0745	5.20	5.35

Table B.4: CHEMICAL PROPERTIES

Chemical	Gram Molecular Weight (g/mol)	Air Diffusion Coefficient (cm^2/s)	Pure Chemical Vapor Pressure (mm Hg)	Gas Viscosity (μ P)
Chemical	78.11	0.0905	75.20	72.22
Toluene	92.14	0.0824	21.84	74.01
Xylene	106.17	0.0668	6.16	53.72

Table B.5: PHYSICAL PROPERTIES OF WATER

Temperature (°C)	Specific Weight (N/m^3)	Density (kg/m^3)	Visocity (N · s/m^2)
0	9806	999.9	1.792
10	9804	999.7	1.308
20	9789	998.2	1.005
30	9764	995.7	0.801
40	9730	992.2	0.656
50	9690	988.1	0.549

APPENDIX C

Table C.1: MONTHLY VALUES OF HEAT INDEX CORRESPONDING TO MONTHLY MEAN TEMPERATURES

T°F	.0	.1	.2	.3	.4	.5	.6	.7	.8	.9
32	.00	.00	.00	.00	.01	.01	.02	.02	.03	.03
33	.04	.04	.05	.05	.06	.06	.07	.08	.09	.09
34	.10	.10	.11	.12	.13	.14	.15	.16	.17	.18
35	.19	.20	.21	.22	.23	.24	.25	.26	.27	.28
36	.29	.30	.32	.33	.34	.35	.36	.37	.39	.40
37	.41	.42	.43	.44	.46	.47	.48	.50	.51	.52
38	.54	.55	.56	.58	.59	.60	.62	.63	.65	.66
39	.68	.70	.71	.73	.74	.76	.77	.79	.80	.82
40	.83	.85	.86	.88	.90	.91	.93	.95	.96	.98
41	1	1.01	1.03	1.05	1.07	1.08	1.10	1.12	1.14	1.16
42	1.17	1.19	1.21	1.23	1.24	1.26	1.28	1.30	1.32	1.33
43	1.35	1.37	1.39	1.41	1.43	1.45	1.47	1.49	1.50	1.52
44	1.54	1.56	1.58	1.60	1.62	1.64	1.66	1.68	1.70	1.72
45	1.74	1.76	1.78	1.80	1.82	1.85	1.87	1.89	1.91	1.93
46	1.95	1.97	2.00	2.02	2.04	2.06	2.08	2.10	2.13	2.15
47	2.17	2.19	2.21	2.23	2.26	2.28	2.30	2.32	2.34	2.37
48	2.39	2.41	2.43	2.46	2.48	2.50	2.53	2.55	2.57	2.60
49	2.62	2.64	2.67	2.69	2.71	2.74	2.76	2.79	2.81	2.84
50	2.86	2.89	2.91	2.93	2.96	2.98	3.01	3.03	3.06	3.08
51	3.11	3.13	3.16	3.18	3.21	3.23	3.25	3.28	3.30	3.33
52	3.35	3.38	3.40	3.43	3.45	3.48	3.50	3.53	3.55	3.58
53	3.6	3.63	3.65	3.68	3.71	3.73	3.76	3.79	3.81	3.84
54	3.87	3.89	3.92	3.95	3.97	4.00	4.03	4.06	4.08	4.11
55	4.14	4.16	4.19	4.22	4.25	4.27	4.30	4.33	4.35	4.38
56	4.41	4.44	4.47	4.50	4.52	4.55	4.57	4.60	4.63	4.66
57	4.69	4.72	4.75	4.77	4.80	4.83	4.86	4.89	4.92	4.95
58	4.98	5.01	5.04	5.07	5.10	5.13	5.15	5.19	5.22	5.25
59	5.28	5.31	5.34	5.37	5.40	5.43	5.46	5.49	5.52	5.55
60	5.58	5.61	5.64	5.67	5.70	5.73	5.76	5.79	5.82	5.85
61	5.88	5.91	5.94	5.97	6.00	6.03	6.06	6.10	6.13	6.16
62	6.19	6.22	6.25	6.28	6.31	6.34	6.38	6.41	6.44	6.47
63	6.50	6.53	6.56	6.59	6.62	6.66	6.69	6.72	6.75	6.79
64	6.82	7.85	6.88	6.92	6.95	6.98	7.02	7.05	7.08	7.12
65	7.15	7.18	7.22	7.25	7.28	7.32	7.35	7.38	7.42	7.45
66	7.48	7.52	7.55	7.58	7.62	7.65	7.68	7.72	7.75	7.78
67	7.82	7.85	7.89	7.92	7.95	7.99	8.02	8.05	8.09	8.12
68	8.16	8.19	8.23	8.26	8.30	8.33	8.37	8.40	8.44	8.47
69	8.51	8.54	8.57	8.61	8.64	8.68	8.71	8.75	8.78	8.82
70	8.85	8.89	8.92	8.96	8.99	9.03	9.06	9.10	9.13	9.17

Table C.1: MONTHLY VALUES OF HEAT INDEX CORRESPONDING TO MONTHLY MEAN TEMPERATURES

T°F	.0	.1	.2	.3	.4	.5	.6	.7	.8	.9
71	9.2	9.24	9.27	9.31	9.34	9.38	9.42	9.45	9.49	9.53
72	9.57	9.60	9.64	9.67	9.71	9.75	9.78	9.82	9.85	9.89
73	9.93	9.97	10.01	10.04	10.08	10.12	10.15	10.19	10.22	10.26
74	10.30	10.34	10.37	10.41	10.45	10.48	10.52	10.56	10.60	10.64
75	10.67	10.71	10.75	10.78	10.82	10.86	10.89	10.93	10.97	11.01
76	11.05	11.09	11.13	11.17	11.20	11.24	11.28	11.31	11.35	11.39
77	11.43	11.47	11.51	11.54	11.58	11.62	11.66	11.70	11.74	11.76
78	11.82	11.85	11.89	11.93	11.97	12.01	12.05	12.09	12.13	12.17
79	12.21	12.25	12.29	12.33	12.37	12.41	12.45	12.49	12.53	12.57
80	12.61	12.65	12.69	12.73	12.77	12.81	12.85	12.89	12.93	12.97
81	13.01	13.05	13.09	13.13	13.17	13.21	13.25	13.29	13.33	13.37
82	13.41	13.45	13.49	13.53	13.57	13.61	13.65	13.69	13.73	13.77
83	13.81	13.85	13.89	13.94	13.98	14.02	14.06	14.10	14.14	14.18
84	14.22	14.26	14.31	14.35	14.39	14.43	14.47	14.52	14.56	14.60
85	14.64	14.69	14.73	14.77	14.81	14.85	14.90	14.94	14.98	15.02
86	15.07	15.11	15.15	15.19	15.23	15.28	15.32	15.36	15.40	15.45
87	15.49	15.53	15.58	15.62	15.66	15.71	15.75	15.79	15.84	15.88
88	15.92	15.97	16.01	16.05	16.10	16.14	16.18	16.23	16.27	16.31
89	16.36	16.40	16.44	16.49	16.53	16.57	16.62	16.66	16.70	16.75
90	16.79	16.83	16.88	16.92	16.96	17.01	17.05	17.09	17.14	17.18
91	17.23	17.27	17.32	17.36	17.41	17.45	17.49	17.54	17.58	17.63
92	17.67	17.72	17.76	17.81	17.85	17.89	17.94	17.98	18.03	18.07
93	18.12	18.16	18.21	18.25	18.30	18.34	18.39	18.43	18.48	18.52
94	18.57	18.62	18.66	18.71	18.75	18.80	18.84	18.89	18.93	18.98
95	19.03	19.07	19.12	19.16	19.21	19.25	19.30	19.34	19.39	19.44
96	19.48	19.53	19.58	19.62	19.67	19.71	19.76	19.81	19.86	19.90
97	19.95	20.00	20.04	20.09	20.14	20.18	20.23	20.28	20.32	20.37
98	20.42	20.46	20.51	20.56	20.60	20.65	20.70	20.74	20.79	20.84
99	20.88	20.93	20.98	21.03	21.08	21.13	21.17	21.22	21.27	21.32
100	21.36	21.41	21.46	21.51	21.56	21.60	21.65	21.70	21.75	21.79
101	21.84	21.89	21.94	21.99	22.03	22.08	22.13	22.18	22.23	22.29
102	22.33	22.38	22.42	22.47	22.52	22.57	22.62	22.67	22.71	22.76
103	22.81	22.86	22.91	22.96	23.00	23.05	23.10	23.15	23.20	23.25
104	23.30									

"Example - for a temperature of 77.5°F, I = 11.62"

Table C.2 VALUES OF UNADJUSTED DAILY POTENTIAL EVAPOTRANSPIRATION (IN.) FOR DIFFERENT MEAN TEMPERATURES AND I VALUES

T°F	I Value (25.0–80)																							T°F
	25	27.5	30	32.5	35	37.5	40	42.5	45	47.5	50	52.5	55	57.5	60	62.5	65	67.5	70	72.5	75	77.5	80	
32	0	0	0	0	0	0	0	0	0	0	0	0	0	0	0	0	0	0	0	0	0	0	0	32
32.5	0	0	0	0	0	0	0	0	0	0	0	0	0	0	0	0	0	0	0	0	0	0	0	32.5
33	0	0	0	0	0	0	0	0	0	0	0	0	0	0	0	0	0	0	0	0	0	0	0	33
33.5	0.01	0.01	0	0	0	0	0	0	0	0	0	0	0	0	0	0	0	0	0	0	0	0	0	33.5
34	0.01	0.01	0.01	0.01	0.01	0	0	0	0	0	0	0	0	0	0	0	0	0	0	0	0	0	0	34
34.5	0.01	0.01	0.01	0.01	0.01	0.01	0.01	0.01	0.01	0	0	0	0	0	0	0	0	0	0	0	0	0	0	34.5
35	0.02	0.02	0.01	0.01	0.01	0.01	0.01	0.01	0.01	0.01	0.01	0	0	0	0	0	0	0	0	0	0	0	0	35
35.5	0.02	0.02	0.01	0.01	0.01	0.01	0.01	0.01	0.01	0.01	0.01	0.01	0	0	0	0	0	0	0	0	0	0	0	35.5
36	0.02	0.02	0.02	0.01	0.01	0.01	0.01	0.01	0.01	0.01	0.01	0.01	0.01	0	0	0	0	0	0	0	0	0	0	36
36.5	0.02	0.02	0.02	0.02	0.02	0.01	0.01	0.01	0.01	0.01	0.01	0.01	0.01	0.01	0.01	0	0	0	0	0	0	0	0	36.5
37	0.02	0.02	0.02	0.02	0.02	0.02	0.02	0.02	0.02	0.01	0.01	0.01	0.01	0.01	0.01	0	0	0	0	0	0	0	0	37
37.5	0.02	0.02	0.02	0.02	0.02	0.02	0.02	0.02	0.02	0.02	0.02	0.01	0.01	0.01	0.01	0.01	0	0	0	0	0	0	0	37.5
38	0.03	0.03	0.02	0.02	0.02	0.02	0.02	0.02	0.02	0.02	0.02	0.01	0.01	0.01	0.01	0.01	0.01	0	0	0	0	0	0	38
38.5	0.03	0.03	0.02	0.02	0.02	0.02	0.02	0.02	0.02	0.02	0.02	0.02	0.01	0.01	0.01	0.01	0.01	0.01	0	0	0	0	0	38.5
39	0.03	0.03	0.03	0.02	0.02	0.02	0.02	0.02	0.02	0.02	0.02	0.02	0.02	0.01	0.01	0.01	0.01	0.01	0.01	0.01	0.01	0.01	0	39
39.5	0.04	0.03	0.03	0.03	0.02	0.02	0.02	0.02	0.02	0.02	0.02	0.02	0.02	0.01	0.01	0.01	0.01	0.01	0.01	0.01	0.01	0.01	0.01	39.5

Table C.2: VALUES OF UNADJUSTED DAILY POTENTIAL EVAPOTRANSPIRATION (IN.) FOR DIFFERENT MEAN TEMPERATURES AND I VALUES

I VALUE (82.5 – 40)

T°F	82.5	85	87.5	90	92.5	95	97.5	100	102.5	105	107.5	110	112.5	115	117.5	120	122.5	125	127.5	130	132.5	135	137.5	140	T°F
32	0	0	0	0	0	0	0	0	0	0	0	0	0	0	0	0	0	0	0	0	0	0	0	0	32
32.5	0	0	0	0	0	0	0	0	0	0	0	0	0	0	0	0	0	0	0	0	0	0	0	0	32.5
33	0	0	0	0	0	0	0	0	0	0	0	0	0	0	0	0	0	0	0	0	0	0	0	0	33
33.5	0	0	0	0	0	0	0	0	0	0	0	0	0	0	0	0	0	0	0	0	0	0	0	0	33.5
34	0	0	0	0	0	0	0	0	0	0	0	0	0	0	0	0	0	0	0	0	0	0	0	0	34
34.5	0	0	0	0	0	0	0	0	0	0	0	0	0	0	0	0	0	0	0	0	0	0	0	0	34.5
35	0	0	0	0	0	0	0	0	0	0	0	0	0	0	0	0	0	0	0	0	0	0	0	0	35
35.5	0	0	0	0	0	0	0	0	0	0	0	0	0	0	0	0	0	0	0	0	0	0	0	0	35.5
36	0	0	0	0	0	0	0	0	0	0	0	0	0	0	0	0	0	0	0	0	0	0	0	0	36
36.5	0	0	0	0	0	0	0	0	0	0	0	0	0	0	0	0	0	0	0	0	0	0	0	0	36.5
37	0	0	0	0	0	0	0	0	0	0	0	0	0	0	0	0	0	0	0	0	0	0	0	0	37
37.5	0	0	0	0	0	0	0	0	0	0	0	0	0	0	0	0	0	0	0	0	0	0	0	0	37.5
38	0	0	0	0	0	0	0	0	0	0	0	0	0	0	0	0	0	0	0	0	0	0	0	0	38
38.5	0	0	0	0	0	0	0	0	0	0	0	0	0	0	0	0	0	0	0	0	0	0	0	0	38.5
39	0	0	0	0	0	0	0	0	0	0	0	0	0	0	0	0	0	0	0	0	0	0	0	0	39
39.5	0	0	0	0	0	0	0	0	0	0	0	0	0	0	0	0	0	0	0	0	0	0	0	0	39.5

Table C.2: VALUES OF UNADJUSTED DAILY POTENTIAL EVAPOTRANSPIRATION (IN.) FOR DIFFERENT MEAN TEMPERATURES AND I VALUES

I Value (25.0–80)

T°F	25	27.5	30	32.5	35	37.5	40	42.5	45	47.5	50	52.5	55	57.5	60	62.5	65	67.5	70	72.5	75	77.5	80	T°F
40	0.04	0.04	0.03	0.03	0.03	0.02	0.02	0.02	0.02	0.02	0.02	0.02	0.02	0.02	0.01	0.01	0.01	0.01	0.01	0.01	0.01	0.01	0.01	40
40.5	0.04	0.04	0.03	0.03	0.03	0.03	0.02	0.02	0.02	0.02	0.02	0.02	0.02	0.02	0.02	0.01	0.01	0.01	0.01	0.01	0.01	0.01	0.01	40.5
41	0.04	0.04	0.04	0.03	0.03	0.03	0.03	0.02	0.02	0.02	0.02	0.02	0.02	0.02	0.02	0.02	0.01	0.01	0.01	0.01	0.01	0.01	0.01	41
41.5	0.04	0.04	0.04	0.04	0.03	0.03	0.03	0.03	0.03	0.02	0.02	0.02	0.02	0.02	0.02	0.02	0.02	0.01	0.01	0.01	0.01	0.01	0.01	41.5
42	0.04	0.04	0.04	0.04	0.04	0.03	0.03	0.03	0.03	0.02	0.02	0.02	0.02	0.02	0.02	0.02	0.02	0.02	0.01	0.01	0.01	0.01	0.01	42
42.5	0.05	0.04	0.04	0.04	0.04	0.04	0.03	0.03	0.03	0.03	0.03	0.02	0.02	0.02	0.02	0.02	0.02	0.02	0.02	0.01	0.02	0.01	0.01	42.5
43	0.05	0.05	0.04	0.04	0.04	0.04	0.04	0.03	0.03	0.03	0.03	0.03	0.02	0.02	0.02	0.02	0.02	0.02	0.02	0.01	0.01	0.01	0.01	43
43.5	0.05	0.05	0.04	0.04	0.04	0.04	0.04	0.04	0.03	0.03	0.03	0.03	0.03	0.02	0.02	0.02	0.02	0.02	0.02	0.02	0.02	0.02	0.02	43.5
44	0.05	0.05	0.05	0.04	0.04	0.04	0.04	0.04	0.04	0.03	0.03	0.03	0.03	0.03	0.02	0.02	0.02	0.02	0.02	0.02	0.02	0.02	0.02	44
44.5	0.06	0.05	0.05	0.04	0.04	0.04	0.04	0.04	0.04	0.04	0.03	0.03	0.03	0.03	0.03	0.02	0.02	0.02	0.02	0.02	0.02	0.02	0.02	44.5
45	0.06	0.06	0.05	0.05	0.04	0.04	0.04	0.04	0.04	0.04	0.04	0.03	0.03	0.03	0.03	0.03	0.02	0.02	0.02	0.02	0.02	0.02	0.02	45
45.5	0.06	0.06	0.05	0.05	0.05	0.04	0.04	0.04	0.04	0.04	0.04	0.04	0.03	0.03	0.03	0.03	0.03	0.02	0.02	0.02	0.02	0.02	0.02	45.5
46	0.06	0.06	0.06	0.05	0.05	0.05	0.04	0.04	0.04	0.04	0.04	0.04	0.04	0.03	0.03	0.03	0.03	0.02	0.02	0.02	0.02	0.02	0.02	46
46.5	0.06	0.06	0.06	0.05	0.05	0.05	0.05	0.04	0.04	0.04	0.04	0.04	0.04	0.04	0.03	0.03	0.03	0.03	0.02	0.02	0.02	0.02	0.02	46.5
47	0.06	0.06	0.06	0.06	0.05	0.05	0.05	0.05	0.04	0.04	0.04	0.04	0.04	0.04	0.03	0.03	0.03	0.03	0.03	0.02	0.02	0.02	0.02	47
47.5	0.06	0.06	0.06	0.06	0.06	0.05	0.05	0.05	0.05	0.04	0.04	0.04	0.04	0.04	0.04	0.04	0.03	0.03	0.03	0.03	0.03	0.03	0.02	47.5

Table C.2: VALUES OF UNADJUSTED DAILY POTENTIAL EVAPOTRANSPIRATION (IN.) FOR DIFFERENT MEAN TEMPERATURES AND I VALUES

I VALUE (87.5 – 140)

T°F	82.5	85	87.5	90	92.5	95	97.5	100	102.5	105	107.5	110	112.5	115	117.5	120	122.5	125	127.5	130	132.5	135	137.5	140	T°F
40	0.01	0.01	0	0	0	0	0	0	0	0	0	0	0	0	0	0	0	0	0	0	0	0	0	0	40
40.5	0.01	0.01	0.01	0	0	0	0	0	0	0	0	0	0	0	0	0	0	0	0	0	0	0	0	0	40.5
41	0.01	0.01	0.01	0.01	0	0	0	0	0	0	0	0	0	0	0	0	0	0	0	0	0	0	0	0	41
41.5	0.01	0.01	0.01	0.01	0.01	0.01	0	0	0	0	0	0	0	0	0	0	0	0	0	0	0	0	0	0	41.5
42	0.01	0.01	0.01	0.01	0.01	0.01	0.01	0	0	0	0	0	0	0	0	0	0	0	0	0	0	0	0	0	42
42.5	0.01	0.01	0.01	0.01	0.01	0.01	0.01	0.01	0	0	0	0	0	0	0	0	0	0	0	0	0	0	0	0	42.5
43	0.01	0.01	0.01	0.01	0.01	0.01	0.01	0.01	0	0	0	0	0	0	0	0	0	0	0	0	0	0	0	0	43
43.5	0.01	0.01	0.01	0.01	0.01	0.01	0.01	0.01	0.01	0.01	0.01	0	0	0	0	0	0	0	0	0	0	0	0	0	43.5
44	0.02	0.01	0.01	0.01	0.01	0.01	0.01	0.01	0.01	0.01	0.01	0.01	0	0	0	0	0	0	0	0	0	0	0	0	44
44.5	0.02	0.02	0.01	0.01	0.01	0.01	0.01	0.01	0.01	0.01	0.01	0.01	0.01	0	0	0	0	0	0	0	0	0	0	0	44.5
45	0.02	0.02	0.02	0.01	0.01	0.01	0.01	0.01	0.01	0.01	0.01	0.01	0.01	0.01	0	0	0	0	0	0	0	0	0	0	45
45.5	0.02	0.02	0.02	0.02	0.01	0.01	0.01	0.01	0.01	0.01	0.01	0.01	0.01	0.01	0	0	0	0	0	0	0	0	0	0	45.5
46	0.02	0.02	0.02	0.02	0.02	0.01	0.01	0.01	0.01	0.01	0.01	0.01	0.01	0.01	0.01	0.01	0	0	0	0	0	0	0	0	46
46.5	0.02	0.02	0.02	0.02	0.02	0.02	0.01	0.01	0.01	0.01	0.01	0.01	0.01	0.01	0.01	0.01	0.01	0	0	0	0	0	0	0	46.5
47	0.02	0.02	0.02	0.02	0.02	0.02	0.02	0.02	0.01	0.01	0.01	0.01	0.01	0.01	0.01	0.01	0.01	0.01	0.01	0	0	0	0	0	47
47.5	0.02	0.02	0.02	0.02	0.02	0.02	0.02	0.02	0.01	0.01	0.01	0.01	0.01	0.01	0.01	0.01	0.01	0.01	0.01	0	0	0	0	0	47.5

Table C.2: VALUES OF UNADJUSTED DAILY POTENTIAL EVAPOTRANSPIRATION (IN.) FOR DIFFERENT MEAN TEMPERATURES AND I VALUES

T°F	I Value (25.0–80)																							140
	25	27.5	30	32.5	35	37.5	40	42.5	45	47.5	50	52.5	55	57.5	60	62.5	65	67.5	70	72.5	75	77.5	80	
48.5	0.07	0.07	0.06	0.06	0.06	0.06	0.06	0.05	0.05	0.05	0.05	0.04	0.04	0.04	0.04	0.04	0.04	0.04	0.03	0.03	0.03	0.03	0.03	48.5
49	0.07	0.07	0.06	0.06	0.06	0.06	0.06	0.06	0.05	0.05	0.05	0.05	0.04	0.04	0.04	0.04	0.04	0.04	0.04	0.03	0.03	0.03	0.03	49
49.5	0.07	0.07	0.07	0.06	0.06	0.06	0.06	0.06	0.06	0.05	0.05	0.05	0.05	0.04	0.04	0.04	0.04	0.04	0.04	0.03	0.03	0.03	0.03	49.5
50	0.07	0.07	0.07	0.07	0.06	0.06	0.06	0.06	0.06	0.06	0.05	0.05	0.05	0.05	0.04	0.04	0.04	0.04	0.04	0.04	0.04	0.03	0.03	50
50.5	0.07	0.07	0.07	0.07	0.07	0.06	0.06	0.06	0.06	0.06	0.06	0.05	0.05	0.05	0.05	0.04	0.04	0.04	0.04	0.04	0.04	0.04	0.03	50.5
51	0.08	0.07	0.07	0.07	0.07	0.07	0.06	0.06	0.06	0.06	0.06	0.06	0.05	0.05	0.05	0.05	0.04	0.04	0.04	0.04	0.04	0.04	0.04	51
51.5	0.08	0.08	0.07	0.07	0.07	0.07	0.06	0.06	0.05	0.06	0.06	0.06	0.06	0.05	0.05	0.05	0.05	0.04	0.04	0.04	0.04	0.04	0.04	51.5
52	0.08	0.08	0.07	0.07	0.07	0.07	0.07	0.06	0.06	0.06	0.06	0.06	0.06	0.05	0.05	0.05	0.04	0.04	0.04	0.04	0.04	0.04	0.04	52
52.5	0.08	0.08	0.08	0.07	0.07	0.07	0.07	0.07	0.06	0.06	0.06	0.06	0.06	0.06	0.05	0.05	0.05	0.05	0.04	0.04	0.04	0.04	0.04	52.5
53	0.09	0.08	0.08	0.08	0.07	0.07	0.07	0.07	0.07	0.06	0.06	0.06	0.06	0.06	0.06	0.05	0.05	0.05	0.05	0.04	0.04	0.04	0.04	53
53.5	0.09	0.09	0.08	0.08	0.07	0.07	0.07	0.07	0.07	0.07	0.07	0.06	0.06	0.06	0.06	0.06	0.06	0.05	0.05	0.05	0.05	0.04	0.04	53.5
54	0.09	0.09	0.08	0.08	0.08	0.07	0.07	0.07	0.07	0.07	0.07	0.06	0.06	0.06	0.06	0.06	0.06	0.06	0.05	0.05	0.05	0.05	0.05	54
54.5	0.09	0.09	0.09	0.08	0.08	0.08	0.07	0.07	0.07	0.07	0.07	0.07	0.06	0.06	0.06	0.06	0.06	0.06	0.06	0.05	0.05	0.05	0.05	54.5
55	0.09	0.09	0.09	0.08	0.08	0.08	0.08	0.07	0.07	0.07	0.07	0.07	0.07	0.06	0.06	0.06	0.06	0.06	0.06	0.06	0.05	0.05	0.05	55
55.5	0.09	0.09	0.09	0.08	0.08	0.08	0.08	0.08	0.07	0.07	0.07	0.07	0.07	0.07	0.06	0.06	0.06	0.06	0.06	0.06	0.05	0.05	0.05	55.5

Table C.2: VALUES OF UNADJUSTED DAILY POTENTIAL EVAPOTRANSPIRATION (IN.) FOR DIFFERENT MEAN TEMPERATURES AND I VALUES

I VALUE (82.5 – 140)

T°F	82.5	85	87.5	90	92.5	95	97.5	100	102.5	105	107.5	110	112.5	115	117.5	120	122.5	125	127.5	130	132.5	135	137.5	140	T°F
48	0.02	0.02	0.02	0.02	0.02	0.02	0.02	0.02	0.02	0.01	0.01	0.01	0.01	0.01	0.01	0.01	0.01	0.01	0.01	0.01	0.01	0	0	0	48
48.5	0.02	0.02	0.02	0.02	0.02	0.02	0.02	0.02	0.02	0.02	0.01	0.01	0.01	0.01	0.01	0.01	0.01	0.01	0.01	0.01	0.01	0.01	0	0	48.5
49	0.03	0.03	0.02	0.02	0.02	0.02	0.02	0.02	0.02	0.02	0.02	0.02	0.01	0.01	0.01	0.01	0.01	0.01	0.01	0.01	0.01	0.01	0.01	0	49
49.5	0.03	0.03	0.03	0.02	0.02	0.02	0.02	0.02	0.02	0.02	0.02	0.02	0.02	0.02	0.01	0.01	0.01	0.01	0.01	0.01	0.01	0.01	0.01	0.01	49.5
50	0.03	0.03	0.03	0.03	0.02	0.02	0.02	0.02	0.02	0.02	0.02	0.02	0.02	0.02	0.01	0.01	0.01	0.01	0.01	0.01	0.01	0.01	0.01	0.01	50
50.5	0.03	0.03	0.03	0.03	0.03	0.02	0.02	0.02	0.02	0.02	0.02	0.02	0.02	0.02	0.01	0.01	0.01	0.01	0.01	0.01	0.01	0.01	0.01	0.01	50.5
51	0.03	0.03	0.03	0.03	0.03	0.03	0.03	0.02	0.02	0.02	0.02	0.02	0.02	0.02	0.02	0.02	0.01	0.01	0.01	0.01	0.01	0.01	0.01	0.01	51
51.5	0.04	0.03	0.03	0.03	0.03	0.03	0.03	0.02	0.02	0.02	0.02	0.02	0.02	0.02	0.02	0.02	0.02	0.02	0.01	0.01	0.01	0.01	0.01	0.01	51.5
52	0.04	0.04	0.03	0.03	0.03	0.03	0.03	0.03	0.02	0.02	0.02	0.02	0.02	0.02	0.02	0.02	0.02	0.02	0.02	0.01	0.01	0.01	0.01	0.01	52
52.5	0.04	0.04	0.04	0.03	0.03	0.03	0.03	0.03	0.03	0.02	0.02	0.02	0.02	0.02	0.02	0.02	0.02	0.02	0.02	0.02	0.01	0.01	0.01	0.01	52.5
53	0.04	0.04	0.04	0.04	0.04	0.03	0.03	0.03	0.03	0.03	0.03	0.02	0.02	0.02	0.02	0.02	0.02	0.02	0.02	0.02	0.02	0.01	0.01	0.01	53
53.5	0.04	0.04	0.04	0.04	0.04	0.04	0.03	0.03	0.03	0.03	0.03	0.03	0.02	0.02	0.02	0.02	0.02	0.02	0.02	0.02	0.02	0.02	0.01	0.01	53.5
54	0.04	0.04	0.04	0.04	0.04	0.04	0.04	0.03	0.03	0.03	0.03	0.03	0.03	0.02	0.02	0.02	0.02	0.02	0.02	0.02	0.02	0.02	0.02	0.01	54
54.5	0.05	0.04	0.04	0.04	0.04	0.04	0.04	0.04	0.03	0.03	0.03	0.03	0.03	0.03	0.02	0.02	0.02	0.02	0.02	0.02	0.02	0.02	0.02	0.02	54.5
55	0.05	0.05	0.04	0.04	0.04	0.04	0.04	0.04	0.04	0.03	0.03	0.03	0.03	0.03	0.02	0.02	0.02	0.02	0.02	0.02	0.02	0.02	0.02	0.02	55
55.5	0.05	0.05	0.05	0.04	0.04	0.04	0.04	0.04	0.04	0.04	0.04	0.03	0.03	0.03	0.03	0.03	0.02	0.02	0.02	0.02	0.02	0.02	0.02	0.02	55.5

Table C.2: VALUES OF UNADJUSTED DAILY POTENTIAL EVAPOTRANSPIRATION (IN.) FOR DIFFERENT MEAN TEMPERATURES AND I VALUES

I Value (25.0 – 80)

T°F	25	27.5	30	32.5	35	37.5	40	42.5	45	47.5	50	52.5	55	57.5	60	62.5	65	67.5	70	72.5	75	77.5	80	T°F
56	0.09	0.09	0.09	0.09	0.09	0.08	0.08	0.08	0.08	0.07	0.07	0.07	0.07	0.07	0.07	0.06	0.06	0.06	0.06	0.06	0.06	0.06	0.05	56
56.5	0.1	0.09	0.09	0.09	0.09	0.09	0.08	0.08	0.08	0.08	0.08	0.07	0.07	0.07	0.07	0.07	0.06	0.06	0.06	0.06	0.06	0.06	0.06	56.5
57	0.1	0.1	0.09	0.09	0.09	0.09	0.09	0.08	0.08	0.08	0.08	0.07	0.07	0.07	0.07	0.07	0.07	0.07	0.06	0.06	0.06	0.06	0.06	57
57.5	0.1	0.1	0.09	0.09	0.09	0.09	0.09	0.09	0.07	0.08	0.08	0.08	0.07	0.07	0.07	0.07	0.07	0.07	0.07	0.06	0.06	0.06	0.06	57.5
58	0.1	0.1	0.1	0.09	0.09	0.09	0.09	0.09	0.09	0.08	0.08	0.08	0.08	0.07	0.07	0.07	0.07	0.07	0.07	0.07	0.06	0.06	0.06	58
58.5	0.11	0.1	0.1	0.1	0.1	0.09	0.09	0.09	0.09	0.09	0.09	0.08	0.08	0.08	0.07	0.07	0.07	0.07	0.07	0.07	0.07	0.06	0.06	58.5
59	0.11	0.1	0.1	0.1	0.1	0.09	0.09	0.09	0.09	0.09	0.09	0.08	0.08	0.08	0.08	0.07	0.07	0.07	0.06	0.06	0.06	0.07	0.06	59
59.5	0.11	0.11	0.1	0.1	0.1	0.1	0.09	0.09	0.09	0.09	0.09	0.09	0.09	0.08	0.08	0.08	0.07	0.07	0.07	0.07	0.07	0.07	0.07	59.5
60	0.11	0.11	0.11	0.1	0.1	0.1	0.1	0.09	0.09	0.09	0.09	0.09	0.09	0.08	0.08	0.08	0.08	0.07	0.07	0.07	0.07	0.07	0.07	60
60.5	0.11	0.11	0.11	0.11	0.1	0.1	0.1	0.1	0.09	0.09	0.09	0.09	0.09	0.09	0.09	0.08	0.08	0.08	0.08	0.07	0.07	0.07	0.07	60.5
61	0.11	0.11	0.11	0.11	0.11	0.1	0.1	0.1	0.1	0.1	0.1	0.1	0.1	0.1	0.09	0.09	0.08	0.08	0.08	0.08	0.08	0.07	0.07	61
61.5	0.12	0.11	0.11	0.11	0.11	0.11	0.11	0.11	0.11	0.11	0.11	0.11	0.09	0.09	0.09	0.08	0.08	0.08	0.08	0.08	0.08	0.07	0.07	61.5
62	0.12	0.11	0.11	0.11	0.11	0.11	0.11	0.1	0.1	0.1	0.1	0.1	0.09	0.09	0.09	0.09	0.09	0.08	0.08	0.08	0.08	0.08	0.08	62
62.5	0.12	0.12	0.11	0.11	0.11	0.11	0.11	0.11	0.1	0.1	0.1	0.1	0.1	0.09	0.09	0.09	0.09	0.09	0.09	0.08	0.08	0.08	0.08	62.5
63	0.12	0.12	0.11	0.11	0.11	0.11	0.11	0.11	0.11	0.1	0.1	0.1	0.1	0.1	0.09	0.09	0.09	0.09	0.09	0.09	0.09	0.08	0.08	63
63.5	0.12	0.12	0.12	0.12	0.11	0.11	0.11	0.11	0.11	0.11	0.11	0.1	0.1	0.1	0.1	0.09	0.09	0.09	0.09	0.09	0.09	0.09	0.09	63.5

Table C.2: VALUES OF UNADJUSTED DAILY POTENTIAL EVAPOTRANSPIRATION (IN.) FOR DIFFERENT MEAN TEMPERATURES AND I VALUES

T°F	82.5	85	87.5	90	92.5	95	97.5	100	102.5	105	107.5	110	112.5	115	117.5	120	122.5	125	127.5	130	132.5	135	137.5	140	T°F
56	0.05	0.05	0.05	0.05	0.04	0.04	0.04	0.04	0.04	0.04	0.04	0.04	0.03	0.03	0.03	0.03	0.02	0.02	0.02	0.02	0.02	0.02	0.02	0.02	56
56.5	0.05	0.05	0.05	0.05	0.05	0.04	0.04	0.04	0.04	0.04	0.04	0.04	0.04	0.03	0.03	0.03	0.03	0.03	0.02	0.02	0.02	0.02	0.02	0.02	56.5
57	0.06	0.06	0.05	0.05	0.05	0.05	0.04	0.04	0.04	0.04	0.04	0.04	0.04	0.03	0.03	0.03	0.03	0.03	0.03	0.03	0.02	0.02	0.02	0.02	57
57.5	0.06	0.06	0.05	0.05	0.05	0.05	0.05	0.05	0.04	0.04	0.04	0.04	0.04	0.03	0.03	0.03	0.03	0.03	0.03	0.03	0.03	0.02	0.02	0.02	57.5
58	0.06	0.06	0.06	0.05	0.05	0.05	0.05	0.05	0.04	0.04	0.04	0.04	0.04	0.04	0.03	0.03	0.03	0.03	0.03	0.03	0.03	0.03	0.02	0.02	58
58.5	0.06	0.06	0.06	0.05	0.05	0.05	0.05	0.05	0.05	0.04	0.04	0.04	0.04	0.04	0.04	0.04	0.04	0.03	0.03	0.03	0.03	0.03	0.03	0.02	58.5
59	0.06	0.06	0.06	0.06	0.06	0.05	0.05	0.05	0.05	0.05	0.04	0.04	0.04	0.04	0.04	0.04	0.04	0.04	0.04	0.03	0.03	0.03	0.03	0.03	59
59.5	0.06	0.06	0.06	0.06	0.06	0.06	0.05	0.05	0.05	0.05	0.05	0.05	0.05	0.04	0.04	0.04	0.04	0.04	0.04	0.04	0.03	0.03	0.03	0.03	59.5
60	0.07	0.07	0.06	0.06	0.06	0.06	0.06	0.06	0.05	0.05	0.05	0.05	0.05	0.05	0.04	0.04	0.04	0.04	0.04	0.04	0.04	0.03	0.03	0.03	60
60.5	0.07	0.07	0.07	0.07	0.06	0.06	0.06	0.06	0.06	0.05	0.05	0.05	0.05	0.05	0.04	0.04	0.04	0.04	0.04	0.04	0.04	0.04	0.04	0.03	60.5
61	0.07	0.07	0.07	0.07	0.06	0.06	0.06	0.06	0.06	0.06	0.06	0.06	0.05	0.05	0.05	0.05	0.04	0.04	0.04	0.04	0.04	0.04	0.04	0.04	61
61.5	0.07	0.07	0.07	0.07	0.07	0.07	0.06	0.06	0.06	0.06	0.06	0.06	0.06	0.06	0.05	0.05	0.05	0.05	0.04	0.04	0.04	0.04	0.04	0.04	61.5
62	0.08	0.07	0.07	0.07	0.07	0.07	0.07	0.07	0.06	0.06	0.06	0.06	0.06	0.06	0.05	0.05	0.05	0.05	0.05	0.05	0.04	0.04	0.04	0.04	62
62.5	0.08	0.08	0.07	0.07	0.07	0.07	0.07	0.07	0.06	0.06	0.06	0.06	0.06	0.06	0.06	0.05	0.05	0.05	0.05	0.05	0.05	0.04	0.04	0.04	62.5
63	0.08	0.08	0.08	0.08	0.07	0.07	0.07	0.07	0.07	0.07	0.06	0.06	0.06	0.06	0.06	0.06	0.05	0.05	0.05	0.05	0.05	0.05	0.05	0.04	63
63.5	0.08	0.08	0.08	0.08	0.08	0.07	0.07	0.07	0.07	0.07	0.07	0.07	0.06	0.06	0.06	0.06	0.06	0.06	0.06	0.05	0.05	0.05	0.05	0.05	63.5

I VALUE (82.5 – 140)

Table C.2: VALUES OF UNADJUSTED DAILY POTENTIAL EVAPOTRANSPIRATION (IN.) FOR DIFFERENT MEAN TEMPERATURES AND I VALUES

T°F	25	27.5	30	32.5	35	37.5	40	42.5	45	47.5	50	52.5	55	57.5	60	62.5	65	67.5	70	72.5	75	77.5	80	T°F
64	0.13	0.12	0.12	0.12	0.12	0.11	0.11	0.11	0.11	0.11	0.11	0.11	0.1	0.1	0.1	0.1	0.09	0.09	0.09	0.09	0.09	0.09	0.09	64
64.5	0.13	0.12	0.12	0.12	0.12	0.12	0.11	0.11	0.11	0.11	0.11	0.11	0.11	0.1	0.1	0.1	0.1	0.1	0.09	0.09	0.09	0.09	0.09	64.5
65	0.13	0.13	0.12	0.12	0.12	0.12	0.12	0.11	0.11	0.11	0.11	0.11	0.11	0.11	0.1	0.1	0.1	0.1	0.1	0.09	0.09	0.09	0.09	65
65.5	0.13	0.13	0.13	0.12	0.12	0.12	0.12	0.12	0.12	0.11	0.11	0.11	0.11	0.11	0.11	0.11	0.1	0.1	0.1	0.1	0.1	0.1	0.1	65.5
66	0.13	0.13	0.13	0.13	0.13	0.12	0.12	0.12	0.12	0.12	0.12	0.11	0.11	0.11	0.11	0.11	0.1	0.1	0.1	0.1	0.1	0.1	0.1	66
66.5	0.13	0.13	0.13	0.13	0.13	0.13	0.12	0.12	0.12	0.12	0.12	0.12	0.11	0.11	0.11	0.11	0.11	0.11	0.11	0.1	0.1	0.1	0.1	66.5
67	0.13	0.13	0.13	0.13	0.13	0.13	0.13	0.12	0.12	0.12	0.12	0.12	0.12	0.11	0.11	0.11	0.11	0.11	0.11	0.11	0.11	0.1	0.1	67
67.5	0.14	0.13	0.13	0.13	0.13	0.13	0.13	0.13	0.13	0.12	0.12	0.12	0.12	0.12	0.11	0.11	0.11	0.11	0.11	0.11	0.11	0.11	0.11	67.5
68	0.14	0.14	0.13	0.13	0.13	0.13	0.13	0.13	0.13	0.13	0.13	0.12	0.12	0.12	0.12	0.12	0.11	0.11	0.11	0.11	0.11	0.11	0.11	68
68.5	0.14	0.14	0.14	0.13	0.13	0.13	0.13	0.13	0.13	0.13	0.13	0.13	0.13	0.12	0.12	0.12	0.12	0.12	0.11	0.11	0.11	0.11	0.11	68.5
69	0.14	0.14	0.14	0.14	0.14	0.13	0.13	0.13	0.13	0.13	0.13	0.13	0.13	0.13	0.13	0.12	0.12	0.12	0.12	0.12	0.11	0.11	0.11	69
69.5	0.15	0.14	0.14	0.14	0.14	0.14	0.13	0.13	0.13	0.13	0.13	0.13	0.13	0.13	0.13	0.12	0.12	0.12	0.12	0.12	0.12	0.11	0.12	69.5
70	0.15	0.14	0.14	0.14	0.14	0.14	0.14	0.13	0.13	0.13	0.13	0.13	0.13	0.13	0.13	0.13	0.13	0.12	0.12	0.12	0.12	0.12	0.12	70
70.5	0.15	0.15	0.14	0.14	0.14	0.14	0.14	0.14	0.14	0.14	0.14	0.13	0.13	0.13	0.13	0.13	0.13	0.13	0.13	0.12	0.12	0.12	0.12	70.5
71	0.15	0.15	0.15	0.15	0.14	0.14	0.14	0.14	0.14	0.14	0.14	0.14	0.14	0.13	0.13	0.13	0.13	0.13	0.13	0.13	0.13	0.13	0.13	71
71.5	0.15	0.15	0.15	0.15	0.15	0.15	0.14	0.14	0.14	0.14	0.14	0.14	0.14	0.14	0.14	0.14	0.13	0.13	0.13	0.13	0.13	0.13	0.13	71.5

I Value (25.0–80)

Table C.2: VALUES OF UNADJUSTED DAILY POTENTIAL EVAPOTRANSPIRATION (IN.) FOR DIFFERENT MEAN TEMPERATURES AND I VALUES

T°F	82.5	85	87.5	90	92.5	95	97.5	100	102.5	105	107.5	110	112.5	115	117.5	120	122.5	125	127.5	130	132.5	135	137.5	140	T°F
64	0.09	0.09	0.08	0.08	0.08	0.08	0.07	0.07	0.07	0.07	0.07	0.07	0.07	0.07	0.06	0.06	0.06	0.06	0.06	0.06	0.06	0.05	0.05	0.05	64
64.5	0.09	0.09	0.08	0.08	0.08	0.08	0.08	0.08	0.07	0.07	0.07	0.07	0.07	0.07	0.06	0.06	0.06	0.06	0.06	0.06	0.06	0.06	0.06	0.05	64.5
65	0.09	0.09	0.08	0.09	0.09	0.09	0.08	0.08	0.07	0.07	0.07	0.07	0.07	0.07	0.07	0.06	0.06	0.06	0.06	0.06	0.06	0.06	0.06	0.06	65
65.5	0.1	0.09	0.09	0.09	0.09	0.09	0.09	0.08	0.08	0.08	0.07	0.07	0.07	0.07	0.07	0.07	0.07	0.07	0.06	0.06	0.06	0.06	0.06	0.06	65.5
66	0.1	0.09	0.09	0.09	0.09	0.09	0.09	0.09	0.08	0.08	0.08	0.08	0.08	0.07	0.07	0.07	0.07	0.07	0.06	0.06	0.06	0.06	0.06	0.06	66
66.5	0.1	0.1	0.09	0.09	0.09	0.09	0.09	0.09	0.08	0.08	0.08	0.08	0.08	0.08	0.07	0.07	0.07	0.07	0.07	0.07	0.07	0.07	0.07	0.06	66.5
67	0.1	0.1	0.1	0.1	0.09	0.09	0.09	0.09	0.09	0.09	0.09	0.09	0.08	0.08	0.07	0.07	0.07	0.07	0.07	0.07	0.07	0.07	0.07	0.06	67
67.5	0.11	0.1	0.1	0.1	0.1	0.1	0.09	0.09	0.09	0.09	0.09	0.09	0.09	0.09	0.08	0.08	0.08	0.08	0.08	0.07	0.07	0.07	0.07	0.07	67.5
68	0.11	0.11	0.1	0.1	0.09	0.09	0.09	0.09	0.09	0.09	0.09	0.09	0.09	0.09	0.08	0.08	0.08	0.08	0.08	0.08	0.07	0.07	0.07	0.07	68
68.5	0.11	0.11	0.11	0.1	0.1	0.1	0.1	0.1	0.09	0.09	0.09	0.09	0.09	0.09	0.09	0.08	0.08	0.08	0.08	0.08	0.08	0.08	0.08	0.07	68.5
69	0.11	0.11	0.11	0.1	0.11	0.11	0.11	0.1	0.1	0.1	0.1	0.1	0.1	0.09	0.09	0.09	0.09	0.09	0.09	0.09	0.08	0.08	0.08	0.08	69
69.5	0.12	0.12	0.11	0.11	0.11	0.11	0.11	0.11	0.11	0.11	0.1	0.1	0.1	0.1	0.09	0.09	0.09	0.09	0.09	0.09	0.09	0.09	0.09	0.09	69.5
70	0.12	0.12	0.11	0.11	0.11	0.11	0.11	0.11	0.11	0.11	0.11	0.11	0.11	0.1	0.1	0.09	0.09	0.09	0.09	0.09	0.09	0.09	0.09	0.09	70
70.5	0.12	0.12	0.12	0.12	0.11	0.11	0.11	0.11	0.11	0.11	0.11	0.11	0.11	0.11	0.1	0.1	0.1	0.1	0.1	0.1	0.1	0.09	0.09	0.09	70.5
71	0.13	0.13	0.12	0.12	0.12	0.12	0.12	0.12	0.11	0.11	0.11	0.11	0.11	0.11	0.1	0.1	0.1	0.1	0.1	0.1	0.1	0.1	0.09	0.09	71
71.5	0.13	0.13	0.12	0.12	0.12	0.12	0.12	0.12	0.11	0.11	0.11	0.11	0.11	0.11	0.11	0.11	0.11	0.11	0.11	0.11	0.11	0.1	0.1	0.1	71.5

I VALUE(82.5 – 140)

Table C.2: VALUES OF UNADJUSTED DAILY POTENTIAL EVAPOTRANSPIRATION (IN.) FOR DIFFERENT MEAN TEMPERATURES AND I VALUES

T°F	25	27.5	30	32.5	35	37.5	40	42.5	45	47.5	50	52.5	55	57.5	60	62.5	65	67.5	70	72.5	75	77.5	80	T°F
72	0.15	0.15	0.15	0.15	0.15	0.15	0.15	0.15	0.14	0.14	0.14	0.14	0.14	0.14	0.14	0.13	0.13	0.13	0.13	0.13	0.13	0.13	0.13	72
72.5	0.15	0.15	0.15	0.15	0.15	0.15	0.15	0.15	0.15	0.15	0.15	0.14	0.14	0.14	0.14	0.14	0.14	0.14	0.14	0.14	0.14	0.14	0.14	72.5
73	0.16	0.15	0.15	0.15	0.15	0.15	0.15	0.15	0.15	0.15	0.15	0.15	0.15	0.15	0.14	0.14	0.14	0.14	0.14	0.14	0.14	0.14	0.14	73
73.5	0.16	0.15	0.15	0.15	0.15	0.15	0.15	0.15	0.15	0.15	0.15	0.15	0.15	0.15	0.15	0.14	0.14	0.14	0.14	0.14	0.14	0.14	0.14	73.5
74	0.16	0.16	0.16	0.15	0.15	0.15	0.15	0.15	0.15	0.15	0.15	0.15	0.15	0.15	0.15	0.15	0.15	0.15	0.15	0.15	0.15	0.15	0.15	74
74.5	0.16	0.16	0.16	0.16	0.16	0.16	0.15	0.15	0.15	0.15	0.15	0.15	0.15	0.15	0.15	0.15	0.15	0.15	0.15	0.15	0.15	0.15	0.15	74.5
75	0.17	0.16	0.16	0.16	0.16	0.16	0.16	0.16	0.16	0.16	0.16	0.15	0.15	0.15	0.15	0.15	0.15	0.15	0.15	0.15	0.15	0.15	0.15	75
75.5	0.17	0.16	0.16	0.16	0.16	0.16	0.16	0.16	0.16	0.16	0.16	0.16	0.16	0.16	0.16	0.15	0.15	0.15	0.15	0.15	0.15	0.15	0.15	75.5
76	0.17	0.16	0.16	0.16	0.16	0.16	0.16	0.16	0.16	0.16	0.16	0.16	0.16	0.16	0.16	0.16	0.16	0.16	0.16	0.16	0.16	0.16	0.16	76
76.5	0.17	0.17	0.17	0.17	0.17	0.17	0.17	0.16	0.16	0.16	0.16	0.16	0.16	0.16	0.16	0.16	0.16	0.16	0.16	0.16	0.16	0.16	0.16	76.5
77	0.17	0.17	0.17	0.17	0.17	0.17	0.17	0.17	0.17	0.17	0.17	0.17	0.17	0.16	0.16	0.16	0.16	0.16	0.16	0.16	0.16	0.16	0.16	77
77.5	0.17	0.17	0.17	0.17	0.17	0.17	0.17	0.17	0.17	0.17	0.17	0.17	0.17	0.17	0.17	0.17	0.17	0.17	0.17	0.17	0.17	0.17	0.17	77.5
78	0.17	0.17	0.17	0.17	0.17	0.17	0.17	0.17	0.17	0.17	0.17	0.17	0.17	0.17	0.17	0.17	0.17	0.17	0.17	0.17	0.17	0.17	0.17	78
78.5	0.17	0.17	0.17	0.17	0.17	0.17	0.17	0.17	0.17	0.17	0.17	0.17	0.17	0.17	0.17	0.17	0.17	0.17	0.17	0.17	0.17	0.17	0.17	78.5
79	0.18	0.17	0.17	0.17	0.17	0.17	0.17	0.17	0.17	0.17	0.17	0.17	0.17	0.17	0.17	0.17	0.17	0.17	0.17	0.17	0.17	0.17	0.17	79
79.5	0.18	0.18	0.18	0.18	0.18	0.18	0.18	0.18	0.18	0.18	0.18	0.18	0.18	0.18	0.18	0.18	0.18	0.18	0.18	0.18	0.18	0.18	0.18	79.5
80	0.18	0.18	0.18	0.18	0.18	0.18	0.18	0.18	0.18	0.18	0.18	0.18	0.18	0.18	0.18	0.18	0.18	0.18	0.18	0.18	0.18	0.18	0.18	80

I VALUE (25.0 – 80)

Table C.2: VALUES OF UNADJUSTED DAILY POTENTIAL EVAPOTRANSPIRATION (IN.) FOR DIFFERENT MEAN TEMPERATURES AND I VALUES

I VALUE (82.5 – 140)

T°F	82.5	85	87.5	90	92.5	95	97.5	100	102.5	105	107.5	110	112.5	115	117.5	120	122.5	125	127.5	130	132.5	135	137.5	140	T°F
72	0.13	0.13	0.13	0.13	0.13	0.13	0.13	0.13	0.12	0.12	0.12	0.12	0.12	0.12	0.11	0.11	0.11	0.11	0.11	0.11	0.11	0.11	0.1	0.1	72
72.5	0.14	0.13	0.13	0.13	0.13	0.13	0.13	0.13	0.12	0.12	0.12	0.12	0.12	0.12	0.11	0.11	0.11	0.11	0.11	0.11	0.11	0.11	0.11	0.11	72.5
73	0.14	0.14	0.13	0.13	0.13	0.13	0.13	0.13	0.12	0.12	0.12	0.12	0.12	0.13	0.12	0.12	0.12	0.12	0.12	0.12	0.12	0.11	0.11	0.11	73
73.5	0.14	0.14	0.14	0.14	0.14	0.14	0.13	0.13	0.13	0.13	0.13	0.13	0.13	0.13	0.12	0.12	0.12	0.12	0.12	0.12	0.12	0.12	0.12	0.12	73.5
74	0.15	0.14	0.14	0.14	0.14	0.14	0.14	0.14	0.13	0.13	0.13	0.13	0.13	0.13	0.12	0.12	0.12	0.12	0.12	0.12	0.12	0.12	0.12	0.12	74
74.5	0.15	0.15	0.14	0.14	0.14	0.14	0.14	0.14	0.13	0.13	0.13	0.13	0.13	0.14	0.13	0.13	0.13	0.13	0.13	0.13	0.13	0.13	0.13	0.13	74.5
75	0.15	0.15	0.15	0.15	0.15	0.15	0.15	0.14	0.14	0.14	0.14	0.14	0.14	0.14	0.13	0.13	0.13	0.13	0.13	0.13	0.13	0.13	0.13	0.13	75
75.5	0.15	0.15	0.15	0.15	0.15	0.15	0.15	0.15	0.14	0.14	0.14	0.14	0.14	0.15	0.14	0.14	0.14	0.14	0.14	0.14	0.14	0.14	0.14	0.14	75.5
76	0.16	0.16	0.16	0.16	0.15	0.15	0.15	0.15	0.15	0.15	0.15	0.15	0.15	0.15	0.14	0.14	0.14	0.14	0.14	0.14	0.14	0.14	0.14	0.14	76
76.5	0.16	0.16	0.16	0.16	0.15	0.15	0.15	0.15	0.15	0.15	0.15	0.15	0.15	0.15	0.15	0.15	0.15	0.15	0.15	0.15	0.15	0.15	0.15	0.15	76.5
77	0.16	0.16	0.16	0.16	0.16	0.16	0.16	0.16	0.15	0.15	0.15	0.15	0.15	0.16	0.15	0.15	0.15	0.15	0.15	0.15	0.15	0.15	0.15	0.15	77
77.5	0.17	0.17	0.16	0.16	0.16	0.16	0.16	0.16	0.16	0.16	0.16	0.16	0.16	0.16	0.16	0.16	0.16	0.16	0.16	0.16	0.16	0.16	0.16	0.16	77.5
78	0.17	0.17	0.17	0.17	0.17	0.17	0.17	0.17	0.16	0.16	0.16	0.16	0.16	0.17	0.16	0.16	0.16	0.16	0.16	0.16	0.16	0.16	0.16	0.16	78
78.5	0.17	0.17	0.17	0.17	0.17	0.17	0.17	0.17	0.17	0.17	0.17	0.16	0.16	0.17	0.17	0.17	0.17	0.17	0.17	0.17	0.17	0.17	0.17	0.17	78.5
79	0.17	0.17	0.17	0.17	0.17	0.17	0.17	0.17	0.17	0.17	0.17	0.17	0.18	0.17	0.17	0.17	0.17	0.17	0.17	0.17	0.17	0.17	0.17	0.17	79
79.5	0.18	0.18	0.18	0.18	0.18	0.18	0.18	0.18	0.18	0.18	0.18	0.18	0.18	0.17	0.17	0.17	0.17	0.17	0.17	0.17	0.17	0.17	0.17	0.17	79.5
80	0.18	0.18	0.18	0.18	0.18	0.18	0.18	0.18	0.18	0.18	0.18	0.18	0.18	0.18	0.18	0.18	0.18	0.18	0.18	0.18	0.18	0.18	0.18	0.18	80

Table C.3: MEAN POSSIBLE MONTHLY DURATION OF SUNLIGHT IN THE NORTHERN HEMSPHERE EXPRESSED IN UNITS OF 12 HOURS

Northern Latitudes	January	February	March	April	May	June	July	August	September	October	November	December
0	31.2	28.2	31.2	30.3	31.2	30.3	31.2	31.2	30.3	31.2	33.3	31.2
1	31.2	28.2	31.2	30.3	31.2	30.3	31.2	31.2	30.3	31.2	33.3	31.2
2	31.2	28.2	31.2	30.3	31.5	30.5	31.2	31.2	30.3	31.2	33.3	30.9
3	30.9	28.2	30.9	30.3	31.5	30.5	31.5	31.2	30.3	31.2	30.9	30.9
4	30.9	27.9	30.9	30.6	31.8	30.9	31.5	31.5	30.3	30.9	30.9	30.6
5	30.6	27.9	30.9	30.6	31.8	30.9	31.8	31.5	30.3	30.9	29.7	30.6
6	30.6	27.9	30.9	30.6	31.8	31.2	31.8	31.5	30.3	30.9	29.7	30.3
7	30.3	27.6	30.9	30.6	32.1	31.2	32.1	31.8	30.3	30.9	29.7	30.3
8	30.3	27.6	30.9	30.9	32.1	31.5	32.1	31.8	30.6	30.9	29.4	30.0
9	30.0	27.6	30.9	30.9	32.4	31.5	32.4	31.8	30.6	30.9	29.4	30.0
10	30.0	27.3	30.9	30.9	32.4	31.8	32.4	32.1	30.6	30.8	29.4	29.7
11	29.7	27.3	30.9	30.9	32.7	31.8	32.7	32.1	30.6	30.8	29.1	27.9
12	29.7	27.3	30.9	31.2	32.7	32.1	33.0	32.1	30.6	30.8	29.1	27.4
13	29.4	27.3	30.9	31.2	33.0	32.1	33.0	32.4	30.6	30.8	28.8	27.4
14	29.4	27.3	30.9	31.2	33.0	32.4	33.3	32.4	30.6	30.8	28.8	29.1
15	29.1	27.3	30.9	31.2	33.3	32.4	33.6	32.4	30.6	30.8	28.5	29.1
16	29.1	27.3	30.9	31.2	33.3	32.7	33.6	32.7	30.6	30.8	28.5	28.8
17	28.8	27.3	30.9	31.5	33.9	32.7	33.9	32.7	30.6	30.0	28.2	28.8
18	28.8	27.0	30.9	31.5	33.9	33.0	33.9	33.0	30.6	30.0	28.2	28.5
19	28.5	27.0	30.9	31.5	33.9	33.0	34.2	33.0	30.6	30.0	27.9	28.5
20	28.5	27.0	30.9	31.5	33.0	33.3	34.2	33.3	30.6	30.0	27.9	28.2
21	28.2	27.0	30.9	31.5	33.0	33.3	34.5	33.3	30.6	29.9	27.9	28.2
22	28.2	26.7	30.9	31.8	34.2	33.9	34.5	33.3	30.6	29.7	27.9	27.9
23	27.9	26.7	30.9	31.8	34.2	33.9	34.8	33.6	30.6	29.7	27.9	27.9
24	27.9	26.7	30.9	31.8	34.5	34.2	34.8	33.6	30.6	29.7	27.3	27.9

Table C.3: MEAN POSSIBLE MONTHLY DURATION OF SUNLIGHT IN THE NORTHERN HEMISPHERE EXPRESSED IN UNITS OF 12 HOURS

Northern Latitudes	January	February	March	April	May	June	July	August	September	October	November	December
25	27.9	26.7	30.9	31.8	34.5	34.2	35.1	33.6	30.6	29.7	27.3	27.3
26	27.6	26.4	30.9	32.1	34.8	34.5	35.1	33.6	30.6	29.7	27.3	27.3
27	27.6	26.4	30.9	32.1	34.8	34.5	35.4	33.9	30.9	29.7	27.0	27.3
28	27.3	26.4	30.9	32.1	35.1	34.8	35.4	33.9	30.9	29.4	27.0	27.3
29	27.3	26.1	30.9	32.1	35.1	34.8	35.7	33.9	30.9	29.4	26.7	26.7
30	27.0	26.1	30.9	32.4	35.4	35.1	36.0	34.2	30.9	29.4	26.7	26.4
31	27.0	26.1	30.9	32.4	35.4	35.1	36.0	34.2	30.9	29.4	26.4	26.4
32	26.7	25.8	30.9	32.4	35.7	35.4	36.3	34.5	30.9	29.4	26.4	26.1
33	26.4	25.8	30.9	32.7	35.7	35.7	36.3	34.5	30.9	29.1	26.1	25.9
34	26.4	25.8	30.9	32.7	36.0	36.0	36.6	34.8	30.9	29.1	26.1	25.8
35	26.1	25.5	30.9	32.7	36.3	36.3	36.9	34.8	30.9	29.1	25.8	25.5
36	26.1	25.5	30.9	33.0	36.3	36.6	37.5	34.8	30.9	29.1	25.8	25.2
37	25.8	25.5	30.9	33.0	36.9	36.9	37.5	35.1	30.9	29.1	25.5	24.9
38	25.5	25.2	30.9	33.0	36.9	37.2	37.8	35.1	31.2	28.8	25.2	24.9
39	25.5	25.2	30.9	33.3	36.9	37.2	37.8	35.4	31.2	28.8	25.2	24.6
40	25.2	24.9	30.9	33.3	37.5	37.5	38.1	35.4	31.2	28.8	24.9	24.3
41	24.9	24.9	30.9	33.3	37.5	37.8	38.1	35.7	31.2	28.8	24.9	24.0
42	24.6	24.6	30.9	33.6	37.8	38.1	38.4	35.7	31.2	28.5	24.9	23.7
43	24.3	24.6	30.6	33.6	37.9	38.4	38.7	36.0	31.2	28.5	24.3	23.1
44	24.3	24.3	30.6	33.6	38.1	38.7	38.7	36.3	31.2	28.5	24.3	23.0
45	24.0	24.3	30.6	33.9	38.4	38.7	39.3	36.3	31.2	28.2	23.7	22.5
46	23.7	24.0	30.6	33.9	38.7	39.0	39.6	36.3	31.2	28.2	23.7	22.2
47	23.1	24.0	30.6	34.2	39.0	39.0	39.9	37.0	31.5	27.9	23.4	21.9
48	22.0	23.7	30.6	34.2	39.3	39.6	40.2	37.0	31.5	27.9	23.1	21.9
49	22.9	23.7	30.6	34.5	39.3	41.2	40.8	37.2	31.5	27.9	22.8	21.3
50	22.2	23.4	30.6	34.5	39.9	40.8	41.1	37.5	31.8	27.9	22.8	21.0

Table C.4: SOIL MOISRE RETENTION TABLE FOR VARIOUS AMOUNTS OF POTENTIAL
EVAPOTRANSPIRATION FOR AROOT ZONE WATER-HOLDING CAPACITY OF 4 INCHES

PET	0.00	0.01	0.02	0.03	0.04	0.05	0.06	0.07	0.08	0.09
0.0	4.00	3.99	3.98	3.97	3.96	3.95	3.94	3.93	3.92	3.91
0.1	3.90	3.89	3.88	3.87	3.86	3.85	3.84	3.83	3.82	3.81
0.2	3.80	3.79	3.78	3.77	3.76	3.75	3.74	3.73	3.72	3.71
0.3	3.70	3.69	3.68	3.67	3.66	3.65	3.64	3.63	3.62	3.62
0.4	3.61	3.60	3.59	3.58	3.57	3.56	3.55	3.54	3.54	3.53
0.5	3.52	3.51	3.50	3.49	3.48	3.47	3.46	3.46	3.45	3.44
0.6	3.43	3.42	3.41	3.40	3.39	3.38	3.38	3.37	3.36	3.35
0.7	3.34	3.33	3.32	3.31	3.30	3.30	3.29	3.28	3.27	3.26
0.8	3.26	3.25	3.24	3.23	3.23	3.22	3.21	3.20	3.19	3.19
0.9	3.18	3.17	3.16	3.16	3.15	3.14	3.13	3.12	3.12	3.11
1.0	3.10	3.09	3.09	3.08	3.07	3.06	3.05	3.05	3.04	3.03
1.1	3.02	3.02	3.01	3.00	2.99	2.98	2.98	2.97	2.96	2.95
1.2	2.94	2.94	2.93	2.92	2.91	2.90	2.90	2.89	2.88	2.87
1.3	2.86	2.86	2.85	2.84	2.83	2.82	2.82	2.81	2.80	2.79
1.4	2.79	2.78	2.77	2.76	2.75	2.75	2.74	2.73	2.73	2.72
1.5	2.72	2.71	2.70	2.70	2.69	2.68	2.68	2.67	2.66	2.66
1.6	2.65	2.64	2.64	2.63	2.62	2.62	2.61	2.60	2.60	2.59
1.7	2.58	2.58	2.57	2.57	2.56	2.55	2.54	2.54	2.53	2.52
1.8	2.51	2.51	2.50	2.49	2.49	2.48	2.48	2.47	2.47	2.46
1.9	2.45	2.45	2.44	2.43	2.43	2.42	2.41	2.40	2.40	2.39
2.0	2.39	2.38	2.38	2.37	2.36	2.36	2.35	2.35	2.34	2.34

Table C.4: SOIL MOISRE RETENTION TABLE FOR VARIOUS AMOUNTS OF POTENTIAL
EVAPOTRANSPIRATION FOR AROOT ZONE WATER-HOLDING CAPACITY OF 4 INCHES

PET	0.00	0.01	0.02	0.03	0.04	0.05	0.06	0.07	0.08	0.09
2.1	2.33	2.33	2.32	2.32	2.31	2.30	2.29	2.29	2.28	2.28
2.2	2.27	2.27	2.26	2.25	2.25	2.24	2.24	2.23	2.22	2.22
2.3	2.21	2.21	2.20	2.19	2.19	2.18	2.18	2.17	2.16	2.16
2.4	2.15	2.15	2.14	2.14	2.13	2.13	2.12	2.12	2.11	2.11
2.5	2.10	2.10	2.09	2.09	2.08	2.08	2.07	2.07	2.06	2.06
2.6	2.05	2.05	2.04	2.04	2.03	2.03	2.02	2.02	2.01	2.01
2.7	2.00	2.00	1.99	1.99	1.98	1.98	1.97	1.97	1.96	1.96
2.8	1.95	1.95	1.94	1.94	1.93	1.93	1.92	1.89	1.91	1.91
2.9	1.90	1.90	1.89	1.89	1.88	1.88	1.87	1.87	1.86	1.86
3.0	1.85	1.85	1.84	1.84	1.83	1.83	1.82	1.82	1.81	1.81
3.1	1.80	1.80	1.79	1.79	1.78	1.78	1.78	1.77	1.77	1.76
3.2	1.76	1.75	1.75	1.74	1.73	1.73	1.72	1.72	1.71	1.71
3.3	1.71	1.70	1.70	1.69	1.69	1.69	1.68	1.68	1.67	1.67
3.4	1.67	1.66	1.66	1.65	1.65	1.65	1.64	1.64	1.63	1.63
3.5	1.63	1.62	1.62	1.61	1.61	1.61	1.60	1.60	1.59	1.59
3.6	1.59	1.58	1.58	1.57	1.57	1.57	1.56	1.56	1.55	1.55
3.7	1.55	1.54	1.54	1.53	1.53	1.53	1.52	1.52	1.51	1.51
3.8	1.51	1.50	1.50	1.49	1.49	1.49	1.48	1.48	1.47	1.47
3.9	1.47	1.46	1.46	1.45	1.45	1.45	1.44	1.44	1.43	1.43
4.0	1.43	1.42	1.42	1.41	1.41	1.41	1.40	1.40	1.40	1.39
4.1	1.39	1.39	1.38	1.38	1.38	1.37	1.37	1.37	1.36	1.36
4.2	1.36	1.35	1.35	1.35	1.34	1.34	1.34	1.33	1.33	1.33
4.3	1.32	1.32	1.32	1.31	1.31	1.31	1.30	1.30	1.30	1.29
4.4	1.29	1.29	1.28	1.28	1.28	1.28	1.27	1.27	1.27	1.26
4.5	1.26	1.26	1.25	1.25	1.25	1.25	1.24	1.24	1.24	1.23

Table C.4: SOIL MOISRE RETENTION TABLE FOR VARIOUS AMOUNTS OF POTENTIAL
EVAPOTRANSPIRATION FOR AROOT ZONE WATER-HOLDING CAPACITY OF 4 INCHES

PET	0.00	0.01	0.02	0.03	0.04	0.05	0.06	0.07	0.08	0.09
4.6	1.23	1.23	1.22	1.22	1.22	1.22	1.21	1.21	1.21	1.20
4.7	1.20	1.20	1.19	1.19	1.19	1.19	1.18	1.18	1.18	1.17
4.8	1.17	1.17	1.16	1.16	1.16	1.16	1.15	1.15	1.15	1.14
4.9	1.14	1.14	1.13	1.13	1.13	1.13	1.12	1.12	1.12	1.11
5.0	1.11	1.11	1.10	1.10	1.10	1.10	1.09	1.09	1.09	1.09
5.1	1.08	1.08	1.08	1.07	1.07	1.07	1.07	1.06	1.06	1.06
5.2	1.05	1.05	1.05	1.04	1.04	1.04	1.04	1.03	1.03	1.03
5.3	1.02	1.02	1.02	1.01	1.01	1.01	1.01	1.00	1.00	1.00
5.4	1.00	1.00	0.99	0.99	0.99	0.99	0.98	0.98	0.98	0.98
5.5	0.98	0.97	0.97	0.97	0.97	0.97	0.96	0.96	0.96	0.96
5.6	0.95	0.95	0.95	0.94	0.94	0.94	0.94	0.93	0.93	0.93
5.7	0.92	0.92	0.92	0.92	0.91	0.91	0.91	0.91	0.90	0.90
5.8	0.90	0.90	0.90	0.89	0.89	0.89	0.89	0.89	0.88	0.88
5.9	0.88	0.88	0.88	0.87	0.87	0.87	0.87	0.87	0.86	0.86
6.0	0.86	0.86	0.86	0.85	0.85	0.85	0.85	0.85	0.84	0.84
6.1	0.84	0.84	0.84	0.83	0.83	0.83	0.83	0.83	0.82	0.82
6.2	0.82	0.82	0.82	0.81	0.81	0.81	0.81	0.80	0.80	0.80
6.3	0.80	0.79	0.79	0.79	0.79	0.79	0.78	0.78	0.78	9.78
6.4	0.77	0.77	0.77	0.77	0.77	0.77	0.77	0.77	0.76	0.76
6.5	0.76	0.76	0.76	0.76	0.75	0.75	0.75	0.75	0.74	0.74
6.6	0.74	0.74	0.74	0.73	0.73	0.73	0.73	0.73	0.72	0.72
6.7	0.72	0.72	0.72	0.72	0.71	0.71	0.71	0.71	0.71	0.70
6.8	0.70	0.70	0.70	0.70	0.70	0.69	0.69	0.69	0.68	0.68
6.9	0.68	0.68	0.68	0.68	0.67	0.67	0.67	0.67	0.67	0.67
7.0	0.66	0.66	0.66	0.66	0.66	0.66	0.66	0.66	0.65	0.65

Table C.4: SOIL MOISRE RETENTION TABLE FOR VARIOUS AMOUNTS OF POTENTIAL EVAPOTRANSPIRATION FOR A ROOT ZONE WATER-HOLDING CAPACITY OF 4 INCHES

PET	0.00	0.01	0.02	0.03	0.04	0.05	0.06	0.07	0.08	0.09
7.1	0.65	0.65	0.65	0.64	0.64	0.64	0.64	0.64	0.54	0.63
7.2	0.63	0.63	0.63	0.63	0.63	0.62	0.62	0.62	0.62	0.61
7.3	0.61	0.61	0.61	0.61	0.61	0.61	0.60	0.60	0.60	0.60
7.4	0.60	0.60	0.60	0.59	0.59	0.59	0.59	0.59	0.58	0.58
7.5	0.58	0.58	0.58	0.58	0.58	0.58	0.58	0.57	0.57	0.57
7.6	0.57	0.57	0.57	0.57	0.57	0.56	0.56	0.56	0.56	0.56
7.7	0.56	0.56	0.56	0.55	0.55	0.55	0.55	0.55	0.55	0.55
7.8	0.54	0.54	0.54	0.54	0.54	0.54	0.54	0.54	0.53	0.53
7.9	0.54	0.53	0.53	0.53	0.53	0.52	0.52	0.52	0.52	0.52

PET	0.05	0.05	PET	0.00	0.05	PET	0.00	0.05
8.0	0.52	0.51	9.0	0.40	0.40	10.0	0.31	0.31
8.1	0.50	0.50	9.1	0.39	0.39	10.1	0.30	0.30
8.2	0.49	0.48	9.2	0.38	0.38	10.2	0.30	0.29
8.3	0.48	0.47	9.3	0.37	0.37	10.3	0.29	0.28
8.4	0.47	0.46	9.4	0.36	0.36	10.4	0.28	0.28
8.5	0.45	0.45	9.5	0.35	0.35	10.5	0.27	0.27
8.6	0.44	0.44	9.6	0.34	0.34	10.6	0.27	0.26
8.7	0.43	0.43	9.7	0.34	0.33	10.7	0.26	0.26
8.8	0.42	0.42	9.8	0.33	0.32	10.8	0.25	0.25
8.9	0.41	0.41	9.9	0.32	0.32	10.9	0.25	0.24

Note: A storage ability equal to 4 in. of water is the combination of the ability of a given soil to store water and the thickness of the soil layer that provides the equivalent of 4 in. of water

INDEX

Fluid conductivity, 109,111
Freezing depth,186
French drain,256,259
Fresh Kills landfill, New York,431
Freundlich isotherm,364
Fuzzy–set procedures, 49

Gas conductivity,389
(see also hydraulic conductivity)
Gas extraction wells,417
Gas turbine engines,427
Geomembranes, 180,227
        failure categories,237
        leachate interactions,237
        materials,237
        non–destructive tests,236
        quality assurance/quality control,240
        spray–applied,234
        types,230
Geonets,226
Geosynthetics,225
Geotextiles,180,227
Granular filtration,325
Groundwater monitoring,461
        construction materials,470
        monitoring well nets,472
        monitoring well development,472
        well casing and screen materials
Grows Landfill, Pennsylvnia,335

Halifax landfill, 333
Hazen method (for fluid conductivity),111
HELP, see Hydrologic Evaluation Landfill Program
Henry's Law, 393
High density polyethylene,24
Hyde Park Landfill, New York, 345
Hydraulic conductivity (see also, fluid conductivity)
        field scale, 216
        magnitudes,109
Hydraulic dispersion,360
Hydraulic gradient,113
Hydrodynamic dispersion,119,121
Hydrogen sulfide,380
Hydrologic Evaluation Landfill Program (HELP),128,145–152,281